MW00990337

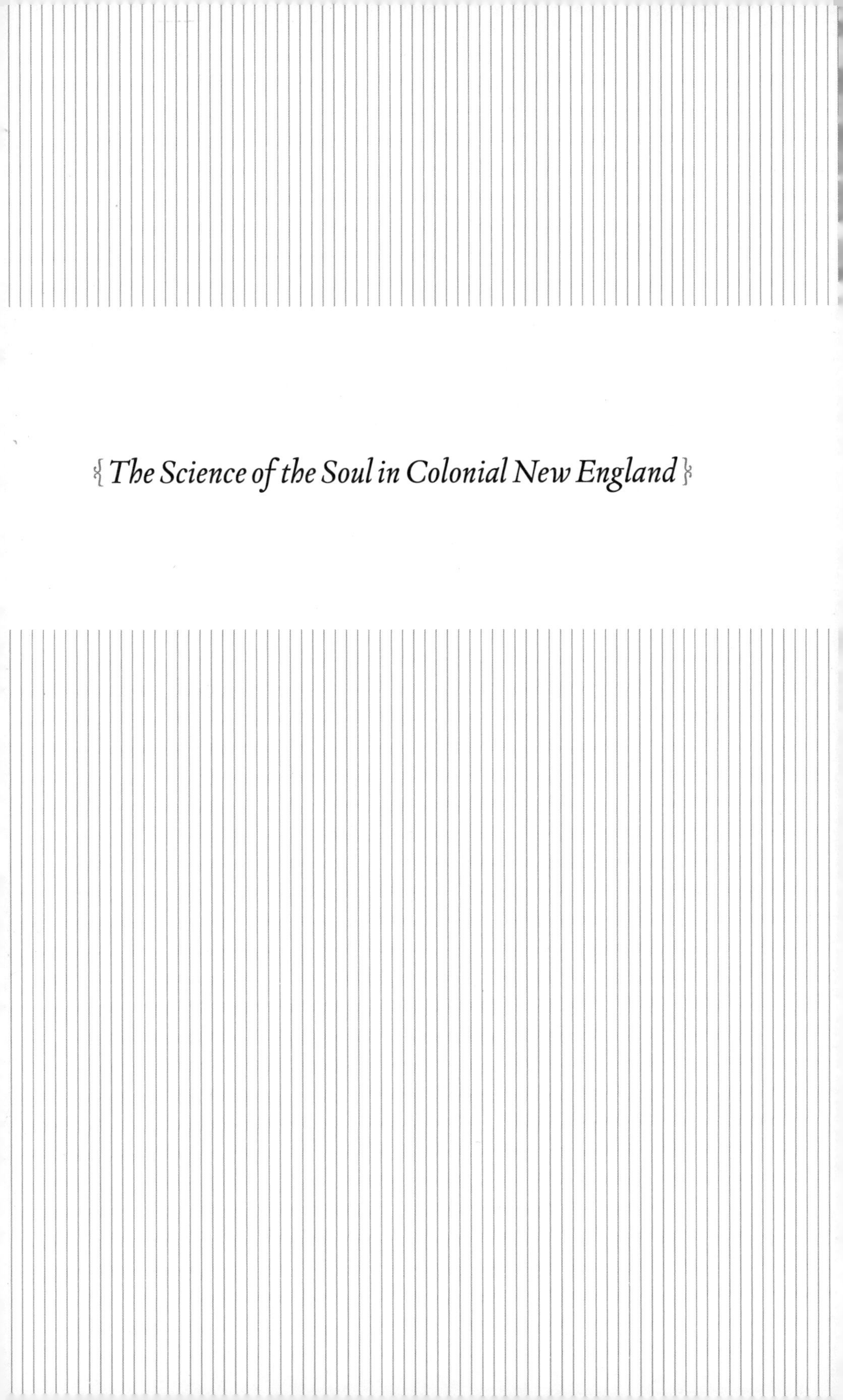

The Science of the Soul in Colonial New England

Published for the Omohundro Institute of Early American History and Culture, Williamsburg, Virginia, by the University of North Carolina Press, Chapel Hill

SARAH RIVETT

The Science of the Soul in Colonial New England

The Omohundro Institute of Early American History and Culture is sponsored jointly by the College of William and Mary and the Colonial Williamsburg Foundation. On November 15, 1996, the Institute adopted the present name in honor of a bequest from Malvern H. Omohundro, Jr.

 Designed and set by Rebecca Evans in Espinosa Nova. Manufactured in the United States of America.

Library of Congress Cataloging-in-Publication Data
Rivett, Sarah.
The science of the soul in colonial New England / Sarah Rivett.
p. cm.
Includes bibliographical references and index.
ISBN 978-0-8078-3524-1 (cloth : alk. paper)
1. Puritans. 2. New England—Church history—17th century.
3. New England—Church history—18th century. I. Omohundro Institute of Early American History & Culture. II. Title.
BX9323.R58 2011
285'.9097409032—dc23
2011028948

The paper in this book meets the guidelines for permanence and durability of the Committee on Production Guidelines of the Council on Library Resources. The University of North Carolina Press has been a member of the Green Press Initiative since 2003.

15 14 13 12 11 5 4 3 2 1

To my parents, Doris and Bob

Acknowledgments

This book is about a quest for knowledge in the early modern Atlantic World. The years that I have spent researching and writing it have in turn taught me a great deal about my own quest for knowledge. For myself as well as for the subjects in this study, the nature of inquiry could be frustrating and even disappointing; new knowledge becomes a fulfilling and rewarding possibility only through communal experiences and shared collaborations. I have benefited immeasurably from the support of my friends, colleagues, collaborators, family, and teachers. Without them, this book would not have been written, and, more important, the process of writing it would not have been nearly as pleasurable.

Two fellowships allowed me to complete the research and revisions for this book: the National Endowment for the Humanities fellowship at the Huntington Library and the Andrew W. Mellon postdoctoral research fellowship at the Omohundro Institute of Early American History and Culture. In addition to the generous financial support that I received from each of these institutions, both also gave me the opportunity to benefit from their thriving scholarly communities.

Princeton University has been an ideal place to complete this project. The early American setting—Nassau Hall, the Revolutionary battlefield, and Jonathan Edwards's grave—have closed the distance between the present and the colonial past. My inspiring colleagues in the English Department and American Studies have offered an ideal balance of encouragement and rigorous feedback. I am appreciative of opportunities to present work at the Eighteenth-Century and Romantics Colloquium and at the American Studies Workshop. In particular, I would like to recognize the comments, feedback, and support that I have received from Eduardo Cadava, Jeff Dolven, Diana Fuss, Sophie Gee, Bill Gleason, Dirk Hartog, Claudia Johnson, Russ Leo, Meredith Martin, Lee Mitchell, Deborah Nord, Jeff Nunokawa, Gayle Salamon, Esther Schor, Nigel Smith, Alex Vazquez, and Susan Wolfson. I would also like to thank the Princeton University Committee on Research in the Humanities and Social Sciences for an award that helped me in the final stages of this book.

Before coming to Princeton, I taught at Washington University in Saint Louis. I wish to thank the university as well as my colleagues there for giving me the time off from teaching to write this book. During my time at Washington University, I benefited from a terrific group of colleagues and had opportunities to present portions of this book at the Eighteenth-Century Salon and the Renaissance Colloquium. For reading sections of the manuscript and engaging my work, I especially thank Guinn Batten, Tili Boon Cuillé, Derek Hirst, Wayne Fields, David Lawton, Joe Loewenstein, Steven Meyer, Bob Milder, Jessica Rosenfeld, Dan Shea, Wolfram Smidgen, and Rafia Zafar.

I owe a special debt of gratitude to Fredrika J. Teute for her editorial brilliance, intellectual rigor, incisive feedback, and keen ability to imagine what is not yet on the page. While I did not live in Williamsburg during my tenure as a Mellon Fellow, I enjoyed the camaraderie and collegial exchange on each visit. Ron Hoffman, director of the Institute, and my fellowship colleagues, Mark Hanna, Robert Parkinson, and Joe Cullon, offered friendly conversation, insightful feedback, and intellectual generosity. I presented a version of my witchcraft chapter for an Institute colloquium that resulted in some very useful questions and comments from everyone present, and particularly Mark Valeri, Doug Winiarski, and Michael McGiffert.

I am especially grateful to Ralph Bauer and Charles L. Cohen, readers, for their extremely lucid and helpful suggestions. Their feedback was invaluable, as they each gave me a fresh perspective on the central contribution of the book as well as helpful techniques on how best to foreground that contribution. Thanks to Gil Kelly and his editorial group for their remarkably thorough revisions and advice about paring down the footnotes and certain sections of the text. During my fellowship tenure, Mendy Gladden offered good advice and sociability while I was visiting Williamsburg. Nadine Zimmerli has given additional support in seeing the book into production. Sally Mason and Beverly Smith's warm, friendly welcome really do make the Institute feel like home.

I completed most of my revisions to this manuscript at the Huntington Library, where I enjoyed not only a first-rate library that has supplied me more than half of the images used in this book but also an ideal community of scholars and friends. I depended on Greg Jackson's expertise in religious history, his thoughtful as well as extensive editorial revisions, and especially his friendship and good humor. Sharon Oster invited me to present a chapter at the Southern California Americanist Group, and Chris Looby invited me to present at the Americanist Research Colloquium at UCLA. My warm thanks to everyone in attendance on each of these occasions. Roy Ritchie, director of research at the Huntington, set the tone for a vibrant and enjoyable intellectual life. I am thankful to him for helping to make my time at the Huntington so productive. The Huntington would not be the warm and welcoming place that everyone returns to without its amazing staff.

David Hall read this manuscript in full. I am indebted to his close, careful reading, his feedback, and his corrections. Kenneth Minkema has helped me a great deal with my work on Jonathan Edwards over the years. In addition to reading my Edwards chapter, he has been my guide through the Edwards Collection at the Beinecke Library, even sharing his own transcriptions and discoveries. My thanks to Rhodri Lewis for his wit and friendship as well as his nuanced and insightful feedback on many sections of this book.

Matt Cohen, Christina Malcolmson, and Laura Stevens read the John Eliot material at an earlier stage and offered stimulating conversation and important insights. I have benefited greatly from exchanging work and conversations with Sally Promey on the topic of deathbed practices and confessions. Wai Chee Dimock and Teresa Toulouse thoughtfully engaged my conference presentations on this material. Many thanks to Francesca Simkin for reading and editing this entire manuscript and her friendship. Stephanie Kirk, my collaborator, dear friend, and coauthor, has taught me so much about colonial religious life outside New England.

I remain grateful to the friends and colleagues that I met during my year as a Barra dissertation fellow at the McNeil Center for Early American studies. Dan Richter, the McNeil Center director, read this manuscript in full at an early stage and offered extremely helpful feedback and suggestions for revision. I am indebted to Dan for his continued support over the years and for cultivating such a vivacious community of young scholars. In particular, I would like to thank Julie Kim, Kyle Farley, Aaron Wunsch, Jen Manion, Justine Murison, and Matthew Osborne for their support, acumen, good humor, and collegiality.

Over the years, the staffs at several libraries have been a great help to me. In particular, I wish to acknowledge my gratitude to librarians and archivists at the Royal Society of London, the American Antiquarian Society, the Beinecke Library, the New England Historic and Genealogical Society, the Massachusetts Historical Society, the Connecticut Historical Society, the Huntington Library, and the Rare Books and Special Collections

Library at Princeton University. A version of Chapter 4 was published in *Early American Literature*, and a portion of Chapter 3 appeared in *Early American Studies*. I thank both journals for permission to reprint.

For my professional achievements, my scholarly development, and—most important—my true passion for teaching and studying the field of early American literature, I owe my greatest debt to my dissertation director, Janice Knight. Janice taught me how to read theological texts as literary texts, to be receptive to the intricacies of doctrine as a malleable rather than fixed set of principles, and to see the Puritan errand as a particularly poignant version of human striving. Another invaluable mentor to me, Lauren Berlant, pushed my thinking into entirely new domains of possibility, offering rigorous yet constructive criticism that made me feel that I was writing a better dissertation than I could have imagined when I started.

I was very fortunate that Eric Slauter joined the Chicago faculty just as I was starting to think about my dissertation and agreed to be on the committee. My conversations with Eric contributed so much to this project. Catherine Brekus joined my dissertation committee once I started writing, and provided detailed and extremely useful feedback. Clark Gilpin was a wonderful reader of sections of the dissertation, and I learned a great deal about theology from him. Richard Strier productively gave me a hard time on several occasions. I greatly appreciate Jim Chandler's friendship and support over the years.

I would also like to express a debt of gratitude to Jenny Franchot for my own conversion to early American literature several years ago. Her intensity, acumen, and brilliant readings of religious genres still inspire my research and teaching. In honor of the memory of Professor Franchot as my own undergraduate teacher, I wish to thank my students for the opportunity to teach and for challenging me to think in new ways.

For the love and support that was always there for me no matter how often or seldom I came up for air as I was writing this book—I owe my greatest debt to my family. This book would not have been possible without the constant encouragement that my parents, Doris and Bob, have given to me over the years. I dedicate this book to them. I am lucky to have amazing siblings, Katharine and William, who are fun, thoughtful, loving, and always willing to make time to spend together. Finally, my deep thanks to my nonacademic friends, Katie, Amanda, Sarah, Alex, Josh, and Darlene, for their abundance of laughter, love, and support.

Contents

Illustrations

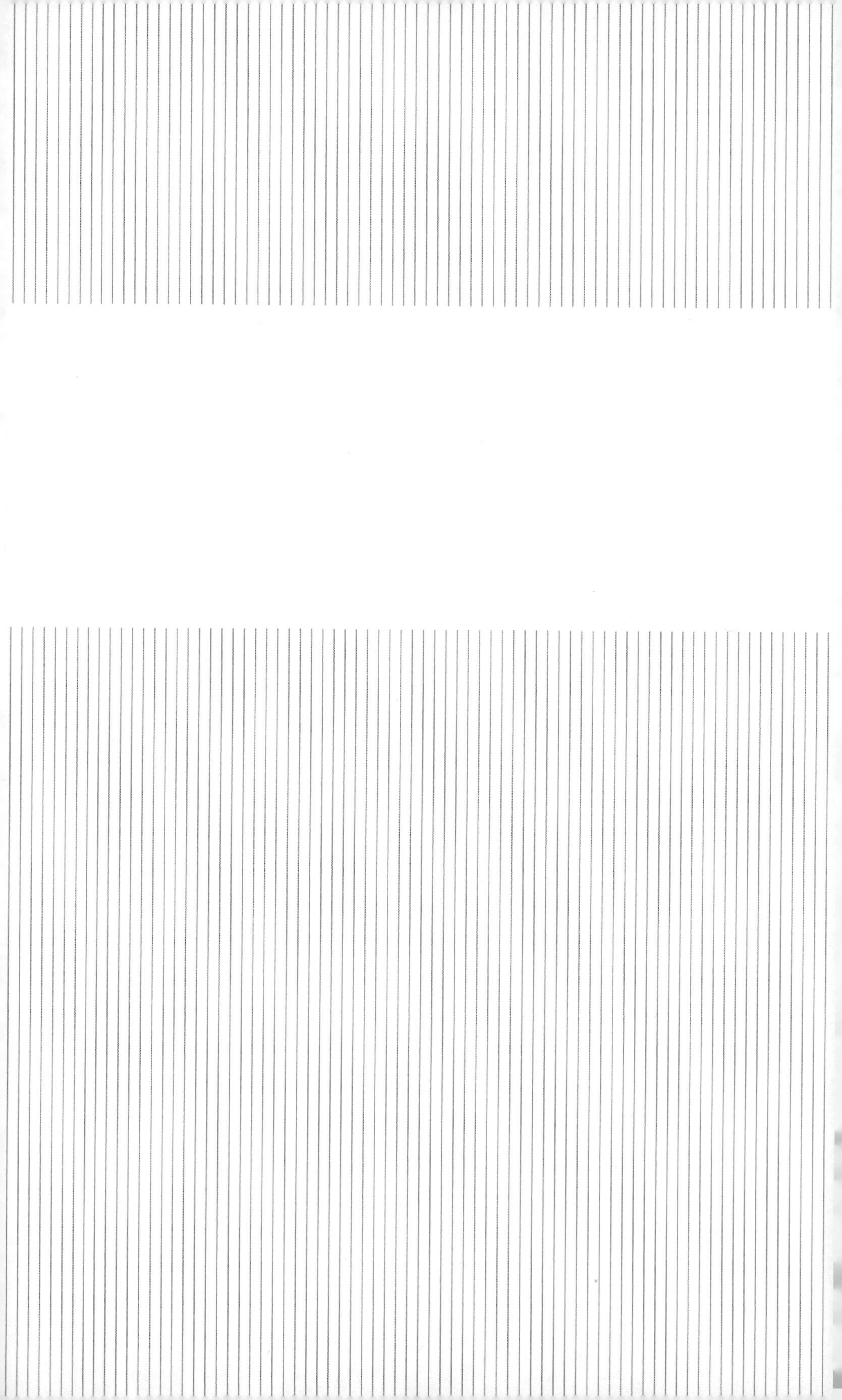

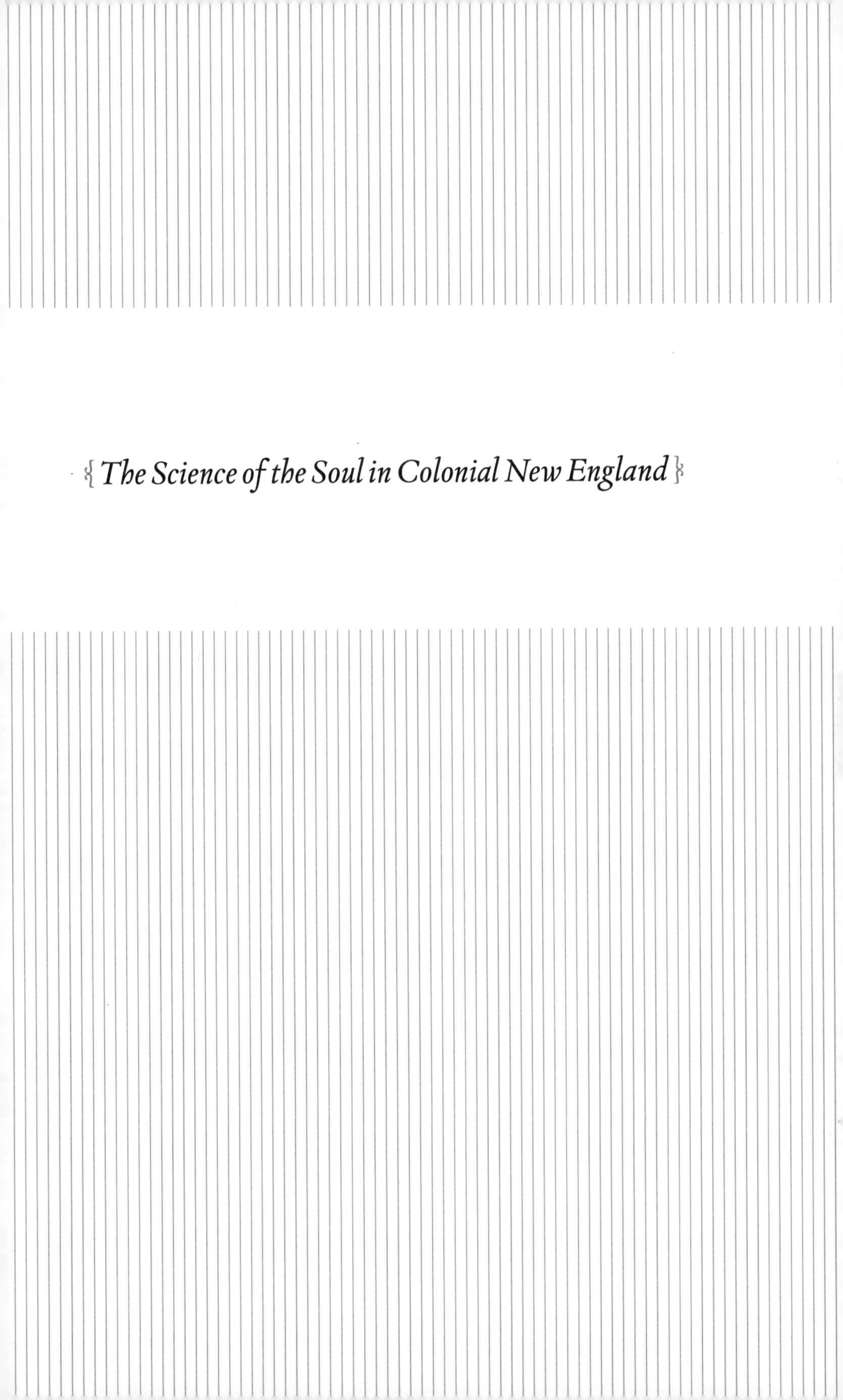

The Science of the Soul in Colonial New England

Introduction
Adam's Perfection Redeemed

John Milton's Eve in *Paradise Lost* (1667) is not built on biblical precedent. She finds knowledge more seductive than flattery as the serpent tempts her through the power of speech and his capacity to rationalize. Listening to the serpent, Eve imagines how she too might possess greater knowledge, augmenting her own "inward powers." She discovers the "virtues" of a fruit that makes her mind "capacious," suddenly capable of discerning "things" erstwhile "visible" only "in heaven." While recognizing how her actions violate God's command, she nonetheless momentarily relishes the reward. From what seems a mixed motive—the desire to share her new authority and implicate another in her shame—she tempts Adam with the desired object, ventriloquizing the serpent's casuistry.

> This tree is not as we are told, a tree
> Of danger tasted, nor to evil unknown
> Opening the way, but of divine effect
> To open eyes, and make them gods who taste.[1]

She tells Adam of the knowledge her new "experience" (9.807) brings. As the serpent has promised Eve that the fruit will open the workings of the universe, disclosing the celestial secrets and expanding her powers of reason, she promises Adam the same. In this rendering of Genesis, Milton breaks with other retellings, focusing less on women's alleged sexual weakness and more on a desire for knowledge. That desire here supersedes the display of sexual seduction narrated by Milton's contemporaries, such as Jakob Böhme, who based his interpretation of Genesis in *Mysterium Magnum* (1654) almost entirely on the rabbinical tradition. In Milton's story, Eve stands in not only for the frailty of women but also for the weakness of clergy and natural philosophers who were unable to contain their ambitions for knowledge, unable to curb their overweening grasp for what lay

1. John Milton, *Paradise Lost,* ed. Alastair Fowler (New York, 1998), 9.600, 603–604, 863–866.

beyond human ken. We see the powers of the serpent through Eve's eyes. The act of disobedience takes on a contradictory double resonance that signals the inevitability of God's displeasure. Milton's Eve posits an originary myth for both the expansion of human knowledge and humanity's frustration at its inability to break beyond the confines of human perception.[2]

Milton's *Paradise Lost* explores the consequences of the Fall for the human intellect, the full measure of which Christian theologians from Augustine of Hippo and Thomas Aquinas to John Calvin and Jonathan Edwards sought to grasp. By the mid-seventeenth century, however, attempts to understand the Fall as an impediment to human progress (and thus humanism) reached a high mark. When Milton completed his Christian epic, the story of the Fall had already become a point of intersection for numerous knowledge traditions as natural philosophers, empiricists, humanists, ministers, and theologians—like Eve—struggled to reconcile their desire for forbidden knowledge with stern prohibitions against human arrogance as personified in Faustian tales about the consequences of overreaching. In a postlapsarian world, this struggle was not without irony: according to Genesis, part of the knowledge gained by the Fall was the knowledge of humanity's shrunken intellectual capacities. The seventeenth century witnessed the Scientific Revolution, New World discoveries, and the spread of Christianity throughout the Atlantic World. Yet those central to these transformations continually confronted the fear of transgression. Transgression involved overstepping boundaries forbidden by scriptural precedent, tradition, and theological law. Philosophers, explorers, and ministers opened onto new and distant frontiers in the the seventeenth century, but not without caution toward the primal and divine proscription at the core of Milton's epic rendering of both Satan and humanity's fall from grace.

2. Philip C. Almond, *Adam and Eve in Seventeenth-Century Thought* (Cambridge, 1999).

Perry Miller explains of the Fall: "In the Puritan view of man, the fall had wrought many melancholy effects, but none so terrible as upon his intellect . . . 'that most efficacious instrument for arriving at deeply hidden truth, for asserting it, vindicating it and eliminating all confusion'—that instrument was warped and twisted. . . .

"Considered therefore in the light of logic, the fall of man had amounted in effect to a lapse from dialectic; the loss of God's image, reduced to the most concrete terms, was simply the loss of an ability to use the syllogism, and innate depravity might most accurately be defined as a congenital incapacity for discursive reasoning" (*The New England Mind: The Seventeenth Century* [New York, 1939], 111). Since then, Peter Harrison has explained the influence of the story of the Fall in early modern science (*The Fall of Man and the Foundations of Science* [Cambridge, 2007]). Part of the task of this book is to bring these conversations together, to show that the story of the Fall was integral to Puritanism and the new science alike and that it in fact generated a continuity of shared methods and goals between the two.

Milton's retelling of the Fall is a retrospective meditation on the career of this story since the Reformation as much as it is a Christian theodicy. The justification of God's ways to man focused on both the universal condition of knowledge and the advance in natural philosophy and empiricism. The paradoxical risk of displeasing God while gaining knowledge, so poignantly rendered in book 9, presents a dilemma that faced New England Puritans as well as empiricists and natural philosophers comprising the early membership of the Royal Society. Despite the different spiritual ends to which these groups sought to advance learning, both were confronted by the specter of the Fall, an inescapable condition of early modern knowledge. This study shows that the story of the Fall was integral to Puritanism and the new science alike and that it in fact generated a continuing of shared methods and goals between the two.

The Reformation made the story of the Fall central to theology and also transferred the condition of knowledge represented by Milton's tree to the human soul. According to Calvin, faith supplies the soul with "new eyes for the contemplation of heavenly mysteries." Yet, as a direct consequence of the Fall, the senses are "deluded in estimating the powers of the soul." Thus, Calvin cautioned, "heavenly mysteries" cannot be fully discerned while on earth. An individual might be able to cultivate a limited awareness of the intricacies of his or her own soul, but this introspective knowledge could not be communicated to others. To presume clear knowledge of divine mysteries was heretical, even blasphemous, an act commensurate with original sin. Yet the soul was a compelling spiritual organ; like the forbidden fruit, its inexplicable nature was tempting to a range of metaphysicians. As a repository of "heavenly mysteries," the soul promised a kind of "opening" similar to the one that Eve experiences upon tasting the fruit. Suspended in the nebulous "middle" between heaven and earth, the invisible and the visible, the soul promised new knowledge if only one could avoid delusion. Moreover, Calvin explained that, in their postlapsarian condition, humans have a limited capacity to resist the allure of divine mystery recorded on human souls: "Because of our own imperfection, we must constantly keep at learning" even while knowing that "the immeasurable cannot be comprehended by our inadequate measure and with our narrow capacities."[3]

3. John Calvin, *Institutes of the Christian Religion,* ed. John T. McNeill, trans. Ford Lewis Battles (Philadelphia, 1960), I, 38, 565; Milton, *Paradise Lost,* 9.605.

Reformed theologians and Protestant ministers such as William Perkins, Thomas Shepard, Richard Baxter, and Jonathan Edwards struggled to resist overreaching the limits of human knowledge. They usually erred on the side of caution by warning their congregants not to rely too heavily on knowledge about the soul gained through religious experience. Yet, despite its forbidden status, the soul often proved too tempting to resist. In the New England colony, religious leaders, in fact, found a particularly promising window into the inner sanctum of the soul through moments of revelation. During the seventeenth and eighteenth centuries, Anglo-American and Native American men, women, and children were called upon by the thousands to give public testimony to the evidence of these mysteries, manifest as the effects of God's grace. In listening to, witnessing, and recording the oral accounts of individuals at the moment of their conversion or in their dying hour, ministers and congregations sought to experience the visual opening that came from small, incremental, and deeply cautious tastes of the forbidden fruit. Collectively, ministers and lay converts pushed against the boundaries of forbidden knowledge, violating their own theological belief in God's law so subtly as to be almost imperceptible.

Performed orally in congregations, missionary towns, deathbeds, witchcraft trials, and revivals, the Puritan testimony of faith was an elusive and continually evolving genre. Each convert struggled to respond to the central question of the Protestant Reformation: "How do I know if I am saved?" There were no rules or conventions to guide the reply. Repeating what had already been said risked producing a form with a deeply suspect content; a replicable experience was a dangerous one, for it meant that conversion could be faked or imagined without divine sanction. Each testimony had to be unique yet intelligible, individual yet commensurate with a communal identity, reflecting God's grace as well as the convert's self-knowledge. Conversion was often difficult to feel and even more difficult to express. Testimonies contained descriptions of what was often called "melting" and "tender" hearts that could not by any theological definition or metaphysical paradigm be systematically charted, yet accuracy was of utmost importance. Religious leaders tried to parse true and false emotions, spiritual insight and human error. Gradually a testimonial idiom emerged from this practice as converts frequently recounted such expressions as, "I felt my heart melting," or, "I found the presence of God in my soul." The members of the visible church called on testifiers to supply the evidence of grace in particular linguistic formulations that ministers could not in the pulpit or in print account for in any formal, theological way. The evidence revealed in such testimonials was, I argue, as much a part of seventeenth- and eighteenth-century philosophy, metaphysics, and empiricism as it was a part of an evolving post-Reformation theological tradition. From the evidence produced by the convert's experience, ministers developed a particular spiritual science for discerning, authenticating, collecting, and recording invisible knowledge of God as it became manifest in the human soul.

What did a seventeenth-century woman know on her deathbed? What did Salem's magistrates learn about the invisible world from tortured adolescent girls? What interest did natural philosophers such as Robert Boyle, Joseph Glanvill, and John Webster have in Native American conversion? Puritan religious communities looked to testimonial forms to manage the central problem imposed by original sin: the postlapsarian limitations that God placed on humanity's rational capacity severely limited the converts' ability to discern their spiritual standing. Because no purely theological resolution to this problem emerged, ministers adapted methods from empiricism and natural philosophy to study evidence of God on human souls. New England Puritans attempted not only to resolve the problem of election through such lay testimonies but also to respond to the philosophical problem of human knowledge circumscribed by the Fall.

Inductive reasoning, recourse to discoveries, the compilation of data, and the testing of a scientific theory through experiment were among the new measurements applied to metaphysics and spiritual study. Each method was integral to the testimonies that constituted the basis of experimental philosophy in the Royal Society as well as to the Puritan testimonies practiced in New England. The experimental religion formulated by Martin Luther, Calvin, and Perkins foreshadowed experimental philosophy while the Scientific Revolution offered Puritan ministers the tools to collect evidence of grace from the souls of testifiers. Ministers learned to study the human soul through the same methods that seventeenth-century scientists used to study nature, in an effort to complete the Reformation project. The science of the soul thus presents an epistemology that was deeply influential in seventeenth- and eighteenth-century religion, science, and philosophy and thus, broadly speaking, in colonial and early modern life.

Religion and Science

Linking Puritanism and empiricism in this way places the former within a historical tradition that is commensurate and contemporaneous with, rather than antithetical and prior to, the Enlightenment. Like their Baconian counterparts, radical Protestants applied the experimental method to witness, observe, and record the manifestations of grace on the souls of others. The science of the soul represents a complex history of epistemological continuity that, while having an end point in the early modern world, nonetheless shaped the dialectical formation and enduring structures of Enlightenment thought and evangelical practice over the long eighteenth century.

Continuity between science and religion can be especially difficult to perceive from a twenty-first-century perspective because we have generally accepted these categories as opposites, organized according to a kind of binary logic. In the early modern period, the opposite was true. Theologians and natural philosophers shared a commitment to pursue knowledge of God as the highest attainable form of truth. Conjoined terms such as "natural theologian," "experimental divine," and "physico-theology" reflect a philosophical culture in which Aristotle's highest form of knowledge was applied to the pursuit of knowledge of God within the natural world. However, this commitment was always tempered by uncertainty and, consequently, hesitancy to make certain truth claims. Early modern Christianity and natural philosophy remained mutually aware that some forms of knowledge resided beyond human ken because of postlapsarian limitations. Ministers and natural philosophers nonetheless grappled with this problem of knowledge, even while remaining wedded to the search for evidence of an impossible truth.[4]

The conjoined goal of spiritual and natural philosophical pursuits for new knowledge of God, the hesitancy surrounding this quest, and the link between new philosophies and new geographic discoveries extended well beyond the Anglo-Protestant world. Since the Spanish Empire began the age of travel and exploration to the Americas, the natural world revealed a host of secrets in distant places. Writings by Christopher Columbus, Hernando Cortez, and José de Acosta promise riches from nature's hidden treasures at the same time that their reports expose the potential forbidden or dangerous content of the natural world. For these and other early Spanish explorers, missionaries, and naturalists, the distant lands of the Americas

4. Describing one of the prevailing assumptions of our secular age, Talal Asad identified a series of oppositions between "belief and knowledge . . . natural and supernatural, sacred and profane in *Formations of the Secular: Christianity, Islam, Modernity* (Stanford, Calif., 2003), 23. Charles Taylor elaborates on the tensions produced between a transcendent and an immanent ontological frame in *A Secular Age* (Cambridge, Mass., 2007).

were both foreign and sacred. Travel, observation, and experience increased the desire for knowledge of undiscovered territory even as this knowledge remained subordinated to the higher purpose of Christianity's providential design. Over the course of sixteenth-century Christendom, nature's secrets became commercial, spiritual, and philosophical goods that soon begged an adequate system of comprehension and organization.[5]

Even though the validation of knowledge gained through travel and observation extends to the ancient world, the early modern discovery of the Americas made firsthand experience more central to knowledge production, foregrounding an important paradigm in empirical modes of inquiry. Broadly defined as a practice based on the inductive accumulation of observations and personal experiences, empiricism emerged and transformed alongside the imperial, commercial, and evangelical goals of settler communities and their correspondent European nations in the Old World. Over the seventeenth century, empiricism became an increasingly institutionalized form of organizing knowledge of the New World, even as its development was gradual, uneven, and sometimes contradictory. Caught up in the transit of Christianity, the intersection of new philosophies and new worlds affected the effort to discover God in American nature as well as to bring the gospel to America's indigenous populations.[6]

By the early seventeenth century, Jesuit colleges sprang up in Catholic territories throughout New France and New Spain as part of the Counter-Reformation. These colleges armed themselves against Protestantism through a curriculum rooted in rational theology, missionary skills, and academic training. In the Jesuit college system, empirical principles differed substantially from the Baconian and Royal Society models. Empirical inquiry also derived from sensory experience, but in the Jesuit tradition the senses were guided by Aristotelian universal principles. This use of experiential knowledge was directly opposed to the Baconian injunction to

5. Edward Peters summarizes this connection between new knowledge and New World exploration in the Middle Ages, in "The Desire to Know the Secrets of the World," *Journal of the History of Ideas,* LXII (2001), 593–610. J. H. Elliot's *Old World and the New, 1492–1650* (Cambridge, 1970) contains an eloquent and compelling reading of José de Acosta's *Natural and Moral History of the Indies,* in which he explains the relationship between knowledge and providence in the sixteenth century. In *The Witness and the Other World: Exotic European Travel Writing, 400–1600* (Ithaca, N.Y., 1988), Mary B. Campbell develops the argument that in Christianity sacred territory is always located elsewhere.

6. For a clear, succinct account of how empirical practice changed in response to New World exploration, see Antonio Barrera-Osorio, "Empiricism in the Spanish Atlantic," in James Delbourgo and Nicolas Dew, eds., *Science and Empire in the Atlantic World* (New York, 2008), 177–202.

seek the anomaly in nature. While differing subtly in methodology and more obviously in doctrine, Catholic Jesuits and Anglo Protestants both relied increasingly on the new philosophy to reveal divine truths. Marin Mersenne and Pierre Gassendi were among the more famous French Jesuits of the mid-seventeenth century to combine theology, philosophy, and mathematics. On the Spanish side, Mexico City's Carlos de Sigüenza y Góngora practiced the same art of intellectual integration, viewing Catholicism through the lens of modern philosophy. Working under a similar culture of the Baroque Enlightenment in Spain, Gabriel Álvarez de Toledo explored the concept of the human soul as further evidence of the book of Genesis. Álvarez explained that the soul was invested with the capacity to discern truth and developed a version of soul science founded on Cartesian rationalism that was quite different from the Anglo-Protestant tradition.[7]

Puritanism and the Calvinist-influenced new science of Bacon and Robert Boyle stand out amid this broader cultural pattern because of the way that both groups intensified the problem of unknowable knowledge. It is difficult to imagine a source of experiential evidence more contested or fraught than the nebulous process of Puritan conversion. Thomas Shepard's well-known proclamation that the greatest part of assurance lies in the "mourning for the want of it" poignantly summarizes this condition. The expression of grace in human terms always involves an elusive movement from the abstract to the concrete, from a numinous to an earthly realm. Yet one of the Puritans' trademarks was to concern themselves intensely with this condition of religious practice.[8]

Uncertainty fundamentally bound Puritan theology and English natural philosophy tightly together. Each movement struggled with uncertainty as an obstacle to increased knowledge of God. Each sought knowledge of God while aware of its forbidden nature. Natural philosophy turned to nature

7. Peter Dear develops a subtle and thorough account of empiricism in the Jesuit colleges in *Discipline and Experience: The Mathematical Way in the Scientific Revolution* (Chicago, 1995), 42–43. On the Baroque Enlightenment in Spain, see Ruth Hill, *Sceptres and Sciences in the Spains: Four Humanists and the New Philosophy (ca. 1680–1740)* (Cambridge, 2000), 119.

8. Michael McGiffert, ed., *God's Plot: Puritan Spirituality in Thomas Shepard's Cambridge,* rev. ed. (Amherst, Mass., 1994), 20, 123. Connections between Puritanism and the new science have long been recognized by philosophers and historians of science who have supplied compelling accounts of the continuities between the Baconian and Protestant Reformations. J. R. Jacob, "Restoration, Reformation and the Origins of the Royal Society," *History of Science,* XIII (1975), 155–176; Charles Webster, *The Great Instauration: Science, Medicine, and Reform, 1626–1660* (New York, 1975); Barbara J. Shapiro, *Probability and Certainty in Seventeenth-Century England: A Study of the Relationships between Natural Science, Religion, History, Law, and Literature* (Princeton, N.J., 1983).

as its principal target of investigation, and the clergy focused on the human soul as a window into the spiritual realm. Despite these different lines of inquiry, the problem of certainty for both groups fostered a complex pattern of convergence; as a result of the anxiety inherent in the proscribed nature of both lines of inquiry, each side developed a sympathy for the other's investigative methodologies. More exemplary than exceptional, the Anglo-Protestant science of the soul emerged at the nexus of a range of philosophical threads, including physico-theology, Comenian linguistics, and antinomianism; it claimed the New England Puritan testimony of faith as the most radical experiment.

Place is essential to the production of scientific knowledge. Such scholars as Ralph Bauer, Jim Egan, Denise Albanese, Jorge Cañizares-Esguerra, Julie Solomon, and Susan Scott Parrish have mapped the new science throughout the New World as well as within early American literature. One thesis unifying this body of work is the centrality of the New World's role as a catalyst for the scientific revolution that occupied the Western world throughout the sixteenth, seventeenth, and eighteenth centuries. The discovery, observation, and need for systems of cataloging and analysis of data produced new, often revolutionary scientific paradigms. The circulation of data between various colonial outposts linked natural philosophy and colonialism. For early modern science, knowledge produced locally in specific places throughout the Atlantic world gradually began to supersede—though not entirely displace—older scientific paradigms that sought first and foremost to establish universally intelligible laws.[9]

9. One method within the philosophy of science subordinates understandings of science as composed of universal laws to an analysis of the production of scientific knowledge in specific geographical and temporal spaces. Seminal studies include Henri Lefebvre, *The Production of Space*, trans. Donald Nicholson-Smith (Cambridge, Mass., 1991); and Bruno Latour, *Science in Action: How to Follow Scientists and Engineers through Society* (Cambridge, Mass., 1987), on information networks. Scholarship on the Enlightenment has integrated this philosophical perspective into a new methodological approach to the history of science. For example, see David N. Livingstone, *Putting Science in Its Place: Geographies of Scientific Knowledge* (Chicago, 2003); Jonathan I. Israel, *Radical Enlightenment: Philosophy and the Making of Modernity* (Oxford, 2002). The new method consists of a reformulation of the Enlightenment as a system of knowledge and philosophical transformation composed, not of universal laws, but rather of the diverse implementation of these laws across a disparate geographical range, across which knowledge varies according to its spatial specificity.

Studies of the intersection between the new science and the New World include Mary Louise Pratt, *Imperial Eyes: Travel Writing and Transculturation* (New York, 1992); Raymond Phineas Stearns, *Science in the British Colonies of America* (Chicago, 1970). See also Jorge Cañizares-Esguerra, *How to Write the History of the New World: Histories, Epistemologies, and Identities in the Eighteenth-Century Atlantic World* (Stanford, Calif., 2001); Susan Scott Parrish, *American Curiosity: Cultures of Natural History in the Colonial British Atlantic World* (Chapel Hill, N.C., 2006).

The geographic expanse of Reformation theology has also long been recognized by scholars who focus on Luther's Germany or Calvin's Geneva and then chart the proliferation of these theological systems throughout Europe. One conceptual frame is the idea of a Calvinist frontier, espousing the view that Calvinist theology depended on an evangelist impetus. Millennial beliefs (principally in the global impact of Christ's return) and the quest for divine truth were so inextricably tied to the spread of the gospel as to make Calvinism global. This paradigm in Reformation studies has much in common with scholarship that parses the Enlightenment geographically rather than as a phenomenon understood in a strictly temporal and universalizing frame. Collectively, scholars have posited a relationship between Christianity and the New World that is analogous to the new science–New World paradigm. Like the new science, Protestantism and Catholicism are no longer considered universal, but rather are seen as composed of symbolic systems that adapted in response to local contexts. Additionally, the seventeenth-century conversion efforts reveal that Christianity was as dependent on New World resources as the new science and that in each case philosophical and theological epistemologies were not simply imposed on a new world of nature but also locally produced and transformed.[10]

My focus on the circulation of ideas throughout the Atlantic world partakes of these complementary theories. Experimental religion not only preceded experimental philosophy, but the practice of Puritan testimony, as it developed most fully in New England, transformed the course of each. Originating as an attempt to resolve the problem of knowledge described by Calvin, the evidence culled from human souls also appealed to the natural philosophical arc that developed through the Enlightenment and that, at least in its early years, was as invested in discovering divine truth as it was aware of the strict limitations of this endeavor.

10. For example of Restoration theology's geographical proliferation, see George P. Fisher, *The Reformation* (New York, 1887); H. A. Enno Van Gelder, *The Two Reformations in the Sixteenth Century: A Study of the Religious Aspects and Consequences of Renaissance and Humanism,* trans. Jan F. Finlay and Alison Hanham (The Hague, 1964); Philip Benedict, *Christ's Churches Purely Reformed: A Social History of Calvinism* (New Haven, Conn., 2002).

David Boruchoff sees the spread of the gospel in pursuit of divine truth as equally integral to the formation of empire in sixteenth-century New Spain and seventeenth-century New England; see "New Spain, New England, and the New Jerusalem: The 'Translation' of Empire, Faith, and Learning *(translatio, imperii, fidei, ac scientiae)* in the Colonial Missionary Project," *Early American Literature,* XLIII (2008), 5–34. See also Graeme Murdock, *Calvinism on the Frontier, 1600–1660: International Calvinism and the Reformed Church in Hungary and Transylvania* (Oxford, 2000); Webb Keane, *Christian Moderns: Freedom and Fetish in the Mission Encounter* (Berkeley, Calif., 2007).

Our narrative of Atlantic crossings and epistemic transformation is incomplete without an integrated account of the science of the soul, the Enlightenment's concomitant effort to systematize the senses, the authority of divine perception, and the representation of religious truth. New England in particular engendered the political, social, theological, and environmental conditions that led to the rise of the testimony of faith. This practice survives as a point of analysis in an effort to understand the range of intersecting Reformed theological and natural philosophical interests, each a way out of the looming shadow of doubt that encased the promise of new knowledge in its own purported impenetrability.

Thomas Shepard and Jonathan Edwards bookend a historical period of investigation into the social problem of hypocrisy. Their theology is, to a large degree, dedicated to discerning true evidence of grace despite the paradox they each acknowledge: if articulating a science of the soul could assist countless converts in understanding their true salvific status, it could also provide countless others with a paradigm and language for misreading (either intentional or self-deluding) that status. This contradiction is not accidental, but speaks to the paradox encoded within early modern philosophy: desired knowledge was also forbidden knowledge; inquiry could reveal new truths that would come twinned with new doubt. To feel doubt was, simultaneously, to feel hope that the knowing heart might also be right. This paradox was as true of the new science as it was of religion. The unending quest for evidence was predicated on doubt and ultimately rewarded with contradictions that undermined any certainty. The contingencies of new experience coupled with doubt, of anxiety read as assurance, of uncertainty transformed into unseen evidence span a range of seventeenth-century philosophical movements. The testimony of faith marked the implementation of these contingencies into religious practice.

Protestantism and natural philosophy did not simply go from a relationship of compatibility to one of opposition with different epistemological goals. Rather, their modern versions emerged through an intricate process of borrowing and differentiating one from the other in order to grapple with the fundamental problem of humanity's limited intellect, a consequence of original sin. This pattern of borrowing was particular to the early modern period, yet it had lasting implications. As much as we tend to think of religion and science as opposites, structured according to the binary logic of secular discourse, along with faith and reason, transcendence and immanence, belief and truth, each carries with it the epistemological underpinnings of its opposite as an intrinsic feature of modernity's formation.

The Science of the Soul

From the ascetic Roman Catholic spiritual practices of the medieval period to the Inquisition, from the Reformation to its later, post-Elizabethan phase, knowledge, faith, and confession were intertwined. The Roman Catholic Church had long used confessional practices as a means of measuring the faith of its converts as well as of regulating sin. By 1215, the Lateran Council had made an annual auricular confession mandatory, a decision that garnered much controversy. As a required practice, the confession was not reliable. How could a required confession be used as an authentic test of faith? Alongside the institutionalization of auricular confession, the Inquisition elevated the mandatory confession of sin to unprecedented heights as a forced means of suppressing heresy. In Spanish and Roman Inquisition practices, forced confession became a means of disciplining behavior against the Roman Catholic Church. The Protestant Reformation both augmented the need for the forced discovery and elimination of heresy and increased Protestant criticisms of Catholic abuses of clerical power. Confession, as a genre and as a mechanism for securing ecclesiastical control, stood at the center of the Reformation and the Counter-Reformation.

As it emerged in 1633, out of the Elizabethan phase of the Reformation, the testimony of faith at once drew on this long history of acquiring spiritual knowledge from confessional practices and inverted it. The testimony of faith was always voluntary. Puritan ministers both felt that this made the information that testifiers produced more accurate and viewed the voluntary aspect of the practice as a corrective to the abuses of the Catholic Church. However, the testimony still stood as a mechanism for maintaining social control. In a community whose safety depended on regulating the spiritual lives of its members, an empirical approach to verifying the experience of grace held great promise. As a consequence of the Reformation idea of *sola scriptura*—scripture alone—converts were believed to have specific moments of intercourse with their souls, unmediated by clerical power. Yet this idea brought with it the danger of enthusiasm, marked by the possibility of misreading signs of deep, often overwhelming emotional outpouring. The testimony of faith both made new use of the data collected from the experience of conversion and made the conundrum of

Calvinist covenant theology a social problem: if the individual could never know the state of his or her own soul with full certainty, he or she most assuredly could not know the state of another's soul with any certainty at all.

To address this dilemma, theologians developed the testimony of faith through a formula that inverted the Catholic tradition of accumulating spiritual knowledge from the senses. In contrast to the visions seen by medieval mystics such as Margery Kempe, evidence of grace worked backward from the soul to the senses in the Puritan tradition, the transformative effects of which the convert reported to a witnessing audience. As the Dublin Puritan minister John Rogers wrote: "*Assurance* is a *reflecting act* of the *soule*, by which a *Saint* sees clearly he is in the *state* of *grace*." In Rogers's published collection of spiritual testimonies and in the many other instances of the practice compiled and recorded in the mid-seventeenth century on both sides of the Atlantic, the task of the saint following this act of spiritual self-reflection was to "clearly" explain what he or she saw in the state of grace. Like Milton's Eve, experience had guided the saint to new knowledge, a knowledge that ministers like Rogers sought, albeit cautiously, to glean through this experimental practice.[11]

Between 1638 and 1649 Thomas Shepard recorded the conversion testimonies of more than sixty of his Cambridge church congregants. Similarly, John Fiske observed and transcribed the aural evidence of grace in his Wenham and Chelmsford congregations between 1644 and 1675. Michael Wigglesworth recorded the testimonies of six converts, which were delivered in Cambridge under the ministry of Shepard's successor, Jonathan Mitchell. Collectively, these texts constitute a representative sampling of a particular genre of spiritual testimony that began in New England congregations—the core of a conversion process leading to membership in the visible church that became known as the New England Way—but soon thereafter appeared in print in London in an almost serial form in 1653 and 1654. John Rogers's *Tabernacle for the Sun* (1653), Samuel Petto's *Roses from Sharon* (1654), Vavasor Powel's *Spirituall Experiences, of Sundry Beleevers* (1653), and John Eliot and Thomas Mayhew's *Tears of Repentance* (1653) each contain collections of exemplary testimonies that record the experiential process of grace in an attempt to catalog the evidence of the soul.[12]

11. John Rogers, *Ohel; or, Beth-shemesh: A Tabernacle for the Sun . . .* (London, 1653), 376.

12. All of these conversion records, 1638–1648 and 1644–1675, were in manuscript in the seventeenth century and have since been published. Several more collections of testimonies from the eighteenth century have survived, but most are still unpublished. Seventeenth-century collec-

New England Puritans implemented the testimony of faith only a few years before Samuel Hartlib's circle initiated the conversation that would bring about the formation of the Royal Society in 1660. As Jan Comenius and John Saltmarsh developed schemas for restoring human perceptive capacities to see divine order in nature, theologians in Old and New England delineated methods for discerning evidence of grace in human souls. All struggled to one degree or another to ameliorate the consequences of the Fall. One method, preferred by linguists, philosophers, and ministers alike, was literally to reconstruct the relationship between seeing and perceiving. Philosophers began experiencing and writing about nature differently. Rather than seeing nature as materiality or even a tabula rasa, they began describing how they perceived it as infused with a sense of divine presence. Ministers took the newly converted soul as their frontier. They began soliciting the spiritual testimonies of converts, ignorant and learned, women and men, young and old, rich and poor, encouraging them to narrate the contents of their souls. Like the Baconian injunction to seek the anomaly in nature, ministers wanted to bear witness to souls unlike their own—the souls of the uneducated, the young, women, or Native Americans—in order to expand the sampling of data and increase knowledge of God.

The senses were elevated to a new ontological status at the same time that linguists and natural philosophers still hoped that language might be redeemed to the state of semiotic perfection that preceded the Fall. The senses supplied a means of achieving this goal, for, as the senses became "spiritualized," the "grammar" spoken by spiritualized subjects reclaimed some of its representational capacity. Over the long historical period through which these transformations occurred, the senses not only opened new avenues of knowledge, but they also reinforced the essential problem of knowledge facing humanity in a postlapsarian world. The senses could deceive, subtly and insidiously, so that it was impossible to know whether it was God, the devil, or the convert's own imperfect mind interpreting the sensory encounter as evidence of grace. The possibility of "trickery" or

---

tions of testimonies include McGiffert, ed., *God's Plot;* George Selement and Bruce C. Woolley, eds., *Thomas Shepard's "Confessions,"* Colonial Society of Massachusetts, *Collections,* LVIII (Boston, 1981); Mary Rhinelander McCarl, "Thomas Shepard's Record of Relations of Religious Experience, 1648–1649," *William and Mary Quarterly,* 3d Ser., LXVIII (1991), 432–466; Robert G. Pope, ed., *The Notebook of the Reverend John Fiske, 1644–1675,* CSM, *Collections,* LXVII (Boston, 1974); Edmund S. Morgan, ed., *The Diary of Michael Wigglesworth, 1653–1657: The Conscience of a Puritan* (1946; Gloucester, Mass., 1970).

"deceit" placed a seemingly impenetrable "mist" upon human understanding. This "mist" is a repeated trope throughout the Bible, functioning as a continual reminder of the barriers to revelation in a fallen world. God's grace could redeem the human senses only partially, just enough to allow for self-knowledge and self-revelation. Protestantism's drive for the individual's unmediated access to God fueled the desire for a deeper awareness of the divine, even while serving as a staunch reminder of the limitations divinely imposed upon human knowledge.[13]

Neither material in its composition nor fully outside the material world, the human soul occupied an enticing space between the material and the spiritual. Consequently, its ontology spawned considerable debate between the sixteenth and eighteenth centuries. In both *The Immortality of the Soul* (1659), his spirited refutation of Thomas Hobbes, and *A Platonick Song of the Soul* (1647), Henry More supplied what he termed "palpable evidence of the souls immortality" based upon knowledge derived "from an inward sense." This was important for More's theology, because he believed that "the very nerves and sinews of Religion is hope of immortality." In other words, the fate of religious belief depended on the fate of the soul; to prove the existence of the soul was to prove the existence of God.[14]

This debate over the soul landed some philosophers, most notably Thomas Hobbes, in a great deal of trouble. Hobbes caused such a stir not simply because *The Leviathan* (1651) highlighted the ascendancy of atheism. Rather, *The Leviathan* threatened to disenchant the world through its fundamental denial of a cosmic perfection that could be only fleetingly grasped. By contrast, knowledge gleaned during the small windows of observation and sensation when converts interacted with their souls suggested how much more knowledge lay beyond. To abandon this hope was to land in a world that was worse than the one that Adam and Eve encountered

13. Ezekias Woodward, *A Light to Grammar, and All Other Arts and Sciences* (London, 1641), 31, 34, 38; Theodore Dwight Bozeman, *The Precisianist Strain: Disciplinary Religion and Antinomian Backlash in Puritanism to 1638* (Chapel Hill, N.C., 2004), 158.

14. Henry More, *A Platonick Song of the Soul,* B4, and *Psychathanasia; or, The Second Part of the Song of the Soul,* H1, both bound with More, *Philosophicall Poems* (Cambridge, 1647). Thomas Willis promoted such an anatomical theory in his treatises on the intercostal nerves. Samuel Parker countered this proposal by using the nervous system as a materially based site for tracking empirical evidence of human souls. For a discussion of the relationship between these two thinkers as their work pertains to an early modern debate about the mind-soul division, see Rina Knoeff, "The Reins of the Soul: The Centrality of the Intercostal Nerves to the Neurology of Thomas Willis and to Samuel Parker's Theology," *Journal of the History of Medicine and Allied Sciences,* LIX (2004), 413–440.

upon their expulsion from Eden. A strictly material world was a world without mystery. And, despite the attendant frustration of not having access to the rich repositories of spiritual knowledge, the fleeting glances beyond the veil and the certainty of the mystery beyond were preferable, for they gave hope for a time when one might receive greater clarity. Because seventeenth- and eighteenth-century Protestant typology, like the fragmented signs in the natural world through which God communicated with his chosen, was predicated on a cosmic totality, the chosen could believe in completeness or perfection of all knowledge, despite the fact that humanity had access to little more than fragments of the invisible world.

Testimony from the New World

The testimony of faith is an example of Puritan hope and striving to know the world beyond. It was much more than the genre of church membership it was once imagined to be. The archive takes us to the home of Mrs. Forbusch, where community members gathered around her bed to hear her dying words in 1727, and to the paper amulet containing "Memoranda" of her husband's dying words that Mrs. Pearce wore around her wrist several weeks after his death. It takes us to the prayer closet of five-year-old Cataret Rede, where her mother heard her exclaim that the "Word doth enlighten my Soul" according to "the Knowledge of Christ." It takes us to the wigwam of Abigail Kesoehtaut on Martha's Vineyard, who recounted a dream in which she sees visions of Christ, and to the Sandwich Islands, just off the coast of Plymouth Colony, where John Eliot went in 1666 to carefully record the testimonies of seven Mashepog Indians and send them to the Royal Society of London.[15]

In each case, a Puritan convert stands before a witnessing audience to publicly, orally, and hesitantly recount the signs of grace upon his or her soul. The audience listens to verify the authenticity of the account. A minister or family member often, though not always, transcribes the evidence supplied in the oral account. Testifiers struggle to find experiential knowledge in spite of its specious nature, exhibiting a historically unprecedented

15. "Extract from the Diary of the Reverend Ebenezer Parkman (Westborough, Massachusetts, 1727)," in John Demos, ed., *Remarkable Providences, 1600–1760* (New York, 1972), 158–162; Jonathan Edwards, "Memoranda by Mrs. Pearce within 4 or 5 weeks of Mr. Pearce's Death," Jonathan Edwards Collection, box 24, fol. 1376, Beinecke Rare Book Library, Yale University, New Haven, Conn.; Sarah Rede, *A Token for Youth . . .* (Boston, 1729), 15; Experience Mayhew, *Indian Converts . . .* (London, 1727), 148.

attempt to not only know the status of one's own soul but in fact to communicate that knowledge to others. These are remarkable documents, born of a deep paradox that a quest for knowledge emerges out of the stark Calvinist reminder of the limits of what could be known.

Conversion testifiers came from diverse segments of the population and were accorded spiritual authority even while the social hierarchy was maintained. Apart from the testimonies of faith initially practiced in Puritan congregations, which aligned spiritual and political agency and from which women were rhetorically and institutionally excluded, the testimony of faith decoupled spiritual knowledge from social and political agency. Deathbed confessors, Praying Indians, possessed girls who became Salem's accusers, and such exemplary Great Awakening figures as Sarah Edwards offered tantalizing new spiritual evidence through pious, expressive modes that represent the dramatic splintering of authority following the Reformation. The evidence supplied by their souls did not change their social place. Within the time span of an hour, deathbed confessors died, of course, supplying the most obvious and literal example of this inverse relationship between spiritual authority and social hierarchy. Salem's accusers lost their relevancy once the ministers and magistrates discarded the skewed empirical system established in the Court of Oyer and Terminer. The Christian testimonies of Native Americans were intelligible only through a colonial discourse, created and sustained by those who also regulated this power.

Ministers worried that those most socially empowered, that is, privy to degrees of education and rudimentary literacy, exposed to the law and legal discourse, and permitted to speak in religious and public forums, might not be the best candidates for confession. English men, with any social authority such as landowners, merchants, religious leaders, elders, deacons, and teachers, were likely to allow their own spiritual expectations to cloud or shape their confessions. Men most attuned to custom, law, or theological study were the least likely to reveal fresh expressions, the least likely to risk accusations of heresy. Thus, it was easier to credit Abigail Kesoehtaut's claim to see a vision of Christ than to hear of a parallel phenomenon experienced by Cotton Mather or Experience Mayhew. Visions of Christ were not theologically sanctioned. Even as late as the publication of Jonathan Edwards's *Treatise concerning Religious Affections* (1746), such evidence was deemed the most specious sort. A minister knew better, whereas a Native American woman did not.

The seventeenth-century theorization of sensory apprehension required an education in Aristotelian and Ramist logic, which at the same time corrupted and distanced the individual from pure and direct access to grace. The problem of hypocrisy, potentially caused by the capacity of the senses to supply corrupt and delusive information, was partially resolved through this focus on Anglo women, children, and American Indians rather than ministers and magistrates. Such groups were, as we have seen, the "least corrupted by Custom, or borrowed Opinions," according to Locke: "Learning, and Education, having not cast their Native thoughts into new Moulds." Scientists of the soul grew increasingly invested in studying the effects of grace on a diversity of populations where divine grace was believed to be channeled through alternative routes, in folkways not as likely reshaped through the conventions of fully established religious practices or normalized through religious and political institutions whose access was largely reserved for enfranchised men. For New England ministers trying to parse true from false affections across a diverse population of religious testifiers, Peter Lipton's maxim holds true: "The central question about testimony is not just whom to trust, but what to believe."[16]

In contrast to the ministerial investment in the Praying Indian as the purveyor of ancient and sacred wisdom, the African was seen as having little, if anything, to offer to this endeavor. The Royal Society invested in the Royal African Company at the time of its formation. Prominent members of the society, most notably Robert Boyle, also invested in the New England Company for the Propagation of the Gospel, but toward very different ends. The Royal African Company was primarily directed toward the transportation of bodies, not the salvation of souls, so that the shareholders might benefit financially from the nascent British Empire. By 1660, the year of the Royal Society's own formation, England had colonies in Barbados, Jamaica, Maryland, and Virginia where the importation of slaves was growing and generating wealth. Englishmen in these colonial outposts, political elites in the Old World, clergy, and plantation owners believed that baptism led to manumission and thus opposed it. One document, the "Conversion of the Negroes in Barbados, 1670," acknowledges that the plantations of the West Indies contain "many thousands of Negros and Blacks being Infidels and without the knowledge of the true God or the means of salvation by Jesus Christ." The author states that these slave populations are unlikely to be converted to Christianity, for this would make them "free and

16. John Locke, *Essay concerning Human Understanding,* ed. Peter H. Nidditch (New York, 1979), 64; Peter Lipton, "The Epistemology of Testimony," *Studies in History and Philosophy of Science,* XXIX (1998), 14.

their several masters and owners loose property in them, it being against the grounds and rules of Christianity that one Christian should be a slave to another." The profits gained from the forced enslavement of African bodies meant that their souls did not offer the same window into the divine that the souls of women, children, and Indians offered.[17]

What justified this profound and unsettling racism remains a historical question. We learn from Benjamin Braude that the myth of the Curse of Ham gained credibility in England between 1589 and 1625, along with the rise of the English slave trade and plantation system. While scholars debate the origins and contingencies of racial categories in the medieval and early modern world, the rise of slave-based economies and plantations is generally understood as consolidating, codifying, and ultimately modernizing notions of race. During the first hundred years of this historical process, debates about Christianity and slavery—and specifically whether or not to convert enslaved Africans—worked to solidify the racist logic of slavery rather than to call the institution into question as such debates would in the late-eighteenth century. By the 1700s and 1710s, many elites began to deny that baptism conferred manumission and turned their attention toward ensuring the spiritual welfare of slaves as a means of reforming them in order to increase their productivity. Efforts to convert Africans happened gradually and with much resistance. Cotton Mather's *Negro Christianized* (1706) is the first printed text in New England that argues that conversion to Christianity may make Africans better slaves. Similar arguments appeared before the British Parliament as proposed legislation. All cases emphasized moderate religious instruction. If conversion to Christianity was encouraged at all, it was to make Africans more effective slaves: accepting of their lot, judicious in their duty, and virtuous in their conduct.[18]

The native peoples of America fitted more easily and necessarily into narratives of God's design. "Nativeness" itself had to be accounted for by the typologies imposed on the New World, its discovery, the success of

17. Mark Govier, "The Royal Society, Slavery, and the Island of Jamaica: 1660–1700," *Notes and Records of the Royal Society of London,* LIII (1999), 203–217; "Conversion of the Negroes in Barbados (West Indies) and St. Hellena, 1670," Boyle Papers, Theology, IV, fols. 1–38, Royal Society, London.

18. Benjamin Braude, "The Sons of Noah and the Construction of Ethnic and Geographical Identities in the Medieval and Early Modern Periods," *WMQ,* 3d Ser., LIV (1997), 138; Ruth Paley, Cristina Malcolmson, and Michael Hunter, "Parliament and Slavery, 1660–c. 1710," *Slavery and Abolition,* XXXI (2010), 257–281. Lara Bovilsky eloquently summarizes this historiography of race studies in *Barbarous Play: Race on the English Renaissance Stage* (Minneapolis, Minn., 2008), 14–27.

colonialism, and the like. Ministers in the first generation of the Great Migration, such as John Eliot, offered what became a widely accepted interpretation of New World discovery and the success of his people there. During the formation of the Society for the Propagation of the Gospel (1644–1648), Eliot developed a theory of the dispersal and degeneration of the lost tribes in North America, basing his argument on Deuteronomy 28:64. His doctrine of simultaneous conversion among the American and Asian peoples connected the immanence of a millennium in Natick to its correspondent immanence in England. In collaboration with Thomas Thorowgood, Eliot wrote *Jews in America* (1660), claiming that "the Judaical badge of circumcision is found upon them." Signs of the Native Americans' "Judaical" status corresponded to eschatological signs of the English nation: God willed that Indians be brought to Christ.[19]

The Native American soul marks a clear case of contrast to that of the African's by ultimately emerging as the most promising site of spiritual potential and ultimately of "true religion" discovered by such luminous ministers as Jonathan Edwards and David Brainerd. Milton also seems to recognize the religious promise of the American Indian in *Paradise Lost* when, immediately following the Fall, Adam and Eve descend "together" "into the thickest wood" where they come upon a "fig-tree" and learn from the "Indians" how to "hide / Their guilt and dreaded shame" with the leaves. Surprisingly, the "Indians" in this scene are already fallen, and they already know to cover their shame. Adam and Eve emulate their model but discover that even once "their shame [is] in part covered," they are "not at rest or ease of mind." Milton has the New World in mind in this passage. He remarks, "Such of late / Columbus found the American so girt / With feathered cincture, naked else and wild / Among the trees on isles and woody shores." Adam and Eve's discovery implicates American Indians in the story of the Fall; they are already living in a state of sin and in need

19. Richard Cogley, *John Eliot's Mission to the Indians before King Philip's War* (Cambridge, Mass., 1999), 87–90; Thomas Thorowgood, *Jews in America; or, Probabilities, That Those Indians Are Judaical, Made More Probable by Some Additionals to the Former Conjectures* (London, 1660), epistle dedicatory. Kristina Bross argues that the occasion of the publication of the Eliot tracts in London (1643–1670) was a scene of metropolitan politics, instigating what the Puritans hoped would be a commensurate and complementary effort to spread the gospel in New England (*Dry Bones and Indian Sermons: Praying Indians in Colonial America* [Ithaca, N.Y., 2004]). James Holstun makes the argument for a connection between Eliot's millennialism and empiricism in *A Rational Millennium: Puritan Utopias of Seventeenth-Century England and America* (New York, 1987). He explains Eliot's mission during the Interregnum as connecting a biblical apocalypse with "contemporary events" such that he could "predict the end of secular history and the beginning of the end time" (45).

of the redemption being supplied by Roger Williams and John Eliot. Yet Adam and Eve also learn that the fig tree is already "to Indians known," acquiring from them a kind of spiritual knowledge about how to exist in this postlapsarian condition. The mention of Columbus's discovery in Milton's epic entwines New World exploration and settlement within the paradoxical desire for forbidden knowledge.[20]

Intrinsic to the optimism purchased through New World discovery, the soul inculcated and augmented empirical desire, the counterpart to the problem of the unknown. If Calvin warned repeatedly against pursuing inaccessible truths, he also recognized the necessity to do so. The need to look was not simply a profitable study in the justice of the consequences of original sin. As Baxter explained in a 1656 letter to Morgan Llwyd, a minister in Scotland or Wales, "It is so naturall to man to desire to know, that I take it for no boast to tell you, that I earnestly long to be acquainted with so lovely a thinge as Truth . . . in this I am still a seeker." Calvin, Baxter, Milton, and others recognized "hope" in humanity's tenacious quest for what it could not achieve; the attempt stood as a testimony to the strength of human striving, a powerful panacea to human suffering and despair. These opposing yet complementary sides of a single condition paralleled for Calvin and his followers the divine capacity to forgive as well as to punish. This was the condition of faith so aptly summarized in the popular Protestant phrase "merciful affliction."[21]

What we learn from the cultural resonance to which *Paradise Lost* so powerfully speaks is the many ways in which Protestants after the Reformation viewed the Fall as fortunate, not only in the conventional, popular way Protestants have understood the Fall as a felix culpa but in the range of more subtle ways examined here. If the wages of original sin had been devastating, God had, in his infinite wisdom and grace freely given, subtly and mercifully mitigated the consequences. This is the point at which the science of the soul enters the contours of modern thought. We too commit to inquiry despite our awareness of what cannot be known and desire knowledge despite our own distance from the object of that knowledge. From the New England Puritans, we learn not only the destructive effects of the Fall but also that the Fall was a fortunate one in the history of human knowledge and perseverance. The great lesson of the Reformation as it prefigured

20. Milton, *Paradise Lost*, 9.1100–1120

21. Calvin, *Institutes*, I, 564–565; N. H. Keeble and Geoffrey F. Nuttall, *Calendar of the Correspondence of Richard Baxter*, I, *1638–1660* (Oxford, 1991), 217.

and was refigured through the Scientific Revolution was that uncertainty is a condition of modern knowledge; the best we can do to circumvent it is to try to supply human answers. We continue to both desire and pursue knowledge despite this, replicating a human condition originally explained through the myth of the Fall.

{1} Evidence of Grace

The sweeping theological transformations that took place over the sixteenth and seventeenth centuries drove bands of people to migrate to diverse locations throughout the early modern Atlantic world. Demonstrating God's grace became of paramount importance to the justification of each New World mission. Radical Protestant sects such as Independents, Congregationalists, and Baptists developed the test of faith, an oral and public testimony designed to mitigate the uncertainty surrounding signs of election. Before a group of discerning witnesses, testifiers declared adequate (albeit uncertain) proof of their own election while also providing data in response to a metaphysical problem, namely, the uncertainty involved in any quest for divine knowledge. This problem of knowledge arose from a largely Calvinist-instantiated Reformation. John Calvin brought new philosophical and theological attention to the quest to know elements of God and the spiritual realm more fully than by what revelation or scripture could reveal. His followers also reiterated in unequivocal terms the proscriptions against certain forms of human inquiry that had been in place since the Fall.

This chapter elucidates the connection between the evidence of grace supplied in the test of faith and the intellectual history that authorized such evidence. The writings of Calvin are an inaugural foray into a specific confrontation between the desire for knowledge of divinity and the limitations of human access to that knowledge. Calvin identifies human limitations through his widely accepted explanation that Adam's fall evacuated all "primal and simple knowledge" of God from earth. But, along with this, Calvin also explicates a form of optimism: the desire to know and the accompanying yearning for access to the spiritual essence of things residing just beyond our grasp. Calvin's intellectual contribution transformed the parameters of human knowledge so substantially that subsequent philosophical changes, including Baconian empiricism, natural theology, and mechanical philosophy, would also bear its mark.[1]

1. John Calvin, *Institutes of the Christian Religion*, ed. John T. McNeill, trans. Ford Lewis Battles (Philadelphia, 1960), I, 40. All citations of the *Institutes* are to this edition.

This contribution solved a problem that Calvin established and codified in new ways but that the sixteenth-century Christian humanist tradition, most fully reflected in the writings of Desiderius Erasmus, also grappled with. If divorced from the revelatory power of the divine word, how could experiential and experimental practices lead to metaphysical truths? What could reconcile the competing frameworks of human and divine learning? The theology of Calvin, like the Christian humanist philosophy of Erasmus, illuminated a discrepancy between these frames that marked a decided break from the traditional belief that the human mind was more or less capable of knowing what God knows. By the late-sixteenth century, the end goals of knowledge had been perceived as serving different purposes; to focus on things human was to forget the indisputable authority and scope of divine, revelatory powers. Clergy and philosophers adhered to the boundaries surrounding their respective domains of knowledge as stipulated by Calvin while also acquiring a desire to transgress these boundaries if only to advance their own knowledge of God.[2]

The coupling of a deep awareness of the kinds of knowledge that could not be known and the compulsion to expand human knowledge nonetheless is integral to the Calvinism that Puritans in Old and New England as well as Baconian empiricists and Royal Society fellows would inherit. This condition of knowledge replaced the seamless continuity imagined in Aristotelian philosophy between what humans could see and perceive and their access to the full scope of divine knowledge. Connections between Calvin and the modern world have been well established by scholars. However, the link that I am making is quite different from traditional narratives of Protestantism as paving the way to capitalism, the Scientific Revolution, and the rise of a democratic society that have long been supplied to us by Max Weber, Christopher Hill, R. H. Tawney, Jürgen Habermas, and others. In these accounts, Calvinism contributes the constitutive and easily recognized features of modernity: Puritans and Baconians represent the shattering of old certainties and the rise of a belief in progress through contested authority, a new stress on disciplined labor, a utilitarian spirit, and a belief in free inquiry. In each case, Calvin anticipates a highly visible, highly recognizable feature of the modern era; the movement from Calvin to capitalism or the new philosophy is also teleological and involves the gradual dissolution of a religious era. These well-established narratives of Protestantism as fostering the rise of the scientific and capitalist revolu-

2. William J. Bouwsma, *John Calvin: A Sixteenth-Century Portrait* (Oxford, 1988), 69.

tions chart Calvin's influence in positivist terms: self-denial engenders a capitalist work ethic, and the contestation and splintering of ecclesiastical authority perpetuates a paradigm of new learning that also discards the presuppositions of ancient philosophy.

I propose that Calvin replaces the optimism of classical knowledge, not exclusively with doubt and harsh reminders of the epistemological challenges of a fallen world, but rather with a new kind of optimism to push the boundaries of what we know, of what we are comfortable knowing, and of how we see and perceive as well as how this perception matches an external reality. Calvinist optimism is unique, for it comes inextricably linked to its twin: doubt. This doubling of inquiry and doubt, this interplay between the limits of knowledge and the transgression of these limits, is what makes this study of Calvinism and modernity different from the traditional narratives of the Protestant Reformation as leading to the secularizing forces of enlightened modernity.

This is not the story of the triumph of reason over revelation, of the replacement of religious beliefs with scientific truths, or of the movement into an increasingly secure methodological frame to buttress the advancement of secular pursuits. Rather, the thread of modern knowledge traceable to Calvin comes out of the complex contestation of authority that in fact crosses—though it takes different forms—the Protestant and Catholic divide. This contestation produced parallel problems along an axis of knowledge and power whereby philosophers and theologians of both faiths sought to delineate the boundaries of what could be known at the same time that they attempted to make a case for the superiority of their way of thinking. Rather than a grand narrative of the triumph of human learning in a Protestant and then secular age, the science of the soul reveals the limiting factors of human knowledge, the boundaries drawn and redrawn around what humans could know and what they could not know.

Francis Bacon inherited the cautious and tentative tone of *The Advancement of Learning* (1605) from theologians who warned against the blasphemous consequences of inquiry into forms of truth that had been cordoned off by God. His writings partake in a classical tradition of ancient Greek skepticism, revived among Reformation theologians who wished to contest what they perceived as Catholic dogmatism. Bacon derived his method of advancing knowledge by studying novelties in the natural world in part through the writings of William of Ockham, who proposed that the human mind could accrue accurate knowledge by isolating singular truths. Ockham also encountered what a long line of English empiricists would take up as a central problem of knowledge: because the human intellect comes into contact with only the material realm and nothing that resembles an immortal spirit, proving the existence of God was impossible. In his turn away from the idealism of Platonic form to a concrete material realm, Ockham represents the first instance in the problem of causality that would consume British empiricism through the mid-eighteenth century. If the movement of material objects could be studied and observed as singular entities in nature, what invisible force controlled the movement of disparate particles?[3]

Through Bacon's work, what Ockham first identified as the problem of causality became a central dilemma of British empiricism. Bacon subordinated physics to metaphysics, explaining that physics dealt only with material and efficient causes while metaphysics dealt with final causes. Metaphysical understanding could not be derived through the methods of empirical observation employed by Bacon to further study the material world. Bacon was also suspicious of mathematics, a discipline remote from the examination of particulars that his method focused on. However, his philosophy did not propose an alternative method for arriving at a conception of a mathematical or metaphysical law about the governing of physical events. As such, Baconian empiricism anticipated the limitations of mechanical philosophy that would emerge most fully by the late-seventeenth century. Mechanical philosophers replicated this hierarchy of metaphysical and physical knowledge and then confronted the frustrations of developing a system for understanding the physical that simply could not lead to the metaphysical.

A complex and varied philosophical system, empiricism is irreducible to any one definition, philosophical lineage, or intellectual culture. Yet the problem of causality sketched above bore a particular mark on British empiricism, producing specific tensions between the limits of knowledge, the desire to circumvent those limits, and the methodologies available to natural philosophers invested in advancing inquiry into the metaphysical realm. In an attempt to thwart the impasse created by causal considerations, the inability to consider physical laws as an avenue to the metaphysical or to fully understand the invisible agent structuring the laws of the material world, Thomas Hobbes shifted the ontological consideration of

3. Harry R. Klocker, *God and the Empiricists* (Milwaukee, Wis., 1968). Richard H. Popkin explains that "in the battle to establish which criterion of faith was true, a sceptical attitude arose among certain thinkers, primarily as a defense of Catholicism." *The History of Scepticism: From Savonarola to Bayle* (Oxford, 2003), 7.

nature to a consideration of the succession of thoughts within the human mind.[4]

In doing so, however, Hobbes replaced one philosophical puzzle with another as his philosophy highlighted the discrepancy between the human perception of nature and the world represented through that perception. This discrepancy required a theory of language to show how words signify concepts in our mind and would take John Locke's *Essay concerning Human Understanding* (1690) to unfold in all of its complexity. British empiricism bore the distinct mark of the two dilemmas characterized by the philosophy of Bacon and Hobbes. Within the Baconian tradition, this dilemma consisted of a persistent frustration with efforts directed toward the study of matter, despite the fact that these methods did not culminate in a deeper awareness of divine truths. Hobbes redirected the problem of knowledge, ruminating on the perceptual gap between the material world and the human perception of it. From the 1640s, when Protestant and empirical perspectives converged through the fervor of the English Civil War, to the 1690s, when Newtonian mechanical philosophy presented an increasingly totalizing view of the natural world, to the rise of radical skepticism in the 1740s, political and religious conditions conjoined with and augmented these problems of knowledge.

These boundaries surrounding knowledge, of the divine as well as of human sensory capacity, come directly out of a Reformed theological tradition. Calvinist cosmologies also registered this twofold limitation. In a fallen world, divisions between the natural and metaphysical, or the visible and invisible world, made knowledge of the former a tenuous avenue to knowledge of the latter. The depleted state of the human senses led to a theological understanding of a deep and potentially irreconcilable gap between external reality and a human understanding of that reality. Within the broader comparative context of overlap and intersection between an Anglo line of empiricist philosophy and Protestantism, Puritanism, particularly in New England, emerged to produce a unique way of reflecting on and grappling with doubt. The Puritans knew that doubt constituted a necessary and inextricable component of knowledge acquisition. The senses bore new information, yet they could also deceive. Assurance only partially resolved this dilemma, for assurance itself was uncertain. For followers of Calvin, the coupling of assurance, however hesitant or tentative, with new

4. Robert L. Armstrong, *Metaphysics and British Empiricism* (Lincoln, Nebr., 1970), 9–35.

paths of empirical inquiry became the only means of managing the anxiety surrounding uncertainty.

This was a radical and unprecedented innovation, for in Calvin's writing individual knowledge of God was not an intellectual achievement. Rather, it came through conversion. Calvin explains succinctly and clearly in the *Institutes* that God's "essence is incomprehensible; hence, his divineness far escapes all human perception." Calvin's theology separates secular learning from the dispensational transformation of divine wisdom to the human soul in the moment of conversion, a primary mark of distinction between his perspective and that of the Christian humanist Erasmus, to whom he is indebted. But, even in Calvin, this separation is tenuous. It divided knowledge in a way that theologians and natural philosophers, from William Perkins to Francis Bacon to the Cambridge Platonists of the 1640s who would seek to direct empiricism toward metaphysical concerns, could not sustain. One of the great architects of Anglo-American Calvinism, William Perkins, found Calvin's explanation unsatisfying. In 1590, he set the terms for an inquiry that would overturn Calvin's explanation by asking how the "unmistakable" yet deeply personal marks of grace might be married to knowledge of Christ in the "very places of learning." How, Perkins's hypothesis implicitly asked, could the evidence of grace imprinted on human souls be translated into knowable knowledge of Christ for the convert and a deeper comprehension of divine essence for the natural philosopher?[5]

By conjoining divine assurance and human learning, experimental theologians and philosophers elevated the inquiry into the divine to new heights. Assurance purified the knowledge acquired through the human senses, cautiously carving out an avenue to divinity that presented the potential to exceed the boundaries of what could be known exclusively through revelation. Greatly reducing the dangerous phenomena that natural philosophers came to call phantasms and theologians called hypocrisy, theologians and natural philosophers turned to human souls to acquire new knowledge of the divine. Throughout the Atlantic world, experimental religion and experimental philosophy increasingly became a means of affirming an otherwise elusive external reality.

5. Calvin, *Institutes,* I, 52; Bouwsma, *John Calvin,* 50–54; William Perkins, *A Declaration of the True Manner of Knowing Christ Crucified* (Cambridge, 1596), A3.

Calvin in Science

Francis Bacon's *Advancement of Learning* uses the separation between divine and human knowledge as a way of protecting the new science against charges of heresy. Defending the new learning against heretical or even blasphemous charges was important not only to seventeenth-century natural philosophers who would promote this new era of discovery and experimental learning but also to theologians. For theologians also directed Baconian methods toward advancing knowledge, though with the aim of furthering the spiritual welfare of the elect rather than employing new Baconian attributes to expand knowledge of the material world.

Isaac Casaubon, a French Huguenot who fled France with his family in the late-sixteenth century and ended up training for the ministry at the Academy of Geneva, owned a first edition of the *Advancement of Learning.* Autographed by Casaubon, this particular copy contains numerous marginal annotations, including a telling note at the beginning of the first book recording Casaubon's own sense that he was reading *"Calvin in scientia."*[6]

More than fifty years after Calvin's *Institutes* codified the central tenets of the Protestant Reformation, Bacon begins his *Advancement of Learning,* the text that is perhaps most singularly and canonically associated with the beginning of the Scientific Revolution, with an identical statement of the paradoxes of learning and the dangers of seeking to know too much. Bacon attributes the cause of the "aspiring to overmuch knowledge" to the "originall temptation and sinne" that caused the serpent to "entreth into a man." Bacon notes the biblical proscriptions against the repetition of this form of heretical inquiry, including the Pauline caveat that "learned men" may be *"spoyled through vaine Philosophie."* Divine proscription must be obeyed as well as carefully transgressed; this is not a repeated act of willful disobedience for Bacon any more than it is for Calvin. Rather, for both, the quest for knowledge beyond the scope of revelation or philosophical study is part of the human condition. The challenge for natural philosophers and theologians alike is to figure out how to manage this condition—to seek new knowledge that skirted the boundaries of heresy without actually violating them.[7]

6. Francis Bacon, *The Twoo Bookes of Francis Bacon: Of the Proficience and Advancement of Learning, Divine and Human* (London, 1605), 3–4, copy at the Huntington Library, San Marino, Calif.

7. Ibid.

Bacon's careful separation of divine and human learning as well as his Calvinist justification for this separation is only tentative throughout his oeuvre. In his *Confession of Faith,* Bacon suggests that the natural philosopher could augment his observational capacity through the gift of free grace. Following Bacon, natural philosophers sought increasingly to merge new empirical methods of discovery and experimental learning with the Redeemed observational capacity of assured saints. Robert Boyle, for example, explained the value of "the fuller Discoveries made of those things by the Preachers of the Gospel." Deeply influenced by Bacon, Boyle expanded experimental philosophy into a practice institutionalized by the Royal Society in 1660. He accumulated vast amounts of information about the natural world and also devised experiments that would reveal new information about phenomena that interested him, such as the movement of particles of matter. While deftly employing such technologies as the air-pump, barometers, and hydroscopes, Boyle also demarcated the different kinds of knowledge made available by human reason versus revealed religion. His writings reveal a reluctance to rely too heavily on human reason, given its limited capacity. Boyle maintained that discoveries made through the rational capacity of the human intellect should complement but remain subordinate to the greater knowledge revealed by God.[8]

Oral conversion testimonies recorded by such ministers as John Eliot, whose New England Company for the Propagation of the Gospel Boyle governed, provided empirical data for supernatural phenomena. The oral conversion performances that Eliot's coterie, including James Janeway, Cotton Mather, and Experience Mayhew, transcribed and collected emulated the experimental method as developed by Robert Boyle. Such testimonies were intelligible to natural philosophers such as Boyle for containing valid revealed evidence that, when properly observed and recorded, could advance new knowledge of God. Ministers such as Richard Baxter, Eliot, and Thomas Shepard as well as those in later generations, such as Cotton Mather and Jonathan Edwards, looked increasingly toward the experimental method of the Royal Society to help augment their own methodologies. Empiricism responded to the problem of knowledge formation that theology in itself had been unable to resolve: "How can one be assured

8. Robert Boyle, *The Christian Virtuoso: Shewing That by Being Addicted to Experimental Philosophy, a Man Is Rather Assisted, than Indisposed, to Be a Good Christian* (London, 1690), 76, 79; Michael Hunter, *Boyle: Between God and Science* (New Haven, Conn., 2009). See also Marie Boas Hall, *Promoting Experimental Learning: Experiment and the Royal Society* (Cambridge, 1991).

of salvific status?" Observers in both the religious and philosophical contexts attempted to discipline the senses so that they would accurately convey and perceive information from the external world, managing the gap between the act of observing the object itself and the internal experience of that observation.[9]

Guided by conversion sermons and by watching exemplary models, converts learned to discipline the senses and frame their experience within appropriate testimonial templates in order to interpret and intuit signs of grace correctly. Ministers and members of the elect, or visible church, acted as curators of experiments of grace, imposing uniformity on the conversion experience through the regulatory frame borrowed from the empirical science increasingly standardized by the Royal Society and the broader suffusion of the Baconian method. No less than scientific experiment or empirical observation, the experience of grace had to be intelligible and recognizable to witnessing audiences. Yet the seventeenth-century clergy knew how dangerously close methods of any kind came to ritual. While spontaneity could not guard against hypocrisy, no authentic experience was conceivable without an element of it. Each testimony had to contain something unique and individual; these proofs of authenticity guarded against the perils of religious formalism. Ministers took the originality of experience itself as a snapshot of God's grace upon the soul. Efforts to observe and discern the authenticity of this encounter represent a particular use of empiricism to capture the kernel of Reformation optimism.

The Problem of Doubt and the Possibility of Redemption

The religious version of philosophical doubt is hypocrisy. Ministers from Thomas Shepard to Jonathan Edwards concerned themselves with hypocrisy, the potential for converts to falsely affirm their own election. Theological definitions of hypocrisy change from Shepard to Edwards, with the former locating the problem largely in the deceptiveness of external evidence and the latter in the problem of knowing the heart. In the *Sincere Convert,* for example, Shepard warns that he is "skild in the deceits of mens hearts" and can readily identify the "Formalist" who merely pretends to have found "soule-enriching truths." Whether a consequence of faulty

9. Richard J. Connell, *Empirical Intelligence: The Human Empirical Mode: Philosophy as Originating in Experience* (Lewiston, N.Y., 1988).

external evidence or a heart susceptible to delusion, hypocrisy represented the human attempt to grapple with how to seek knowledge presumed beyond ordinary perception without trespassing into the Fall's justifiable consequences: the loss of divine knowledge. Doubt thus established the limiting condition through which the divine gift of free grace issued. Martin Luther's *Freedom of a Christian* (1520) defined this gift as that which God gives—and it is from this possibility of salvific pardon that Reformation optimism emerges. According to Luther, man has a "twofold nature, a spiritual and a bodily one." Through "faith alone," the Christian soul becomes free while the body is a "dutiful servant of all." The meaning of freedom in Luther's essay is paradoxical. Freedom comes to those perpetually enslaved to their innately depraved selves. They are bound by a debt to God that can never be repaid. Grace contains the possibility of redemption, the potential to reclaim some portion of humanity's perceptive capacities lost in the Fall. Grace cannot be acquired through any human capacity; once given, it engenders a free and blessed state within the recipient. The inner self becomes pure, newly capable of discerning evil, and then the outer self becomes capable of performing benevolent actions.[10]

In *The Institutes,* Calvin reiterates and expands this concept of the paradoxical nature of Christian freedom. Book 2, chapter 2, perhaps comes closest to the Lutheran paradox of freedom. Like Luther, Calvin also insists that individuals lack agency to effect—and to accurately perceive—their own conversion. The fall of Adam destroyed humanity's capacity to perceive the invisible world, leaving all in servitude to the flesh with just enough perception to recognize and be humbled by their undeserving state. The bleakness of the human condition was for Calvin, as for Luther, ameliorated by individuals' sensory capacity to recognize spiritual assurance within themselves. At the same time that *The Institutes* reminds its reader of the limits of knowledge that result from the Fall, it also opens new epistemological frontiers. Following Augustine, Calvin locates these frontiers within the individual's soul, which acts as a window to both God and the natural world. Calvin explains that "full assurance" is possible, with "our truly feeling its sweetness and experiencing it in ourselves."

10. Tho[mas] Shepard, *The Sincere Convert: Discovering the Paucity of True Beleevers; and the Great Difficulty of Saving Conversion* (London, 1641), 9, 34; Martin Luther, *The Freedom of a Christian,* trans. from Latin by W. A. Lambert, rev. Harold J. Grimm, in Jaroslav Pelikan and Helmut T. Lehmann, eds., *Luther's Works* (Saint Louis and Philadelphia, 1955–1986), XXXI, 344–346, excerpted in *On Christian Liberty* (Philadelphia, 1957), 7, 10.

Spiritual self-knowledge promised a virtually unmediated relationship with God, a key to Protestant humanism in both its religious and secular instantiations.[11]

Calvin's emphasis on the knowledge accumulated in each individual soul links his theology to Christian humanism as developed through the writings of Erasmus. Humanists also promoted the splintering of authority though a critique of "externalism," the Erasmian version of Calvinist hypocrisy. Calvin's criticism of the clergy and of the ecclesiastical structures of power for thwarting individual access to new knowledge of God paralleled Erasmus's resistance to social and political structures of knowledge and power, which he also felt distracted individuals from self-realized knowledge. Distinctly Christocentric, Erasmian humanism believed that philosophy was a gift granted to man by God; access to it came through the power and self-knowledge of each individual soul. The Calvinist reformation and Christian humanism thus represent two sixteenth-century English reform movements that perform and espouse a cyclical pattern of history. For Erasmus as well as for Calvin, social and political institutions inevitably deteriorate because of the unreliable condition of human nature to sustain the governing structures that would perpetuate an organic consistency between inward knowledge of the soul and the self and the external realization of that knowledge.[12]

Hypocrisy and externalism describe an ever-widening gap between these two spheres of human knowledge and agency. As each condition becomes manifest and codified as a pervasively apparent social condition, humans become intolerant of the limited sphere of the social institutions governing their behavior and instigate a reformation. The cycle thus begins again. Erasmus and Calvin characterize a philosophical pattern within sixteenth-century England; their philosophies work through various forms of methodological borrowing. They stand at the end of the sixteenth century as Perkins and Bacon do at the start of the seventeenth century as representing a certain kind of reform, which locates the power for knowledge acquisition within the individual and provides a means of gaining access to

11. B. A. Gerrish, *Grace and Reason: A Study in the Theology of Luther* (Oxford, 1962), 15–30; Calvin, *Institutes,* I, 561; George Huntston Williams, *The Radical Reformation* (Philadelphia, 1962).

12. Here I am thinking of Erasmus's contention in his *Enchiridion* (1503) that religion consists primarily of inward love of God rather than outward signs of devotion. For an overview of his thought as it connects to the Reformation, see John C. Olin, ed., *Christian Humanism and the Reformation: Selected Writings of Erasmus* (New York, 1975).

and cultivating this knowledge so as to erode the boundaries imposed by the Fall.[13]

Book 2 of *The Institutes* begins with a discussion of the postlapsarian consequences on the inward soul. Calvin links ignorance of the self to ignorance of nature. If the Fall hid the regenerated individual from him- or herself, it also hid the divine in nature. "But although the Lord represents both himself and his everlasting Kingdom in the mirror of his works with very great clarity, such is our stupidity that we grow increasingly dull toward so manifest testimonies, and they flow away without profiting us." The mistake of all ancient philosophers, including Plato, was that their contributions to knowledge simply "vanishe[d]" in the "round globe." Because of their diminished spiritual capacities, they failed to see nature as part of this mirror configuration, containing "innumerable evidences" of "heaven." Calvin explains that God's manifestations in nature are unintelligible because "we have not the eyes to see this unless they be illumined by the inner revelation of God through faith." Nature and the inner self, faith and knowledge, are thus intimately linked for Calvin.[14]

Books 1 and 2 of *The Institutes* foreground two related directions of Reformed theology and natural philosophy that would, after the Reformation, develop around Luther's and Calvin's concept of the redemptive possibilities of divine grace. Although Calvin believed in the Fall's irreparable consequences, he nonetheless held out hope that the individual could know God. It was this humanism, this optimism, that inevitably led to early modern developments in natural philosophy as individuals from Bacon to Boyle, from Newton to Leibniz, articulated a theory of the Christian Virtuoso, the concept that a philosopher could repair his observational capacities in order to see evidence of God within nature. As the Baconian line of empiricism developed from Thomas Hobbes to Henry More to such influential Cambridge Platonists as Ralph Cudworth, a concern with final

13. For an assessment of these connections between Erasmus and Calvin, see Bouwsma's intellectual biography, *John Calvin,* 15, 54, 82. Bouwsma explains how Erasmus guided both Calvin and Theodore Beza's biblical and patristic studies. There is no record of Calvin's visiting Erasmus toward the end of Erasmus's life, but Bouwsma argues that Calvin reveals his indebtedness to Erasmus in the *Institutes'* discussion of the gospel, which uses Erasmus's term "Christian philosophy."

14. Calvin, *Institutes,* I, 53, 63, 64, 68. Calvin explains that Plato was "the most religious of all and the most circumspect" but that his perspective is still ultimately inadequate. Calvin's reading of Plato as the most Christian of all the ancient philosophers builds upon Augustine's *City of God,* specifically Augustine's comparison of the *Timaeus* to Genesis. This preference for Platonic philosophy persists through the various phases of theological and natural philosophical overlap described in this study.

causes and metaphysical dimensions of the physical world aligned the end goal of the new science with the pursuit of knowledge of God. Social and political reform movements furthered the reformation of knowledge, as the fragmentation of long-standing structures of authority opened up new possibilities for new knowledge.[15]

This intricate relationship between the advancement of knowledge and political revolution reached new heights in the context of the English Civil War, when the theologians and philosophers who made up the circle of Cambridge Platonists preached sermons to the House of Commons. One of the first of these sermons was Peter Sterry's *Spirits Conviction of Sinne Opened* (1645). This sermon conflates the possibility of new government with the possibility of new knowledge, condensing political and epistemological potential in a millennial moment of great promise. The sermon projects the coming of Christ as bringing flesh and visibility "to all alike" as well as "to the eye of sense." Sterry anticipates a time when the fallen status of the human sense will be redeemed, when the limitations of reason will be superseded by a divine light spreading out into nature. This divine light and its redemptive effect upon the fallen human intellect is the reward promised in exchange for the blood and violence of an English revolution bearing witness to "thousands of slain men" in Scotland, "*England* with her ten thousands," and "*Ireland* with her millions." Scripturally projected as an eschatological phase en route to Christ's Second Coming, this blood foreshadows the imminent actions of a Parliament that would behead the king. The redemptive promise for this violent struggle and strife is both an improved government and a state of philosophical epiphany when the "Lower principles of Truth" will come to "comprehend the Higher."[16]

From the gradual erosion of structures of knowing, instigated by Erasmus and Calvin, to the radical reconfiguration of political and governmental structures represented in Sterry's address to the House of Commons, the epistemological transformations in the century following the Reformation promoted a certain kind of optimism for new knowledge at the same time that they also endorsed a profound splintering of traditional authority that led to uncertainty or doubt. The redemption of human perceptive capacities through the gift of God's grace was not intended to eradicate the memory of the Fall. Adam's powers could be redeemed only within an imperfect state: they were forever inhibited by the fallen world that lay

15. Armstrong, *Metaphysics and British Empiricism.*

16. Peter Sterry, *The Spirits Conviction of Sinne Opened* (London, 1645), 5, 13, 15.

outside the regenerate individual. Nature, language, and human intellect were flawed and imperfect resources that could not be fully attuned either to the status of one's inner soul or to the innumerable evidences of divinity in nature. Evidence of both could be hoped for and occasionally seen but never accepted as absolute or complete knowledge of God.

Streams of Nature, Streams of Grace

Francis Bacon's contemporary, William Perkins, attempted to reconcile the tension between doubt and the possibility of redemption through an increased attention to individual experience. His theological tracts and sermons provide a blueprint for how evidence of God might be intuited, conveyed, analyzed, and recorded as it became manifest on the soul. In *A Treatise Tending unto a Declaration Whether a Man Be in the Estate of Damnation or in the Estate of Grace* (1590), one of the most influential works for seventeenth-century Puritanism, Perkins begins placing responsibility for knowing the status of one's soul on the individual. In a theological intervention that would have broader implications for the spiritual authority accorded converts at their conversions by Baxter and other followers of Perkins's theology, Perkins insists that the elect have privileged knowledge to know the state of their own souls. In the preface "To the Christian Reader," however, he warns that such inward study is no mean task: "It is an hard thing for a man to search out his owne hart," so one can make only modest attempts to "discerne." Adding to the difficulty, as Perkins tacitly acknowledges, there was no clear procedure for inward reflection and salvific discernment. Perkins tells the readers that his pamphlet can assist them only in "examining and observing of thine owne heart." As each soul is individualistic, exhibiting grace in variations of a universal pattern, he warns that each person must account and allow for uniqueness within that larger conversion pattern.[17]

So how did one identify evidence of God's grace without clear rules about what counts as evidence? Perkins sets the parameters of the question, not by attempting to explain what grace was, but rather by explaining what it was not. According to Perkins's generation of Puritans, fallen humans had a tendency to misidentify the manifestations of grace, committing hypocrisy, the sin that endangered both the individual and the social, since

17. Williams Perkins, *A Treatise Tending unto a Declaration Whether a Man Be in the Estate of Damnation or in the Estate of Grace*... (London, 1590), 26.

members of the visible church—the privileged few at the core of the larger congregation—could unwittingly misinterpret divine communication and thus mislead other congregants. False grace, Perkins observes, is "when the word rooteth . . . but not with the residue of the affections." That is, false converts have knowledge of scripture but not the feeling of grace—and thus no capacity for inspired exegesis. In contrast, he identifies grace as a "deepe and a lively rooting of the word, when the word is received into the minde and into the heart." Authenticity was marked by a pure and unique feeling; hypocrisy, by an overweening faith in the fallen capacity of the human intellect. To establish a test for distinguishing between authentic and false grace, Perkins elevates experience as a way of knowing conversion. The experience of grace is distinctly different from belief or knowledge of faith, albeit difficult to detect. He describes the potential pitfalls of his epistemological schema: "The graces of God may be buried in [a Christian] and covered for a time, so that he may bee like a man in a traunse, who both by his own sense and by the judgement of the phisition is taken for dead." There is no sure way to bring these graces to light; the detection of grace depends on the convert's and minister-physician's vigilance and patience. A nebulous phenomenon—fleeting and invisible—the experience of grace could be explained, explored, and codified only through figural language and by examples.[18]

Perkins uses both in his tract. To explain to the reader how grace is seeded in the soul, Perkins draws on Christ's parable of the sower (Matt. 13:1–23, Mark 4:1–20, and Luke 8:1–15) to explain how corn, like grace, takes root in the earth:

> Corne as soone as it is cast into this ground, it sprouteth out verie speedilie, but yet the stones will not suffer the corne to be rooted deeply beneath, and therefore when Summer commeth the blade of the corne withereth with roots and all. So it is with these professours: they have in their hearts some good motions by the holy Ghost, to that which is good: they have a kinde of zeale to Gods worde, they have a liking to good things, and they are as forward as any other for a time, and they doe beleeve. But these good motions and graces are not lasting, but like the flame and flashing of strawe and stubble: neither are they sufficient to salvation.[19]

18. Ibid., 23, 24, 103–104.
19. Ibid., "To the Christian Reader."

This opening figure illustrates how grace has temporal as well as phenomenal qualities. It comes like the summer; it cannot be expected to stay. As soon as Christians feel the presence of grace on their souls, they should expect to feel it "wither," inevitably disappearing completely. Even during the height of this fleeting summer of grace, grace does not occupy a continued and palpable presence on the soul. Grace moves continuously, flickering as if exhibiting the constant threat of being extinguished. The would-be convert knows "good," or authentic, grace through its flamelike motions across the soul. The way that one knows grace is also how one knows its insufficiency to rest as a stable sign of "salvation." With certainty comes incertitude. To know is also to doubt. Perkins's figure illustrates what by description alone would be ineffable—the nuanced, affective patterns of grace. It also encodes the paradox of Reformation optimism within this scheme for discernment and doubt. As soon as the convert feels what appears to be the motion of grace on the soul, the flame dies out, leaving a lingering doubt about its legitimacy as a sign of grace at all.

Perkins pairs his figures of grace with examples of how grace feels. In the "Dialogue of the State of a Christian Man," midway through his tract Eusebius, the perfect Christian, and Timotheus, the wise Christian, converse about grace. Eusebius describes the feeling of grace as the "heart began to waxe soft and melt." Like a freshly plowed field, such melting renders the heart particularly susceptible to receive the seed of grace through the preaching of the gospel. God enters the heart and opens Eusebius's "inward eyes" such that he may have a "livelie faith" "wrought" within himself. Perkins's reader is meant to recognize this example as a sign of true grace based on the evidentiary standards he has already laid out in the tract. Importantly, Eusebius does not express an intellectual understanding of faith. Rather, he bears witness to the visceral effects, the feeling, of divine grace flickering in his soul. That grace is "livelie" suggests that it lives and has agency; it does not originate in the heart of the saint but falls on and gestates in the heart as a seed sown by the spirit. It is not static, but moving. It arrives like a fallen spark, is coaxed into flame by the preaching of the word, and dances for a time across the soul.[20]

In response to Eusebius's lesson, Timotheus confesses a desire for his own "assurance" of "Gods speciall goodnes." But Timotheus has a lesson of his own for Eusebius. Through the wisdom denominated by his title (the wise Christian), he conveys to Eusebius the knowledge that divine

20. Ibid., 125–126.

goodness will reform his "judgment" and "hath lightened [his] eies to see and [his] heart to imbrace his sincere truth." The dialogue thus not only explains how grace is supposed to feel on the soul but also illustrates the transformation of humanity's fallen vision through the effects of grace. The dialogue replicates the relationship between orality and hearing the narrative that would characterize the Puritan testimonial form in the next century. Timotheus increases his wisdom through the aural evidence Eusebius offers. Timotheus in turn translates this aural evidence into an enhanced ocular capacity to perceive "sincere truth." Together perfection and wisdom—Eusebius and Timotheus—offer a powerful model for the evangelical impulse that would emerge out of this conversion process. The move from the aural to the visual also foregrounds the sensory relationship structuring the Puritan testimony, where aural evidence would confirm the status of the visible saint. Just as Timotheus imagines his "eies" seeing his "heart" and perceiving "sincere truth," testifiers would cull visible evidence from their souls to correspond to the invisible world.[21]

The taxonomy of lively faith that evolves through Eusebius and Timotheus's dialogue partakes in the Puritan science of the self that comes from the Perkinsian theology of discerning grace. The science of the self involved an inwardly directed soul searching for sensory manifestations of divinity. Perkins, like his New England followers, advocated this inward gaze in order to strengthen individuals' perceptive capacity so they would improve their ability to read the soul's sensory data. Theologians believed that this practice would respond to the uncertainty of election by reducing the mystery and secrecy clouding perception of the heart. They advocated careful use of reason by minds that had been "reborn by grace" to intuit and map the "soul's terrain." Perkins's insistence that the detection of grace was the individual's fundamental responsibility built on the Reformation ideal of sola scriptura. Only the individual could know the status of his or her own heart, because grace moved differently across different human souls and because the heart of one could never truly be known by another. But, even while proposing a science of the self, Perkins recognized its limitations. Not only was the intellect imperfect, but the senses were also fallible.[22]

21. Ibid., 139. Walter Ong has taught us that this move from the aural to the visual characterizes the larger transformation of Christianity from the early Middle Ages into the print focus of modernity.

22. Theodore Dwight Bozeman, *The Precisianist Strain: Disciplinary Religion and Antinomian Backlash in Puritanism to 1638* (Chapel Hill, N.C., 2009), 163.

While the seventeenth century's science of the self worked to resolve the Calvinist dilemma of desired knowledge and fallen human intellect, the senses still posed a problem that Perkins addressed in another dialogue between a minister and a Christian, "Consolations for the Troubled Consciences of Repentant Sinners," that concludes his tract on conversion. In the dialogue, the minister explains the dual promise of knowing and not knowing encoded in the Protestant creed of sola scriptura: "Faith standeth not in the feeling of Gods mercie, but in the apprehending of it . . . for faith is of invisible things." In other words, faith depends on the Christian's ability to apprehend the sensory manifestations of inward affections as the "invisible," phenomenal presence of God rendered visible through the eye of faith. Agreeing, the Christian affirms: "Everie true beleever feeles the assurance of faith: otherwise Paul would not have said, 'Prove your selves whether you are in the faith or not.'"[23]

The response to the minister's lesson seems almost intuitive; faith is a nebulous phenomenon that can be perceived only at the level of the individual. Apprehending this resolves the mystery surrounding Calvin's description of election. Yet the minister cautions the Christian, and thus reader, about oversimplifying. Occasionally, he observes, believers feel the assurance of faith; at other times, they do not. More troubling, the spirit's communication to the heart can be so subtle, so brief, that it might indefinitely escape individuals' perception, no matter how intently they search their souls. Inevitably, the minister explains, doubt is an outcome of the inward examination. Even when individuals feel certain of their redemptive status, they have to question whether assurance is the product of deception. The senses, after all, are hardly infallible in the material world. How much likelier are they to be deceived while probing the dark mysteries of the spiritual, spurred on by desire, troubled by doubt, ebbing faith, and anxiety about damnation? The limits of sensory knowledge seemed, at last, inescapable.

While theologians struggled to find a way out of a diminished epistemology, without circumventing the Fall's just penalty, natural philosophers were taking another tack with the limits of faculty psychology. Published two years after he wrote the *Confession of Faith,* Bacon's *Advancement of Learning* (1605) outlined a program for transforming sensory data into tangible knowledge—though toward a very different end. Like his college-trained contemporaries, Bacon realized that the senses were the only av-

23. Perkins, *A Treatise Tending unto a Declaration,* 271.

enue of knowledge. Philosophers needed the senses to gain access to the truths to be discovered within the natural world. But also, as the New England Puritans did just a few decades later, Bacon ruminated on the fundamental instability of the senses to secure accurate information. Because for Bacon nature coined material truths, he assigned the natural world a secure ontological status within his hierarchy of knowledge. Nature was as fixed for Bacon as God was for Perkins. Those seeking knowledge thus looked to the stability of the external world rather than relied on cognitive structures. The Baconian mind yields to natural truths rather than the reverse. Nature becomes for the philosopher what God had long been for the divine: an ultimate but rarely accessible source of truth.

In the *Advancement of Learning* as well as elsewhere in Bacon's works, the senses do not gather proof of the invisible world, but rather provide evidence of visible matter as it exists in nature. By carefully distinguishing the material world from the spiritual and permitting the senses to address only the material, Bacon purified the senses, making them valid tools for empirical inquiry. This division between material and spiritual explains why this text is often read—not incorrectly from this perspective—as paving the way for modern, secular science. Yet this assessment of Bacon's legacy is not entirely accurate. His caution comes from the heretical potential of his methodology rather than a denial of God's presence in the natural world. If the senses can convey previously unknown data from visible matter, what is to stop them from seeking invisible phenomena? Bacon seems well aware that he stands on a slippery slope and thus mitigates the possibility of excess by positing three limitations of empirical and natural knowledge. First, "We do not so place our felicity in knowledge, as we forget our mortality." Second, "We make application of our knowledge to give ourselves repose and contentment, and not distaste or repining." Last, "We do not presume by the contemplation of nature to attain to the mysteries of God."[24]

24. Francis Bacon, *The Advancement of Learning* (1605), in Bacon, *A Critical Edition of the Major Works*, ed. Brian Vickers (New York, 1996), 124. Readings of the secularizing impact of the *Advancement of Learning* are a long-standing commonplace in Bacon scholarship. For a summary of the history of the debate, see John Channing Briggs, "Bacon's Science and Religion," in Markku Peltonen, ed., *The Cambridge Companion to Bacon* (Cambridge, 1996), 172–199. Julie Robin Solomon contends that Calvin's and Bacon's epistemologies are fundamentally opposed; the latter's therefore mark a turn to objectivity (*Objectivity in the Making: Francis Bacon and the Politics of Inquiry* [Baltimore, 1998]). Denise Albanese reads Bacon as prompting the replacement of Christianity with scientific inquiry as the primary imperialist mode (*New Science, New World* [Durham, N.C., 1996]). Other Bacon scholars have contested some of these claims toward secularization. For example, see Stephen A. McKnight, *The Religious Foundations of Francis Bacon's Thought* (Columbia, Mo., 2006).

As the *Confession of Faith* asserts, however, Bacon is negotiating between a theological tradition and the philosophical traditions he innovates in the *Advancement of Learning*. He constructs an antithetical yet analogical relationship between philosophy and divinity. This separation of knowledge permits him to enumerate the problems and propose some solutions for the unstable relationship between perception and knowledge acquisition.

> The knowledge of man is as the waters, some descending from above, and some springing from beneath; the one informed by the light of nature, the other inspired by divine revelation. The light of nature consisteth in the notions of the mind and the reports of the senses; for as for knowledge which man receiveth by teaching, it is cumulative and not original; as in a water that besides his own spring-head is fed with other springs and streams. So then according to these two differing illuminations or originals, knowledge is first of all divided into Divinity and Philosophy.[25]

A basic division structures knowledge. Some stems from a revealed source; some stems from nature. Divinity and philosophy are separate because the objects of knowledge in each case are different. But Bacon's repeated insistence on this division also indicates the danger of collapse or, in his idiom, the possibility of a confluence between the waters "descending from above" and those "springing from beneath." His insistence that philosophers "do not presume by the contemplation of nature to attain the mysteries of God" is far from a rejection of a religious worldview. Rather, Bacon is aware of the heretical potential of presuming that the pursuit of knowledge can lead to divine discovery. By reducing the object of inquiry to matter, Bacon implicitly acknowledges the almost infinite expandability of his methodology. The experience of nature could, in fact, be pressed too far in an effort to find divine evidence. Such an acknowledgment seems calculated to assuage the concerns of theologians—to whom Bacon addresses most of *The Advancement of Learning*—and to circumscribe the study of nature to matter. His division of the branches of knowledge, here divided according to religious and philosophical conventions, also functions to curtail his empirical program's potential for expansion. Thus, far from exhibiting a secular propensity, Bacon would not need to insist quite so adamantly on this division of knowledge without an unquestioned belief in the presence of God in nature.

25. Bacon, *Advancement of Learning,* in *The Major Works,* ed. Vickers, 189.

The study of the soul and the study of the world emerged in the early seventeenth century as parallel empirical techniques. Philosophers and Protestant divines participated in a project of sensory purification in order to isolate and identify sensory evidence upon these two objects of investigation. The Baconian desire to evacuate natural philosophy of preconceived taxonomies mirrored the Reformation desire to return to the pureness of the connection between the Christian and the Holy Spirit unencumbered by clerical mediation and dramaturgical ritual. The ideal promoted through the Reformation notion of sola scriptura and *sola fides* (faith alone) was to isolate a spiritual self in the hope that diminished religious formalism would foster a more authentic spiritual encounter. Faith, according to Calvin, is a "steady and certain knowledge of the Divine benevolence towards us." This benevolence "is both revealed to our minds and sealed upon our hearts through the Holy Spirit."[26]

Quite literally knowledge, faith can be revealed and confirmed only by the Holy Spirit. Humans experience this revelation and confirmation as sensory data. Proof of God's existence and presence in the saint's life comes through the experience of divine sweetness as sensory data. Likewise, Bacon's antidote to received wisdom in Scholastic philosophy was to insist on fresh encounters with the natural world, divorced from conventional learning and knowledge forms. The senses would absorb nature, gleaning a range of material data that could facilitate the reconstruction of "the image of the universal world." Noting a distinction between appearances and truth—that things are not always what they seem—Bacon advocates relying on the report of the senses in an attempt to achieve epistemological stability. By supplanting received knowledge with experience, Bacon's followers hoped to clarify Scholasticism. Yet elevating the senses created new doubts for Bacon and Perkins. Senses could be misled or misread. Philosophy and divinity thus inherited the Reformation "crisis of knowing" and located it more specifically in the problem of the senses.[27]

In answer to the evidentiary crisis, theologians and natural scientists eventually became students of each other's methodological variations. The distinction that Bacon initially proposed between two streams of knowledge was impossible to maintain. The streams of knowledge would con-

26. Calvin, *Institutes,* I, 551.

27. Bacon, *Advancement of Learning,* in *The Major Works,* ed. Vickers, 123. I am borrowing "crisis of knowing" from William J. Bouwsma, "Calvin and the Renaissance Crisis of Knowing," *Calvin Theological Journal,* XVII (1982), 190–211.

verge most clearly by the 1640s, when the Invisible College was beginning to lay the foundation for the Royal Society and Reformed religion was spreading throughout the Atlantic world, most significantly through the Great Migration to New England. Bound by the paradox of the Calvinist view, post-Perkinsian ministers would look for ways of gaining access to the self-knowledge acquired through introspection. Natural philosophers who, of course, also believed in God found it difficult to resist knowledge forbidden by theology, particularly when the new science might serve as a means of charting evidence of the invisible world. On each side of Bacon's purported divinity and philosophy divide, early moderns were tempted to seek access to the knowledge of the former with the methods of the latter, both in theory and in practice. By the 1640s a program for doing so was under way.

Waters of Knowledge Converge

By the mid-seventeenth century, once-parallel tracks taken up by Bacon and Perkins had merged. Theology and natural philosophy became two sides of the coin of early modern knowledge formation. For both, grace was the Holy Grail for discovering an epistemic bridge that would link the visible and invisible worlds. A particular convergence between Perkins's and Bacon's methodologies, Puritanism and empiricism, had become the principal tool in the search for grace. Less inhibited than Bacon had been by the need to defend the newfound scientific method against the dangers of heresy, natural philosophers following Bacon in the latter half of the seventeenth century turned their new methods on what for them was literally the alpha and omega of knowledge: the domain of God. Simultaneously, theologians turned to what Calvin had described as the renewed observational capacity of the elect in an effort to gain the convert's toehold in the spiritual world. Thus, the convergence of Bacon's double stream came twinned with externally visible and empirically mappable laws of grace.

The seventeenth-century integration of natural philosophy and metaphysical inquiry still raised the vexing problem of how to make something as fleeting and subjective as the experience of grace a unified object of scientific study. Despite Perkins's compelling attempts to shore up techniques of discernment, any synthesis of the religious experiences empiricists collected depended entirely on the validity of each account. Since testifiers were often and easily deceived by their own misreading of divine signs, post-Perkinsian theologians first needed to come up with a way of testing the authenticity of the individual's conversion experience. The earliest of these tests were, like Thomas Shepard's examination of converts in his Cambridge, Massachusetts, congregation, implemented locally and relied almost entirely on the individual's senses for evidence that the experience was authentic. Yet, given the senses' often ambiguous or incorrect account of the external world—all the more problematic when what they were expected to perceive was grace, an invisible, immutable force absent of any material markers—theologians gradually turned their focus from the senses to the soul, what in the Augustinian tradition had long been held as a storehouse for spiritual experience and divine memories.

The soul offered a unique solution to the problem of grace. The soul, like grace, was immaterial, belonging to the spiritual; however, it had also long been theorized in both religion and empirical inquiry as belonging in part to the material world. Unlike grace, the soul, then, was in the world, even if not of it. Luther explains, "So the Word imparts its qualities to the soul." Calvin describes the nimble soul as a repository of memories of heaven and earth and of past and future; the "signs of immortality" are "implanted" upon the soul. For Perkins, the soul was a temporal repository for the hidden secrets of grace. If Bacon meant to draw a clear distinction between the soul and nature, he muddled the distinction by insisting that, while God created nature and left it alone, he continues to create the soul. "The soule of man," Bacon equivocates, "was not produced of heaven, or earth, but was breathed immediately from God, so that the wayes, and proceedings of God with Spirits are not included in nature that is in the lawes of heaven or earth, but are reserved to the lawe of his secret will, and grace whereby God worketh still."[28]

Providing a kind of synthesis of Luther's, Calvin's, and Perkins's overlapping definitions, Bacon suggested that, while the soul was neither fully spiritual nor material, it most certainly was the place to look to decode the mystery of God's creation. The soul opened new domains of empirical promise, while grace, the manifestations of God upon the soul, remained at the heart of the Calvinist problem of incertitude. If indeed the soul was an entity breathed immediately from God—an act configured in the continuous present tense—then theologians and natural philosophers had something relating to humans to conceptualize apart from fallen nature. Grace

28. Luther, *Freedom of a Christian,* trans. Lambert, rev. Grimm, in Pelikan and Lehmann, eds., *Luther's Works,* XXXI, 349, excerpted in *On Christian Liberty,* 15; Calvin, *Institutes,* I, 57; Francis Bacon, *The Confession of Faith* ([London], 1641), 4.

thus came to be conceived of as having a traceable, mappable nature that, were it to be identified upon human souls, could be treated as empirical evidence.

In 1603, Francis Bacon wrote *The Confession of Faith*, a short pamphlet unpublished until 1641, fifteen years after his death. The narrative structure of this twelve-page pamphlet echoes the relationship between books 1 and 2 of *The Institutes*. As with Calvin's treatise, Bacon begins with a lament for prelapsarian nature. He describes how nature encodes the laws of God and how the Fall's irreversible consequence was the "deface[ment]" of these laws: "Heaven and Earth which were made for mans use, were subdued to corruption by his fall." The correspondence between heaven and earth at Creation ceased with original sin. Like Calvin, Bacon does not resign natural theology to such an account. Rather, he reveals God's "promise," which he describes as a continuing potential for the "righteousnesse of God" to be "wrought by faith," despite "the fall of man."[29]

In his lament for an intelligible nature and his recognition that the potential for natural intelligibility comes from divine grace, Bacon signals his debt to Calvin. The recapitulation of that debt in 1603 when the text was written marks the appearance of Reformation optimism in natural philosophy. Following his description of fallen nature, Bacon's *Confession of Faith* explains "free Grace" as God's "Seed incorruptible" that "quickneth the spirit of Man, and conceiveth him a new the sonne of God." The "Imputation" of this seed into the soul redeems the individual. Bacon delineates two consequences of this redemption in his *Confession of Faith*. The redeemed Christian is counted a member of the elect, the "visible Church," which is "distinguished by the outward workes of Gods Covenant," and their repaired faculties made them superior natural philosophers.[30]

Augustine first conceptualized the visible church as Platonic paradigm. Individuals might deem themselves elect and comprise the visible church, but, because no member could know his or her salvific status for certain, those actually to be among the membership at the Judgment comprised the invisible church. For Anglo-American Calvinists Augustine's paradigm comported well with chapter 9 of the Book of Acts, the scriptural authority for predestination and the congregational concept of an elect body, or the New England Way. Membership within the visible church located the

29. Bacon, *Confession of Faith*, 5, [6], 8–9.

30. For this reading of Bacon's Calvinism, see Vickers, Introduction, in Bacon, *The Major Works;* Bacon, *Confession of Faith*, 9, 11.

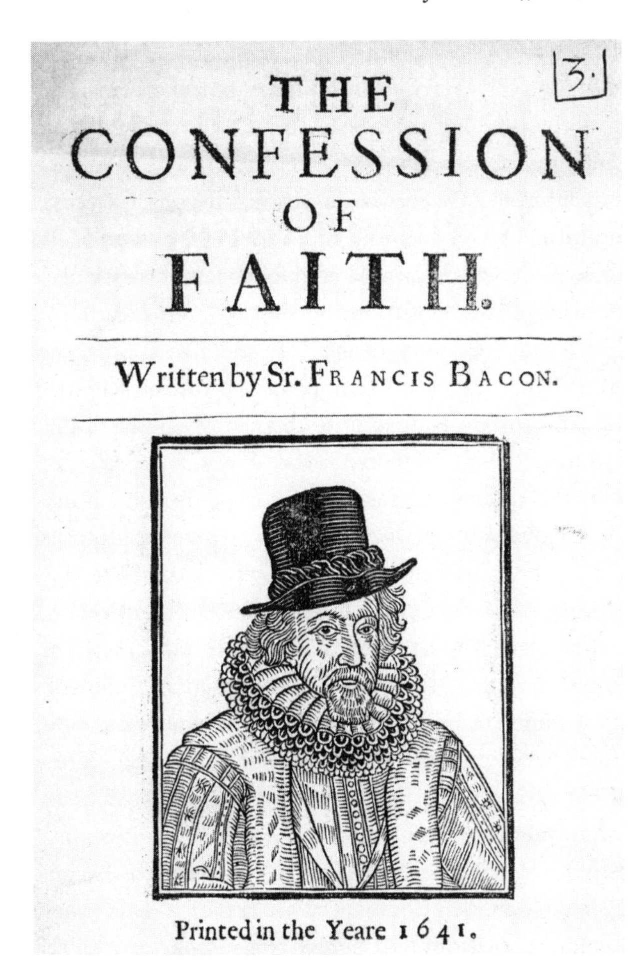

3.

THE CONFESSION OF FAITH.

Written by Sr. FRANCIS BACON.

Printed in the Yeare 1641.

FIGURE 1 • *Francis Bacon,* The Confession of Faith. *Huntington Library, San Marino, Calif.*

redeemed within an unfolding eschatology. Bacon referred to this phase as "the time of the Revelation of the Saints of God; which time is the last, and is everlasting without change." While locating the elect body of saints within Augustine's millennial schema, Bacon would declare that "imputation" theology (Christ's spirit imputed to the sinner) brought the redeemed closer to the prelapsarian correspondence between the divine and the natural world.[31]

Bacon's early pamphlet augurs the direction seventeenth-century natural theology would take. In so doing, it anticipates the modern emergence of what Robert Boyle's and Cotton Mather's writings would, respectively,

31. Bacon, *Confession of Faith,* 9, 12.

term the *Christian Virtuoso* and *Christian Philosopher*. Bacon's treatise transformed a number of theological doctrines into philosophical axioms: revelation precedes reason, the revealed truths of natural philosophy depend on human faculties heightened through salvific regeneration, and natural science and religious metaphysics are convergent truths. Bacon's *Confession of Faith* revealed a commitment to redeeming and studying the laws of God within nature even as his more widely circulated published tracts articulate a division between knowledge and revelation.

Read alongside each other, *The Confession of Faith* and *The Advancement of Learning* speak to the greater purpose Bacon proposed for the scientific method: restoring the philosophical capacity to discern divine law intelligently in an effort to prove the existence of God. The *Confession*'s 1641 publication emerged at the intersection of three complementary philosophical and theological movements: natural theology, Comenian linguistics, and antinomianism. A branch of natural philosophy, natural theology sought to decode evidence of God's existence in nature. In the meetings that ultimately led to the formation of the Royal Society in 1660, the Cambridge Platonists engaged in natural theology while Comenian linguistics also gained popularity. Comenian linguistics imagined a renewed state of semiotic perfection that shared philosophical continuity with the natural theological conception of nature. Additionally, an integral but as yet unexamined link between natural theology, Comenian linguistics, and the antinomian controversies had a significant impact on the reception of Bacon's text. Natural theology and Comenian linguistics took center stage in a new alliance in the antinomian accusations and heresy trials under way by the mid-1630s. No such controversy garnered more attention on both sides of the Atlantic than Anne Hutchinson's trial and expulsion from the Massachusetts Bay Colony. Antinomian controversies focused on the theological legitimacy of developing discernible and recognizable laws of grace. Bacon's *Confession of Faith* appears amid this confluence of philosophical and theological ideas as a testament to how a sequence of epistemological movements might be applied to a renewed study of the human soul.

Bacon begins his account by defining dispensation as the selective revelation of God's "secret will, and grace" through the "Lawes of nature." In so doing, Bacon closely follows Calvin's theology of the Holy Spirit as an "inner teacher by whose effort the promise of salvation penetrates into our minds." For Bacon, like Calvin, religious conversion was a form of divine illumination whereby the passive reception of grace endowed humans with new eyes to see "how secret and lofty the heavenly wisdom is." *The Confes-*

sion of Faith describes the numinous experience of conversion as occurring in the phase known as imputation. In Calvinist-influenced morphologies of conversion, such as the one Perkins developed, imputation is the phase in which the "Image of Christ" enters the "spirit of Man" through an "open passage" in "Man's flesh" to make "a new sonne of God." So ordered, this conversion sequence allowed the individual a brief window into the working of divine grace, permitting the convert to witness and later recount (or "witness") the meaning of Paul's famous encounter with the spiritual realm: "For now we see through a glass darkly, but then face to face. Now I know in part, but then shall I know even as I am known" (1 Cor. 13:12). But Bacon departs from Paul, Calvin, and Perkins in the assertion that the "new eyes" acquired through conversion provide a new form of sight even while serving as a reminder of the limits of human perception. That is, for Bacon, the semiopaque window is transparent enough to offer a greater glimpse into the mechanics of the divine's operation on the soul.[32]

Bacon's metaphysics views God as mysterious, granting only limited spiritual understanding even to the elect. His science of the divine proposes that imputation transforms the "flesh" enough to give the "new sonne of God" privileged access to the "worke of the Spirit" in nature. For Bacon, the window into this divine operation had to function in accordance with the prevalent conception of faculty psychology. For his model of empiricism to be applicable, the temporal and spiritual window into the working of grace on the soul had to account for the time it took for senses to operate—no mere flash or glance would do. Likewise, if the medium between the material and the spiritual were, as many divines interpreted Paul, too dim, it would too firmly restrict an epistemology already starkly circumscribed by the consequences of the Fall. Even though the "world" is "not yet fully revealed," the elect, in Bacon's formulation, acquire the capacity to observe the mysterious workings of grace not only through the dispensation of evidence in their own souls but also through the divine laws operating within the natural world. Here the word "revealed" takes on a double meaning: Bacon supplements the doctrinal limits of the convert's assurance of grace with his or her willed observational capacities. Maintaining a Calvinist model of conversion, *The Confession of Faith* introduces new empirical possibilities for religious knowledge through the methods of natural science. Through these different forms of observation and ab-

32. Calvin, *Institutes,* I, 541, 581; Bacon, *Confession of Faith,* 4, 8–9.

straction, God's elect can begin to discern the subtle forms of divine law in the lived world.[33]

Bacon's 1603 revision of the ocular limits implied in Paul's theory of sight is instructive for understanding mid-seventeenth-century Puritans' negotiating the problem of reading evidence of grace as signs of assurance. Bacon's *Confession of Faith* concerns an effort to establish a continuum between his observational capacity for discerning nature and his capacity for discerning his own soul. New England ministers extrapolated from the relationship between their own inward piety and their ability to discern natural phenomena, to their observation and assessment of conversion testimonies. Whereas Bacon's *Confession of Faith* locates the capacity for discerning grace among the elect in order to apply the "new eyes" acquired through conversion to the study of "things unseen" in nature, New England Puritans take this epistemology of spiritual knowledge one step further, proposing that the elect acquired the capacity for discerning divine law not only within the natural order but also as it affected individuals. The observational capacities of the elect were then applied in a scene of visual witnessing that tracked the aural evidence of religious experience recounted in testimonial form. Ministers differed from Bacon in shifting their focus from God's presence in nature to God's presence upon the souls of others. Just as God becomes newly intelligible in Bacon's reading of a redeemed natural world, God became intelligible for elect Puritan ministers observing the aural and affective evidence as it flickered across the souls of testifying saints.

Bacon's *Confession of Faith,* like the lay testimonies of Puritan converts, functions as a kind of philosophical microcosm through which to view the larger, continuing transformations of the period in which it was published. Walter Charleton, an English Protestant and author of millennial treatises written in the context of the English Revolution, argued in *The Darknes of Atheism Dispelled by the Light of Nature* (1652) that proof of the existence of God and the soul can be attained through "natural theology." While both God and the soul are invisible entities, wholly distinct from corporeal matter, God "imprint[s] . . . upon my *Soul* . . . an indelible *Mark* or *Signature*" of his existence. Philosophers, Charleton observes, are more equipped than

33. Bacon, *Confession of Faith,* 3, 4, 8, 10. Bacon expands upon this method in *Advancement of Learning:* "If we will excite and awake our observation, we shall see in familiar instances what a predominant faculty the subtlity of the spirit hath over variety of matter or form." Bacon, *The Major Works,* ed. Vickers, 210.

divines to perceive this indelible mark. Charleton's notion of a legible sign of grace would have been compatible with New England Puritans' science of visible signs within the testimonial form. After all, Charleton had insisted that God can be established "by conversion . . . since *Faith* is the Gift of God, he that gives Grace sufficient for the stable apprehension of other things contained therein, can also give Grace sufficient to the admission of his Existence." In other words, the evidence that the elect find on their own souls can serve, as we will see, as a set of symbols that natural philosophers traced back to the "fountain" or "Archtype" of the conversion experience.[34]

In *Dawnings of Light* (1644), an even earlier natural theological tract, John Saltmarsh defined his empirical project as a striving to move beyond *"lower* and *more natural interests"* in order to begin to see the divine, "that which is more *hidden* and *secret*." The object of Saltmarsh's spiritual hermeneutic was to dispel such mysteries and unfold the "secrets" of the invisible within the visible domain. Considered a condition of the Fall, secrecy was one of the key categories through which seventeenth-century philosophers viewed nature and divine valences. Such movements as natural theology, the Rosicrucian brotherhood, and sustained projects to develop an artificial language system sought in their disparate ways to restore the semiotic correspondence between nature and human communication, to once again lift—even to a small degree—the veil between the material and spiritual realms. The secrecy encasing the natural world in illegible signs would gradually dissolve through a telos of increasingly intelligible systems of observation, representation, and communication.[35]

Natural philosophical efforts to dispel secrecy through a new way of reading nature corresponded to Puritan efforts to render spiritual secrets visible through the experience of the regeneration. Regeneration allowed a "new man" to emerge from the husk of the old, as Paul laid the transfor-

34. Walter Charleton, *The Darknes of Atheism Dispelled By the Light of Nature* (London, 1652), a3, 14, 20.

35. John Saltmarsh, *Dawnings of Light* . . . (London, 1644), 17–18; William R. Newman and Anthony Grafton, ed., *Astrology and Alchemy in Early Modern Europe* (Cambridge, Mass., 2001); Rhodri Lewis, *Language, Mind, and Nature: Artificial Languages in England from Bacon to Locke* (Cambridge, 2007).

The Rosicrucian movement began in Germany in 1614 through a secret society of alchemists who advocated the reformation of human knowledge in terms similar to Francis Bacon's. Rosicrucian tracts drew from alchemy and mathematics as well as spiritual and millenarian reform. Walter W. Woodward, *Prospero's America: John Winthrop, Jr., Alchemy, and the Creation of New England Culture, 1606–1676* (Chapel Hill, N.C., 2010), 28–29.

mation out in chapter 6 of Romans. The Puritan visible saint was a self-proclaimed and communally affirmed member of the elect. Saints existed as earthly manifestations of the invisible body of the elect that would be revealed only at Judgment. The relationship between visibility and secrecy in the Puritan context thus has two important implications. On the one hand, secrecy could impair the religious community, suggesting the convert might be withholding a crucial component of his or her regeneration that would either substantiate or deny salvific status. On the other hand, secrecy referred to evidence of grace lodged in the hearts of the regenerate, awaiting introspection and public clarification. Perkins spoke of this form of secrecy when he explained that "the graces of God may be buried in [a Christian] and covered for a time." For both Puritans and natural theologians, secrecy marked the veiled presence of the divine. Whether referring to nature or to the soul, rendering evidence visible meant dispelling secrecy and transforming elements of the divine into both theoretical and applied knowledge.[36]

As his text takes on a millennial fervor, Saltmarsh recognizes the continuity of the conception of visibility in natural philosophy with that of Calvinism. Saltmarsh projects a future moment of divine revelation in England, recognizing that God's *"labourers"* had already "harvest[ed]" many "young Prophets" and *"spirituall subsidiaries"* in *"Germany, Geneva, New-England,* the *Netherlands,"* the principal communities of the Reformed Church. Where Saltmarsh looks for visible evidence in nature, these spiritual communities looked for evidence on human souls, but both worked toward a millennial realization of all invisible secrets. Mid-seventeenth-century ministers and natural philosophers might differ in their focus on a particular object of scientific inquiry, but they shared a commitment to using faith to transform sight from a fallen visual sense to a redeemed sense, translated through the working of grace on the soul.[37]

Metaphysics and the emerging sciences borrowed from the methodologies of the other in an effort to compensate for what could not be known through each's own methodological approach to the world. Natural philosophers needed faith to supplement what could not be achieved through reason-based methods, even as Puritan ministers relied upon empiricism to channel revelation into forms of evidence. Visibility assumed an ontologi-

36. Perkins, *Treatise Tending unto a Declaration,* 103–104.
37. Saltmarsh, *Dawnings of Light,* 17, 18, 51.

cal rather than conditional status as ministers and philosophers turned to faith to renew humanity's observational capacities. Faith supplemented the depleted, imperfect state of the human senses, invigorating the study of metaphysics among ministers and natural philosophers alike. From their Reformed theological background, natural philosophers such as Bacon inherited a deep sense of humility that checked the excesses of Scholastic humanism, particularly the Scholastics' claims to know how God made the world. They did not base their empiricism in an optimism born of emerging voluntarism and nascent natural rights doctrines, but rather in the knowledge that only through faith and regeneration could they overcome the inborn limitations. Thus bracketed by faith and humility, the inquiry of ministers and empiricists adopted experimental philosophy in their pursuit of greater knowledge of the divine in an effort to mount the most ambitious theodicy of all: the justification of the ways of God to man not based on received wisdom, the word of God, or piety, but through the empirical evidence of all creation itself.

Controversy

Not all ministers and natural philosophers could agree on the limits of divine inquiry, not even within their respective camps. Debates about the legibility of grace, for instance, culminated in the antinomian controversies of the 1630s and 1640s. In the 1640s, theologians in New and Old England debated the legibility of grace in a contentious, often bitter, series of pamphlet wars. The question of the capacity to read grace through "laws" divided ministers on both sides of the Atlantic. The most contentious question was the most fundamental. Did evidence of grace produced on the soul constitute a legal form of spiritual practice? The use of the term "antinomian"—"against the law"—itself marks the terms of the debate over spiritual legality. Mid-seventeenth-century printed books such as John Graile's *Modest Vindication of the Doctrine of Conditions in the Covenant of Grace* (1655) and Benjamin Woodbridge's *Method of Grace* (1656) promoted a science of divinity through a new relationship between sight and spiritual knowledge. Stricter Calvinists maintained that the soul's union with Christ was "secret and invisible, to be apprehended by Faith, and not by Sense or Reason"; those in favor of a discernible method of grace extended "what is meant by Gods sight" to include, by analogy, "mans sight" and "mans judgment." Woodbridge insists that sensory data from experience could serve as visible evidence of a union that is otherwise mystical, secret, and invisible. Prominent theologians in Old England and New proposed fresh visual and sensory techniques of spiritual discernment that extended scriptural limits of knowledge. Such techniques expose an epistemological rupture in seventeenth-century Calvinist metaphysics, a division opening up in conventional Calvinist exegesis that understood spiritual knowledge as something "neither eye hath seene, nor eare hath heard, nor heart of naturall man can conceive."[38]

Published in London in 1646 (but dating from 1637), John Cotton's *Gospel Conversion* situates New England's own Antinomian Controversy of 1636–1637 within a larger culture of controversy over the new meaning of grace as producing an external sign of one's internal state. Cotton explains methodically and cautiously that the first evidence of grace is assurance from justification, an inwardly felt experience. This first evidence is the convert's own experience of witnessing the evidences of his or her own soul. Only then can the convert "discern, and take his sanctification as a secondary witnesse, or an evident signe or effect of his justification." Justification must become legible to the individual before the evidence can be translated into communal knowledge through testimony, understood here as a form of sanctification. Careful not to suggest that sanctification in the form of testimony or good works is an unequivocal sign of inward grace, Cotton nonetheless concedes to the evidentiary value of external signs for a witnessing community invested in the quest for spiritual knowledge of the invisible world. Conceding sanctification as some evidence of justification, Cotton's retrospective on the Antinomian Controversy is very careful to identify sanctification as secondary evidence. Thomas Shepard, by contrast, makes sanctification crucial to completing the form of the first evidence of grace. Justification becomes a more complete form of assurance when an audience of converted saints confirms its legitimacy and authenticity by bearing witness to the visible evidence of grace upon the soul of the testi-

38. W[illiam] Eyre, *Justification without Conditions; or, The Free Justification of a Sinner* . . . (London, 1695), 11; Benjamin Woodbridge, *The Method of Grace in the Justification of Sinners* (London, 1656), 9. Woodbridge cites 1 Cor. 4:4, Luke 16:15, and Num. 32:22 to establish scriptural grounds for his analogy between what is meant by God's sight and what is visually available to man through human sight.

On epistemological rupture: John Eaton, *The Honey-Combe of Free Justification by Christ Alone* (London, 1642), 226. This is a reference to 1 Cor. 2:19.

fier. Through the discerning capacity of the witnessing audience, testimony becomes evidence of sanctification.[39]

The antinomian controversies express concern over appropriate techniques for discerning grace. Theologians on both sides of the Atlantic struggled to address the Calvinist problem of incertitude, expanded from an individual to a collective phenomenon. The central question about Protestant conversion shifted from how an individual might acquire some inkling of his or her elect status to how experimental knowledge of this status might be communicated to others. Following the antinomian debates, ministers began to make the soul a location for identifying evidence of God within the natural world. Reason-based techniques of observation and experimentation became the means for acquiring this evidence through conversion testimony. Converts would orally narrate the feeling of conversion as an evidentiary and phenomenological feeling that superseded knowledge of faith. Ministers and testifiers developed an expertise of human souls around this practice, transforming grace into an object of scientific inquiry.

God could be observed in nature through special discerning capacities, yet additional techniques were needed to render the soul visible through the testimony of faith. The key difference between the natural theological approach of reading God in nature and the ministerial approach of reading the soul for signs was that the evidence of the soul had to be communicated in words. It could not be seen through direct observation. This raised the possibility for further errors. Witnessing evidence came to have a double meaning—to see, and to speak, registering the hearsay quality for recording evidence of grace on the soul. Yet, words were a problem because the Fall rendered them imperfect vehicles for communicating divine truths. By the time the testimony of faith took hold in New England congregations, linguistic philosophers, influenced by the work of Comenius as popularized and translated into Ezekias Woodward's *Light to Grammar* (1644) and Saltmarsh's *Dawnings of Light,* were working toward a resolution of this crisis of language, specifically by advocating the recovery of the lost languages of Babel, an enterprise promising a rapprochement of natural and revealed knowledge in the recovery of a perfect tongue.

39. John Cotton, *Gospel Conversion* . . . (London, 1646), 7. For the history of this text as a scribal publication preceding its initial printing in London, see David D. Hall, *Ways of Writing: The Practice and Politics of Text-Making in Seventeenth-Century New England* (Philadelphia, 2008), chap. 2, and 132.

Not surprisingly, then, Comenius was reportedly invited to New England, as Cotton Mather explains in his *Magnalia Christi Americana:*

> That brave Old Man *Johannes Amos* COMENIUS . . . was indeed agreed withall, by our Mr. *Winthrop* in his Travels through the *Low Countries,* to come over into *New-England,* and Illuminate this *Colledge* and *Country,* in the Quality of a *President:* But the Solicitations of the *Swedish* Ambassador, diverting him another way, that Incomparable *Moravian* became not an *American.*

This might well be Mather's embellishment of historical fact; scholars have debated whether Winthrop had ever issued an invitation. Apocryphal or not, the story suggests a link between pansophic Comenian linguistics and the community of testifying Puritans seeking to establish a harmony between language and nature that would restore language to its pre-Babel state. The Puritans might thus be placed within a larger movement of linguistic purification that spanned a variety of overlapping circles from the plain-style ministers and the Cambridge Platonists to early Royal Society scientists and historians. As they sought to purify the senses through empiricism, philosophers also sought to purify language. This effort has much in common with the late-sixteenth-century development of plain style, or the Puritan program for linguistic purification that was driven by the central, organizing idea of reducing the distracting elements of rhetorical flourish in order to allow language to more clearly convey divine ideas.[40]

The Cambridge Platonists and Royal Society fellows followed a similar linguistic program, which Thomas Sprat described as the Royal Society's objective of bringing "*Knowledge* back again to our very senses." Attempting to develop a linguistic method that followed this model, society members, according to Sprat, advocated a primitive purity of speech, developing "a close, naked, *natural way* of speaking; positive expressions; clear senses; a *native easiness:* bringing all things as near the Mathematical plainness, as

40. Cotton Mather, *Magnalia Christi Americana* . . . (London, 1702), book 4, 128. See also G. H. Turnbull, *Hartlib, Dury, and Comenius: Gleanings from Hartlib's Papers* (Liverpool, 1947). Robert Fitzgibbon Young confirms this story in *Comenius and the Indians of New England* (London, 1929) without much evidence to back his claim. However, Walter W. Woodward has shown more thoroughly and convincingly that "Comenius believed the pansophic educational program provided an especially effective means for converting the American Indians, who had to be won to Christ as a prerequisite to the millennium" (*Prospero's America,* 58).

they can: and preferring the language of Artizans, Countrymen, and Merchants, before that, of Wits, or Scholars" (my emphasis).[41]

This subordination of wits and scholars to countrymen echoes the philosophy of plain style and the Reformation insistence that the vernacular Bible would secure the most immediate connection between the convert and God. Both proposals for linguistic purification are rooted in an attempt to reduce the problems of hypocrisy endemic in language—Bacon's concern that words are mere appearances and the corresponding concern that rhetoric can mask a larger structure of incertitude. Sprat's method of speaking returns language to the society's commitment to "Nullius in verba" ("Take nobody's word for it"), a saying that marks a turn from reliance on textual language to reliance on a disciplined kind of sensual data discernible through techniques of experiment and witnessing.[42]

Visible Souls

From the convergence of "nature" and "divine revelation," a particular form of religious expertise emerged that ultimately worked against the regnant Protestant concept of sola scriptura. The idea here was not simply to elevate the discerning capacity of regenerate individuals but to mobilize their experiential data for a higher collective purpose. Ministers and natural philosophers each proposed ways of acquiring the authority to see and discern evidence of God, despite the allegedly restricted terrain of such knowledge. Intersections between empiricism and faith transformed the soul into a "vehiculum scientiae" that provided theologians, philosophers, and New England communities with aural and visual sensory data. Testifiers not only proclaimed their faith; they also displayed discernible manifestations before a witnessing audience of that which, according to Paul, the "eye hath not seen, nor the ear heard." Reinterpreting Paul's teachings, conversion testimonies performed sensory impressions of grace as recorded

41. Thomas Sprat, *History of the Royal Society* (1667), ed. Jackson I. Cope and Harold Whitmore Jones (Saint Louis, Mo., 1959), 112, 113.

42. For modern work on the epistemic shifts that accompanied the rise of experimental philosophy in the Royal Society, see Joyce E. Chaplin, *Subject Matter: Technology, the Body, and Science on the Anglo-American Frontier, 1500–1676* (Cambridge, Mass., 2001); Mary Poovey, *A History of the Modern Fact: Problems of Knowledge in the Sciences of Wealth and Society* (Chicago, 1998); Barbara J. Shapiro, *Probability and Certainty in Seventeenth-Century England: A Study of the Relationships between Natural Science, Religion, History, Law, and Literature* (Princeton, N.J., 1983); Peter Dear, *Discipline and Experience: The Mathematical Way in the Scientific Revolution* (Chicago, 1995).

upon the human soul. Represented in the Puritan concept of the visible saint, a newly visible soul became a site for investigating what Bacon called a "science of divine things" that might ultimately lead to the apprehension of things unseen. Testimonies of faith transformed the inward vision that the soul receives through conversion to an outwardly manifest vehicle of the "Holy Spirit" witnessed by others.[43]

In addition to contradicting the conventional exegetical understanding of Pauline law, the Puritan concept of the visible saint challenged a more general history of theological reluctance to use vision as the central tool for acquiring divine knowledge. Early Christian theory rejected ocularity. Augustine cautioned against visual desire, which can distract the mind from the proper spiritual path. Vision tempts Christians to substitute idolatrous forms for spiritual sight. Calvin valued physical blindness, because it led to spiritual truths. In this interpretive framework, the new science replaced a former religious suspicion of visual sense with new possibility. Lucien Febvre postulated the "vision was unleashed" in a seventeenth-century "world of science." Bacon's proclamation—"I admit nothing but on the faith of the eyes"—epitomizes the emerging visual regime in natural philosophy that contested conventional understandings of Paul's proscription against sight.[44]

The Puritan concept of the visible saint aptly illustrates this account of how the eyes gradually became a more stable vehicle for securing material truths over the course of the seventeenth century. Yet, if we move back in time before Paul's era, we see that alternative understandings of sight had been around for some time. The history of Plato's *Timaeus,* for instance, re-

43. Bacon, *Advancement of Learning,* in *The Major Works,* ed. Vickers, 152 (where he also proposes his theory of testimony of faith), and 292. Quotation is from 1 Cor. 2:19: " Eye hath not seen, nor ear heard, neither have entered into the heart of man." Edmund S. Morgan discusses the concept of the "visible saint" as a Puritan approximation of Augustine's visible church, an earthly correspondent to the invisible church of the elect (*Visible Saints: The History of a Puritan Idea* [New York, 1963]).

Much early Americanist and early modernist work focuses on the senses. Peter Charles Hoffer claims that historians have not fully understood the extent to which Puritan religiosity was sensory, dependent on a visible and audible culture that produced a kind of "sensory self-consciousness" (*Sensory Worlds in Early America* [Baltimore, 2003], 79). See also Richard Cullen Rath, *How Early America Sounded* (Ithaca, N.Y., 2003); Bruce R. Smith, *The Acoustic World of Early Modern England: Attending to the O-Factor* (Chicago, 1999); Leigh Eric Schmidt, *Hearing Things: Religion, Illusion, and the American Enlightenment* (Cambridge, Mass., 2000).

44. Martin Jay, *Downcast Eyes: The Denigration of Vision in Twentieth-Century French Thought* (Berkeley, Calif., 1993), 13 (quoting Bacon); Lucien Febvre, *The Problem of Unbelief in the Sixteenth Century: The Religion of Rabelais,* trans. Beatrice Gottlieb (Cambridge, Mass., 1982), 34.

minds us that sight had not always endured the kind of suspicion Paul's theology evinces. Through the character Timaeus, a "leading astronomer and special devotee of natural science," Plato explains the importance of the soul as the central tool for rendering spiritual "things visible." "Nothing without understanding," Timaeus observes, "would ever be more beauteous than with understanding, and further that understanding cannot arise anywhere without soul." In this ancient proscription, the soul enhances a special vision that is not unlike the special vision that Baxter promotes as acquired through faith or that Bacon perceives as a consequence of imputation. Timaeus assigns a specific, divine purpose of vision as an outgrowth of the soul:

> God invented it and bestowed it on us that we might perceive the orbits of understanding in the heavens and apply them to the revolutions of our own thought that are akin to them, the perturbed to the imperturbable, might learn to know them and compute them rightly and truly, and so correct the aberrations of the circles in ourselves by imitating the never erring circles of the god.[45]

In other words, divine vision acts as a supplemental corrective to the depleted state of human vision. Seeing nature becomes a mechanism for seeing the heavens. Given this formulation of redeemed visual capacity, the *Timaeus* was a foundational text in both natural philosophical and theological traditions of the seventeenth and eighteenth centuries. Cotton Mather makes use of Plato's *Timaeus* in *The Christian Philosopher* (1721), exclaiming at one point: "*Oh my Redeemer!* Bestow thou an *Eye* upon me: A *Faculty* to discern the Things that are *Spiritually to be discerned*." This recuperation of the ocular is one Protestant legacy of Plato's *Timaeus*.[46]

The Renaissance elevation of sight as a sense with renewed potential occurred through the prism of Plato's statement: "Sight, then, as I hold, is the cause of our chiefest blessing, inasmuch as no word of our present discourse of the universe could have been uttered, had we never seen stars, sun nor sky." Paradoxically, vision provides access to things unseen. Its preeminent status in the hierarchy of the senses accords the eyes their access—conventionally figured as windows to the soul and hence to a range of erstwhile invisible things newly manifest in the spiritual world. The Renaissance Neoplatonist Leone Ebreo explained, "Sight is the most spiritual of the

45. Plato, *Timaeus and Critias*, ed. A. E. Taylor (London, 1929), 24, 27, 45.
46. Cotton Mather, *The Christian Philosopher*, ed. Winton U. Solberg (Chicago, 1994), 25.

external senses because it 'sees things that are in the last circumference of the world.'" Sight is the most important sense for teaching, learning, and discovery, thus suggesting to Boethius the need to sharpen the hierarchy of the senses by which the higher may judge the lower.[47]

The testimony of faith aided a collective effort to identify Augustine's visible church as an earthly approximation of the invisible body of God's elect. The communal witnessing of testimony in the immediacy of a singular moment marked the exclusive privilege of visible saints. Profane time folded into the sacred in this moment. Through this condensing the temporal into a sacred narrative in which the numinous and evanescent became visible, the aural testimony was a performance designed to recreate the space of Augustine's invisible church. Time stood still within this testimonial space, allowing the words—or visible signs—of the testifying subject to occupy an originary moment within eschatological time. Within this performative space, testimony experientially recuperated the Augustinian ideal of linguistic expression where in the moment of conversion words in fact become signs of the invisible domain. Empirical techniques permitted the transformation of the relentless quotidian search for signs of grace in the Perkinsian model into a moment in which assurance materialized in a form that could contain the oscillation between anxiety and assurance so characteristic of Puritan conversion.[48]

In developing the category of the visible saint, the Puritans adopted two elements of the Renaissance judgment of sense, which exhibited a continuum between theology and natural philosophy rather than a replacement of the former by the latter. First came the compulsion to elevate visual sense and to use it to judge other senses. Augustine taught that we see passively with our material eyes and judge with our soul. Seventeenth-century Puritans wished both to return the discernment of spiritual experience to this formulation and to invert it so that vision could also be used to discern the contents of the soul as divine manifestations rendered intelligible through subtle information from other sensory organs. This reformulation expanded the Renaissance continuity between seeing nature and seeing divinity as vision became an augmented sense—and seeing an augmented

47. Plato, *Timaeus and Critias*, ed. Taylor, 45. Also quoted in David Summers, *The Judgement of Sense: Renaissance Naturalism and the Rise of Aesthetics* (Cambridge, 1987), 32, and see 33.

48. The invisible was the perfect church, consisting of only the predetermined and the elect. The visible church was imperfect because in their fallen human condition not everyone who professed belief would actually possess saving faith. Before the 1640s, Presbyterians, Congregationalists, and Independents agreed on these basic principles of church organization. Morgan, *Visible Saints*, 29.

discipline—for discerning nature in Baconian circles. Puritan ministers transformed sight from the passive and receptive material organ that it had been for Augustine into a muscularized sense endowed with the capacity to open apertures into the souls of others. Far from marking the replacement of religion with science, the emerging status of vision, as Febvre describes it, helped to resolve a doctrinal "crisis of knowing" by offering an empirical link between nature and divinity.[49]

The elect members witnessing the testimony as well as the ministers recording the spoken words attempted to translate a complex sensory scene into a visible taxonomy. Each testifier had to have self-knowledge of divine grace upon his or her soul in order to supply evidence of election. The event of conversion itself consisted of feeling hearts hearing God whispering to another's soul. Testimonies refer to touching, hearing, and taste more frequently than seeing God. As such, the testimonies heed Paul's warning to be suspicious of vision. But the act of translation—from an inwardly directed phenomenal experience into usable knowledge for others—is where rendering these sensory clues as visual signs becomes apparent. Conversion itself remained wholly and fully Calvinist. The testimony, constructed retrospectively and after the conversion, transformed the soul into a scientific vehicle.

Testifying before Thomas Shepard's congregation in 1640, Alice Stedman states, "I had need to know what was in my own heart." Speaking in the past tense, she narrates a desire for an inward quest, implying that she has learned something through the careful examination of her own heart that she will reveal to the congregation. Yet Stedman's testimony is not simply an account of the evidence of grace she found on her heart. She attempts to render her introspective journey visible to a witnessing audience who, by judging the evidence on Stedman's heart to be manifestations of grace, will substantiate her membership in the visible church. Through empirical techniques of observing and discerning experience, the formal properties of testimony transformed the scriptural version of sight from "*f*aith is the grounde of things, which are hoped for, and the evidence of things which are not sene" to the understanding that *"faith is a kind of sight; it is the eye of the soul:* the evidence of things not seen." The testifier provides

49. Febvre, *The Problem of Unbelief*, trans. Gottlieb, 34. Here again, I am borrowing this term, "crisis of knowing," from Bouwsma, *John Calvin*. I am referring primarily to a doctrinal crisis, but others have more broadly applied the concept to Renaissance epistemology: for example, Elizabeth Hanson, *Discovering the Subject in Renaissance England* (Cambridge, 1998).

testimony for others to witness evidence of grace in that testifier's soul. Senses became recalibrated through the oral testimony; the transcription translates an inward justification to outward sanctification. The effort to discipline the senses through a visual taxonomy and to record a system of signs addressed the Puritan problem of hypocrisy. If the senses lied, such a record could help to detect the deception and to tease out the secrecy of the Holy Spirit's playing upon the human soul. The object of the testimonial scene of witnessing was thus to verify the information conveyed to the senses as authentic signs of God.[50]

Unlike modern attempts to explain religion through science, the seventeenth-century science of the soul did nothing to delegitimize the religious experience as purely and wholly transcendent. The empirical science that developed around the mystery of the divine involved taking a snapshot of the transcendent experience. Through testimony, experience became a usable form that could be studied and transformed into divine knowledge in order to further a continuing empirical project. Ministers hoped that data conveyed to the soul from God in the form of subtle feelings and soft whisperings might be translated into usable knowledge for the ministerial mind. The public nature of the scene—a group of visible saints gathered before a testifying individual—literally and metaphorically superimposed a paradigm of discernment and visual intelligibility onto the scene.

This seems an odd way to describe the sensory and taxonomic process taking place, given that the most immediate evidence supplied by the testifier was aural. Audiences listened attentively, and indeed aural evidence was also crucial. The first stage of sanctifying or rendering the saint visible during conversion took place through this interplay between the testifier's words and the absorption of aural evidence on the part of the audience. But the words were also written down and arranged on the page in such a way as to establish order in the evidence communicated. Images display the

50. Michael McGiffert, ed., *God's Plot: Puritan Spirituality in Thomas Shepard's Cambridge,* rev. ed. (Amherst, Mass., 1994), 188; Heb. 11:1, Geneva Bible; Richard Baxter, *A Call to the Unconverted to Turn and Live and Accept Mercy . . .* (London, 1663), 48.

For this transformation from sight to understanding, see Chapter 3 and my reading of Richard Baxter's *Call to the Unconverted* as containing a passage that revises the scriptural epistemology of Heb. 11:1. While Baxter clearly had this passage in mind, his transformation is significant. Whereas Hebrews states that faith gives hope, Baxter claims that faith gives sight. For Baxter faith is no longer simply evidence of things to come; it actually transforms the ontological status of vision in the way displayed in my reading of Bacon's *Confession* above.

chronological arc of visual evidence of the soul recorded through ministers' transcriptions and the authority derived from observation. One image (Figure 2) depicts Shepard's initial attempts to evaluate the testimonies of his congregants in his Cambridge congregation between 1638 and 1645. While the deletions and revisions are not entirely clear, they seem to indicate a retrospective attempt to cull from his testifiers' words the precise nature of divine evidence recounted in their statements. Another image represents Parkman's more advanced phase of this editorial and taxonomic practice (Figure 3). He extracts and enumerates "Evidence of Grace" from the manuscript memoir of Sarah Pierpont. James Pierpont sent the memoir to Parkman after her death. It records Sarah's poignant and sustained effort to grasp the ineffable nature of God's grace upon her soul. Parkman intervenes as a minister-editor, intent on winnowing from her inward journey a visible and usable taxonomy of gracious affections. Through this practice of transcribing the quotidian search and signs of grace discovered by the converts under their pastoral care, both ministers draw from the Renaissance commonplace tradition, which also influenced the advent of experimental knowledge in the Royal Society. Writing down impressions rendered them visible, creating an archive of empirical evidence of divinity from which ministers and congregants developed a particular kind of discerning expertise.[51]

The written record of each convert displays an impression of Adam's perfection, the faint, visible outline of the once-perfect law of nature. He or she (the gendering of this process is complicated) did not acquire a new ontological status as would emerge in later evangelical models of Christianity that adhere to a doctrine of universal salvation. Rather, through conversion, the visible saint regained a capacity to see the law of nature, to tease out patterns of dispensation within the law of nature. This perceptive capacity allows the convert to see "the harmony between the image of God in his heart, and the image of God in nature" in a way that echoes imputation as recounted in Bacon's *Confession of Faith* as a lens for witnessing God's "secret will and grace" through the "light of nature." In partially regaining Adam's innocence, the convert regains Adam's knowledge of the harmony

51. Adrian Johns, "Reading and Experiment in the Early Royal Society," in Kevin Sharpe and Steven N. Zwicker, eds., *Reading, Society, and Politics in Early Modern England* (Cambridge, 2003), 247. Also see William R. Newman and Lawrence M. Principe's discussion of the methodology of George Starkey's notebooks in *Alchemy Tried in the Fire: Starkey, Boyle, and the Fate of Helmontian Chemistry* (Chicago, 2002).

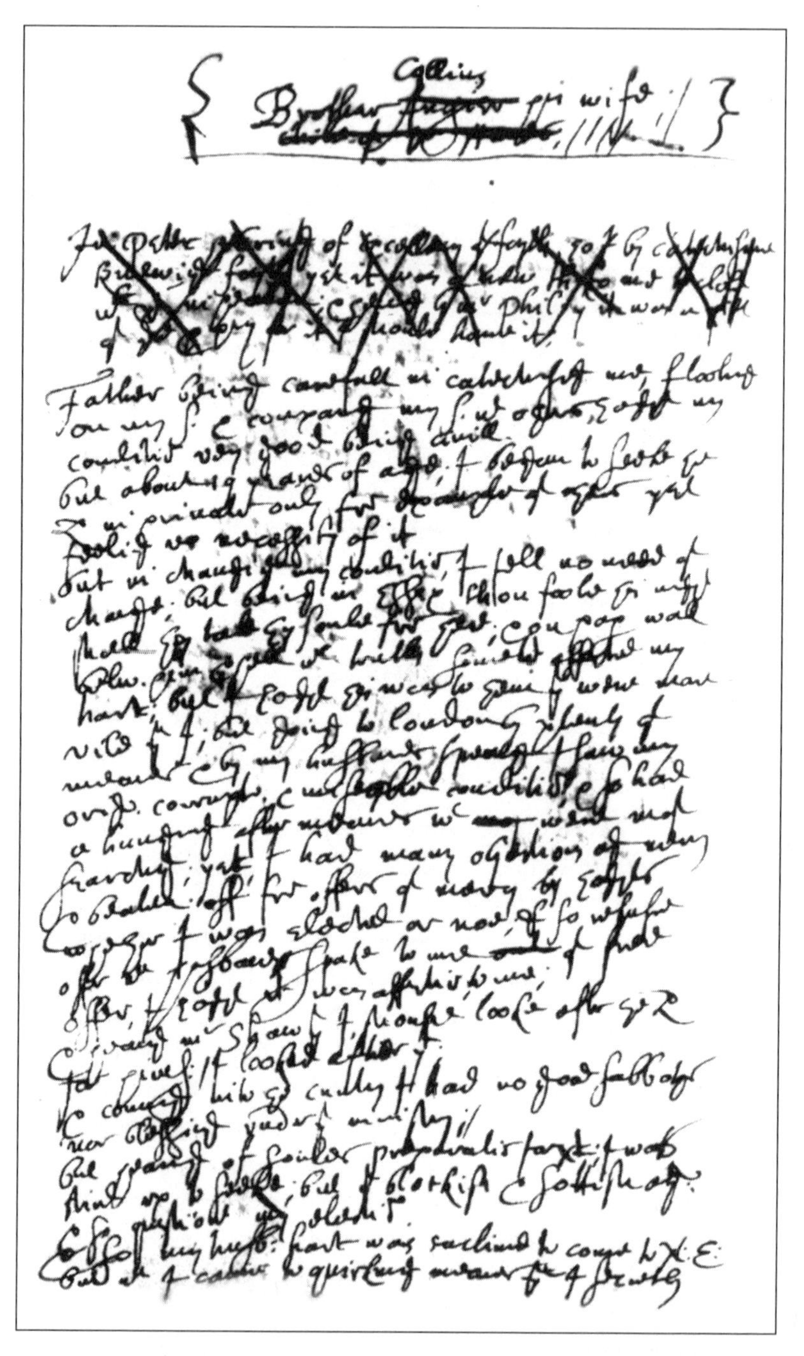

FIGURE 2 • *Confession of Brother Collins, His Wife, and the Confession of John Stedman. In Thomas Shepard, "The Confessions of Diverse Propounded to Be Received and Were Entertained as Members." MS 553, New England Historic Genealogical Society, Boston. Permission New England Historic Genealogical Society*

~~Mr Sanders his Confession~~:

John Stedman his Confession

It pleased God about 15 years since to
move my heart to hearken to God and the
Lord convinced my course was a fearful
sermon of my uncle which ~~was~~ showed me
out of 2 Thes. 1: flaming fire showing the
wrathful estate of men it brought me to
consider of ~~my~~ former course and sadness of
the former course. and another time preaching
out of 1 Thes. 5: be not drunk with
wine. which he showed the fearfulness of the sin of
drunkenness being found guilty of that
fearful sin I was much affected with the former
and it was down it appeared [illegible] I was
continuance in my vain and idle company who
asking me what ailed me I said I was cause
enough to see them walk so idly. after such
manner and so I labored to pray and got a
book but I [illegible] Lord to help me to
pray and so ~~was~~ cast off the book and was
affected with hardness of heart and seeing 1 Pet. 5
God resists the proud and gives [illegible] and I saw a
great want of humility and saw more
of vileness of my hard unsubdued heart
and so seeking the Lord and went to [illegible]
to help me ~~about~~ a hard heart and the
Lord [illegible] I was obedient to the Lord it
was [illegible] and I found and that I was
admitted to private fast[?] of the
which I found much sweetness.

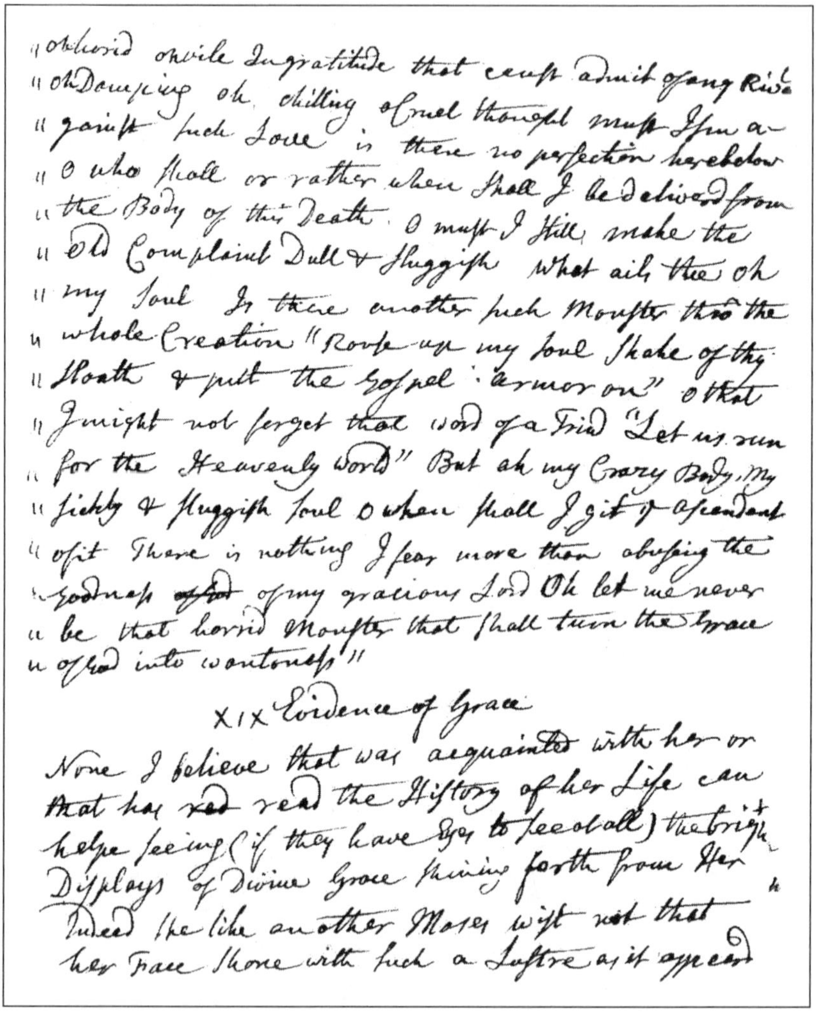

"Oh horrid oh vile Ingratitude that can'st admit of any Rival
"Oh Damping oh chilling oh Cruel thought must I sin a-
"gainst such Love is there no perfection here below
"O who shall or rather when shall I be delivered from
"the Body of this Death O must I still make the
"Old Complaint Dull & sluggish What ails thee oh
"my Soul Is there another such Monster thro the
"whole Creation "Rouse up my soul shake of thy
"sloath & put the Gospel armor on" O that
"I might not forget that word of a Friend "Let us run
"for the Heavenly World" But ah my Crazy Body, My
"sickly & sluggish soul O when shall I git ye Ascendant
"of it There is nothing I fear more than abusing the
"goodness ~~of God~~ of my gracious Lord Oh let me never
"be that horrid Monster that shall turn the Grace
"of God into wantoness"

XIX Evidence of Grace

None I believe that was acquainted with her or
that has ~~red~~ read the History of her Life can
helpe seeing (if they have Eyes to see at all) the bright
Displays of Divine Grace shining forth from Her "
Indeed she like another Moses wist not that
her Face shone with such a Lustre as it appeard

FIGURE 3 • *Sarah Pierpont's Diary: Ebenezer Parkman's Edited Version. American Antiquarian Society, Worcester, Mass.*

between natural and divine law through renewed access to the visibility of God's secrets within the natural world. One of the larger philosophical problems of the seventeenth century—how to negotiate the limits of knowledge that result from the Fall—is thus transferred to the lay testimonies of Shepard's "Confessions."[52]

As a source of this philosophical problem, Reformation optimism structured each confession with the deeply inculcated awareness that grace was always coupled with doubt. In the New World and in other colonial outposts, such as Dublin, the Puritans sought to negotiate this dilemma through their commitment to sola scriptura and to their belief that assurance could be intuited and discerned in limited degrees. Because it disguised a deep awareness of the limits of human knowledge, this individualistic model of conversion created a serious social problem. What could unite religious communities of elect saints if their elect status could be determined only in degrees and if their experiences hewed to individualized expressions and emotions? The Reformation's regnant question—"How can one be assured of one's election?"—had become a collective problem: How can a community of the redeemed cohere if salvific status remains uncertain? As Puritan communities developed disciplinary mechanisms in an effort to check the danger of religious freedom, the Reformation optimism in Calvin's initial proposal paradoxically threatened Puritan unity even as it proffered that optimism as a unifying form. What, ministers asked, if the Puritan convert *could* glean a fuller sense of assurance from the experience of grace? What kind of evidence of God proceeded from this form of experiential knowledge?

The testimony of faith signals the implementation of Reformation optimism. Converts struggled in their testimony to narrate, and thus synthesize, the vexed relationship between original sin, self-knowledge, and knowledge of God proposed by Calvin. The genre responded to a theological and social conundrum by violating theological law. It became the focus of sectarian division, including between Scottish Presbyterians and New England Congregationalists. The crux of this debate was the controversial use of the senses as data of the effects of divinity upon the souls of the elect, for, even though the senses could delude, they also represented tremendous empirical promise. The human senses embodied the problem of knowledge at the heart of both theology and natural philosophy. They paradoxically displayed the consequences of the Fall and the redemptive

52. Bacon, *The Major Works,* ed. Vickers, 210.

The VOICE
of the
SPIRIT.
OR
An Essay towards a Discoverie of the witnessings of the Spirit by opening and answering these following weighty Queries.

Q. 1. *What is the witnessing worke of the Spirit?*
2 *How doth the Spirit witnesse to a soule its Adoption?*
3. *Who are capable of attaining the witnessings of the Spirit?*
4. *How may a soule know its injoyment of them?*
5. *By what meanes may a soule attaine them?*

To which is added.
Roses from Sharon or sweet Experiences reached out by Christ to some of his beloved ones in this Wildernes.

By SAMUEL PETTO Preacher of the Gospell at *Sandcroft* in *Suffolke*.

Isaiah 59. 19. *When the enemy shall come in like a flood, the Spirit of the Lord shall lift up a Standard against him.*

LONDON: Printed for *Livewell Chapman*, at the *Crowne* in *Popes-head*-Alley. 1654.

FIGURE 4 • *Samuel Petto,* The Voice of the Spirit. *Beinecke Rare Book and Manuscript Library, Yale University, New Haven, Conn.*

possibility through the acquisition of new and more empirically secure forms of knowledge. The debates marked the genre's status as a novelty, an emergent object of scientific inquiry that potentially violated certain limits of knowledge as described by Calvin as well as natural philosophers.

Through replicating interior assurance as communal meaning, individual perception as collective judgment, and spiritual sight as visual evidence, conversion testimony opened up a new set of practices in what Paul Ricoeur describes as a "quasi-empirical" space. Thus testimony works to transfer meaning from perception to judgment about that perception, and from the credibility of the testifier to the reliability of the discerning witness. New scientific epistemologies defined the contours of this empirical space by expanding the limits of what could be seen through a dark glass and known about the invisible world while simultaneously enacting limits upon that knowledge. Samuel Petto's *Voice of the Spirit* describes a method of tracking God's "secret whisperings to the soule" and "inward, unspeakeable inspirations" by "mak[ing] report of" and visually "how the Spirit doth witnes to a soule its Adoption." He imagines identifying secret, mysterious, and hidden things through sound and plain observation. Through the introduction of new ways of imagining, discerning, and articulating experience, testimony offers a unique response to what Ricoeur describes as the "unbridgeable chasm . . . between the interiority of original affirmation and the exteriority" of testifying to the absolute.[53]

In Puritan communities, the public, oral testimonies enact these patterns of transference not solely by recounting interior spiritual experience but by actually transforming this experience into a series of empirical signs visible and audible to others. Each collection promotes an empirical approach to cataloging, discovering, and authenticating religious experience. Rendering a numinous experience visible for empirical study, testimonies were arranged and recorded in manuscript and occasionally circulated and published as inductive evidence of God's presence in the souls of his saints. It is to the first set of these testimonies that we now turn.[54]

53. Samuel Petto, *The Voice of the Spirit; or, An Essay toward a Discoverie of the Witnessing of the Spirit* . . . (London, 1654), 4, 6. Testimony has a "quasi-empirical meaning," according to Paul Ricoeur, because it is a narration of an event rather than the reporting of an actual perception. See Ricoeur, "The Hermeneutics of Testimony," in Ricoeur, *Essays on Biblical Interpretation* (London, 1981), 119–154, esp. 142–143.

54. John Rogers, *Ohel; or, Beth-shemesh: A Tabernacle for the Sun* . . . (London, 1653), 392; [John] Eliot and [Thomas] Mayhew, *Tears of Repentance; or, A Further Narrative of the Progress of the Gospel amongst the Indians in New-England* . . . (London, 1653), "To the Reader."

{2} Congregations

Masculine Form and Reluctant Women in Puritan Testimony

We came not here to find gracious hearts, but to see them too. 'Tis not faith, but a visible faith, that must make a visible church, and be the foundation of visible communion; which faith I say, because my weakness could not see in some of them by their profession.—THOMAS SHEPARD TO RICHARD MATHER, Apr. 2, 1636

The year 1633 is one of transformation in John Winthrop's *History of New England*. Nothing bespoke change more than his catalog of the ships arriving in the Massachusetts Bay Colony bearing colonists to the New World settlement. Among the arrivals were the soon-to-be-infamous Anne Hutchinson, the much-looked-for John Cotton, and Thomas Hooker, two ministers whose Old World reputations prompted a small-scale migration to Boston. Taking six to eight weeks on average, voyages from London brought in hundreds of new passengers annually, each eager to settle among the original migrants of 1630. Although most of these Puritans had prepared for years for this occasion, fasting, praying, planning, and saving, they also expected to undergo something of an emotional, if not spiritual, transformation through the journey itself. The Atlantic crossing became a vital allegory for a spiritual journey, one Winthrop himself recorded as the personalized induction of individuals joining the church and the communal pattern exhibited through the formation of new congregations where "divers profane and notorious evil persons came and confessed their sins."[1]

In a new practice, quite possibly one that John Cotton brought to the New World, a group of colonists "joined" in Boston in 1633 to "make confession of their faith." Winthrop defines this confessing as a declaration of

1. John Winthrop, *The History of New England from 1630 to 1649*, ed. James Savage (Boston, 1825), I, 121.

"what work of grace the Lord had wrought in them." As few as six to eight individuals constituting the pillars of this new congregation gathered in the Boston town commons. In a succinct public address lasting not more than ten minutes, they attempted to respond to the central question of the Calvinist Reformation, How did the individual know he or she was saved? This question could not be answered with knowledge of faith or a declaration of individual belief in God. Rather, testifiers had to supply some account of what divine grace felt like on their souls. Individual knowledge then translated into communal knowledge. Following such declarations, the church "covenant was read, and they all gave a solemn assent to it." A formative moment in Winthrop's *History,* this scene records putting the ideal of Christian charity into practice, linking the covenant of grace to the "federal" covenant, which Winthrop famously described in his 1630 lay sermon, "A Modell of Christian Charity," as the diverse members of the elect bound collectively together through "love."[2]

Testifying began in Boston in 1633 as an improvisational form that responded to a practical need to form a New World congregation. The geographic urgency of the migration and the social and political demands of reinventing society upon New World arrival are both integral to explaining why the practice of declaring one's faith emerged in New England when it did. The Puritans arrived in the Massachusetts Bay Colony in 1630 with a vision for a godly community, most famously described in "A Modell of Christian Charity." Although not the first time he delivered the sermon, Winthrop most famously preached it aboard the *Arbella* in 1630; in it he imagines a godly community in which the congregants become the visible earthly members of the invisible body of Christ. Winthrop's ser-

2. Ibid., I, 180. In "A Modell of Christian Charity," Winthrop uses the metaphor of being bound together through Christ's "ligaments" of love throughout to suggest both the cohesiveness of the visible church and to announce the project of the federal covenant, or the enfolding of disparate, individual covenants of grace into a collective, corporate body. Although this is a rather idealized image of a form of spiritual equality, Winthrop's sermon also famously maintains a rigid social hierarchy in which some members of Christ's body form the head and others the feet (Giles Gunn, ed., *Early American Writing* [New York, 1994], 108–112). As the testimony of faith became codified as an official genre of church membership, the Massachusetts General Court applied language to explain the communal function of the test of faith: ". . . as those similitudes hold forth which the Scripture makes use of to shew the nature of particular Churches: as a *Body,* a *Building,* or *House, Hands, Eyes, Feet,* and other members must be united, or else (remaining separate) are not a Body. Stones, Timber, though squared, hewn, and polished, are not an House, until they are compacted and united: so Saints or Believers, in judgement of charity, are not a Church, unless orderly knit together." *A Platform of Church Discipline: Gathered out of the Word of God: And Agreed upon by the Elders: And Messengers of the Churches . . .* (Cambridge, Mass., 1649), 5.

mon resolves a theological and social tension in Puritanism between an individual and a corporate identity. Predicated on the commitment to the individual's autonomous relationship with God, Protestantism from its earliest days proved particularly susceptible to social atomism. Puritan leaders struggled to find a means to unify the Christian community in the absence of traditional unifying forms of Roman Catholic liturgical, dramaturgical, and devotional practices.[3]

Part of this dilemma was theological: the Puritan New World errand required an elaborate plan for suturing the covenant of grace promised to Adam in Genesis to the federal covenant on which its collective nature came to depend. The other part was social and political, encapsulated in Winthrop and other first-generation magistrates' attempts to implement and even mandate this theological idea among a diverse sectarian group of Calvinists fleeing to New England after 1630 to escape the unifying reforms initiated by Archbishop William Laud. Relentless in their commitment to their particular doctrine and sure of the moral and divine sanctioning of their faith, the Puritans nonetheless remained uncertain and anxious about the form the New World governmental and ecclesiastical structures should take. The testimony of faith stems from this sense of uncertainty rather than from the unwavering commitment to doctrinal imperatives. As Winthrop's *History* lays out the architecture of the New World city on a hill—a beacon to all, particularly the Anglican Church that he and others deemed to be standing in the way of true reform—the formation of the first congregation was essentially an experimental and improvisational attempt to come up with a pattern, a usable template, for the pure Reformation church.[4]

The success of the testimony of faith in the Bay Colony did not go unnoticed by Old World reformers. Ministers in Suffolk, Wales, and Dublin

3. Gregory S. Jackson, *The Word and Its Witness: The Spiritualization of American Realism* (Chicago, 2009), 37–88.

The case for the genre of testifying as American exceptionalism has been most strongly made by Patricia Caldwell in *The Puritan Conversion Narrative: The Beginnings of American Expression* (Cambridge, 1983). However, Edmund S. Morgan argued for the theological particularity of New England Augustinianism in relation to the testimony of faith in *Visible Saints: The History of a Puritan Idea* (New York, 1963).

4. Darren Staloff explores the implications of Puritanism as a radical sectarian political movement. He shows how the New England way led to the development of a "thinking class," capable of generating and maintaining a unique form of political authority until 1680, when the crown revoked the colony's charter. *The Making of an American Thinking Class: Intellectuals and Intelligentsia in Puritan Massachusetts* (New York, 1998), 11–16.

began adopting the practice as a means of discerning the visible church, of winnowing God's good seed from among the bad. Testimonial collections from each location as well as from John Eliot's first praying town of Natick, Massachusetts, began appearing in print in London in the 1650s, many passing into multiple editions. This geographic diversity periodically undermines the claim for an exclusive New England particularity, born of a unique theological frame that produced a narrative experience that in turn produced subsequent forms of American religious conversion. Records of these orally and publicly produced mini–spiritual autobiographies cannot be divorced from place, for, within each, the migratory pattern indexes the journey of the soul. Movement across land and sea marks a turning toward Christ; settlement signifies unprecedented spiritual fulfillment; the unknown within the wilderness is the unknown within the human soul. But, rather than a genre of American exceptionalism, the incipient and gradual appearance of testimonies of faith on the peripheries of the Anglo world reflects unresolved tension between theology and experience: the former resists a uniform and universal application of its doctrinal dimensions without clear engagement with the transformative effects of the latter.[5]

One of the first instances of the Puritan testimony of faith, this scene became common, as such testimonials became required for church membership in New England's congregational system. While constituting the body of Christ, the test of faith also built the body politic. Experiential knowledge of election not only verified that New England's settlers were among the chosen people who would soon be the initial recipients of Christ's descendant love but also constructed a viable body politic in which those deciding matters of the state were also members of the church. These complementary communal functions of the testimony of faith reflected its dual purpose as a method of knowledge acquisition that also facilitated the reestablishment of the social hierarchy within the New World.

This link between the individual and the community, the singular and the corporate voice, was tenuous throughout the first generation of New England history, dependent as it had become upon the Puritan effort to suture the theological and the political, particularly since New England

5. These testimonial collections include Samuel Petto, *The Voice of the Spirit; or, An Essay towards a Discoverie of the Witnessing of the Spirit* (London, 1654); Vavasor Powel, *Spirituall Experiences, of Sundry Beleevers . . .* (London, 1653); John Rogers, *Ohel; or, Beth-shemesh: A Tabernacle for the Sun . . .* (London, 1653); and [John] Eliot and [Thomas] Mayhew, *Tears of Repentance; or, A Further Narrative of the Progress of the Gospel amongst the Indians in New-England . . .* (London, 1653).

Calvinist theology was intrinsically focused on the individual. For first-generation New World Puritans, the testimony of faith was at the heart of this effort to conjoin these competing strands; its implementation thus exhibited the perceived need to facilitate the translation of the covenant of grace into the federal covenant whereby elect saints would share what they knew of Christ's love, tightening communal bonds.

The testimony also offered visible sainthood as the preliminary stage to a supplemental form of political membership for those of the elect who were also landowning white males. In declaring their faith, they had to give a short and formulaic freeman's oath in order to become part of the body politic and participate in communal affairs. This state-sanctioned and institutionally enforced exclusion of women from the body politic produced a distinctive formal and rhetorical gender difference in the testimony of faith as a direct result of the genre's communal function of bridging the theological to the political and the individual voice to the corporate. Emerging out of an intricate overlap between experimental religion and the burgeoning use of experimental philosophy, an empiricism of grace thus instituted a masculine norm as the ideal testimonial witness of grace in contrast to the patterns of shyness, reluctance, and silence that characterized the testimonies performed by New England's first generation of women. These socially mediated patterns, I contend, reveal that scientific techniques of discernment not only mitigated doctrinal insecurities over hypocrisy and the condition of unknowable knowledge but also engendered a new form of social invisibility that comes across in the testimonial records as a gender-inflected form of religious affect.

Male testimonies invoke and enact a self-discovering "I" in order to orally perform the evidence of grace recorded on their souls as it fills an empty Adamic form, sermonically constructed for them—initially by Thomas Shepard in what became the pervasive New England pattern—as a way of expressing the ontological status of the visible saint as a redeemed Adam and then using this second Adam as a vehicle for exercising political agency. Women's testimonies, by contrast, neither invoke this self-discovering "I" nor find the Adamic form necessary to collect and make use of evidence of grace. Consequently, their testimonies register a sense of spiritual and political incompleteness that compounds and exacerbates the open-ended narrative structure long recognized as theologically mandated and thus integral to the practice. Patterns of hesitancy and reluctance characterize the testimonies of New England women as they defer consistently to the minister's help in discerning their own signs of grace. Testifying men foreground a self-scrutinizing narrative "I" that displays grace through the repetition of this self-authorizing subject and a catalog of active verbs.

Among the male testifiers recorded by Shepard was Jonathan Mitchell, who states, "I found my heart unwilling to stick to Christ." Captain Gookin similarly attests, "I had no power to believe." Sir Starr, "I saw I knew not God but was ignorant of him." John Shepard, "I saw my inability to prepare." And Robert Browne testifies, "I saw myself such a one, for I was ignorant if Christ knew me." Mr. Collins recalls, "God . . . let me see I was one needed preparation." Following this divinely guided self-realization, evidence of grace then flows into a hollow Adamic frame as a structuring vehicle for translating evidence of religious affections into the status of the visible saint, who, infused with qualified knowledge of election, also becomes qualified to adjudicate affairs of the state. Women testifiers, on the other hand, repeatedly manifest an inability to express commensurate forms of self-authorizing scrutiny and an incapacity to find the Adamic form that would channel spiritual experience into productive knowledge. Narrative and rhetorical patterns in women's testimonies ultimately reflect a deep severance between the individual and corporate voice because of an institutional and political process of exclusion that was unique to New England's congregational system and governing body from 1630 to 1660.[6]

The declaration of faith described by Winthrop in 1633 thus subtly departs from Calvinism not only to meet the political demands of the congregational structure but also to gain access to a new domain of forbidden knowledge. Winthrop bases this illicit knowledge in experience, foregrounding attempts by subsequent ministers and natural philosophers to escape the fallen realm of human reason by basing knowledge on the senses. The confessors knew what "work of grace the Lord had wrought in them" by the evidence of this work as supplied by their senses. As with the epistemic transformations of a Baconian world that privileged experience as an ontological category over classical sources and traditional Scholasticism, the testimony of faith elevated experience over scriptural precedent.[7]

6. These testimonies come from the supplemental collection of sixteen conversion testimonies recorded by Shepard and discovered at the American Antiquarian Society by Mary Rhinelander McCarl, published as "Thomas Shepard's Record of Relations of Religious Experience, 1648–1649," *William and Mary Quarterly*, 3d Ser., XLVIII (1991), 444, 451, 452, 454–456, 462; and from Edmund S. Morgan, ed., *The Diary of Michael Wigglesworth, 1653–1657: The Conscience of a Puritan* (1946; Gloucester, Mass., 1970), 107.

7. Repeatedly throughout the *Institutes*, John Calvin speaks of the individualized, privatized, and inwardly directed nature of grace, whereby knowledge of God comes only from "a feeling that can

Winthrop recorded in his *Journal* in 1633 not only an atavistic turn toward Christian primitivism but also the beginning of an emergent ecclesiastical practice. A new, unfolding terrain of theological possibility converged with a philosophical reworking of the human soul as a site of empirical inquiry into the workings of God throughout the world. Puritans began declaring what work God hath wrought upon their souls just a few years before Comenian linguists and natural theologians looked to nature to recuperate the semiotic relationship between nature and the divine that might both restore the knowledge lost in the Fall and enlarge humanity's understanding of God. Confessional declarations of salvific status also anticipated a late-seventeenth-century theorization of the soul as a terra incognita of untapped potential, unfolding divine evidence. This concept of the soul as an uncharted domain spanned theological and natural philosophical writing from Bacon's *Confession* (1641) and Henry More's *Immortality of the Soul* (1659) to Richard Baxter's *Call to the Unconverted* (1658) and William Derham's *Physico-Theology* (1714).

For each of these disparate writers, the soul was the remaining link between the visible and invisible worlds. The post-Reformation world increasingly separated these domains. Theologians and then natural philosophers insisted that nature could be perceived only through the imperfect capacities of fallen human senses. Adhering to a rigid distinction between nature and an invisible domain of revealed truths protected human inquiry from the inevitable possibility of blasphemy as a result of transgressing divinely proscribed boundaries between the invisible and visible, the sacred and profane. While the 1633 testimony of faith preceded ecclesiastical codification as an official requirement for church membership, the testimonies recorded soon thereafter also preceded the theological and philosophical developments that would inaugurate an empirical quest for certain evidentiary forms of God. The convergent programs of Comenian thought, natural philosophy, and antinomian debate to chart evidences of grace helped crystallize this empirical quest in the 1640s. The testimony of faith, initially described by Winthrop and sustained as a central feature of New England lay practice by such ministers as Thomas Shepard, John Fiske, John Eliot,

be born only of heavenly revelation." He writes: "I speak of nothing other than what each believer experiences within himself—though my words fall far beneath a just explanation of the matter" (*Institutes of the Christian Religion*, ed. John T. McNeill, trans. Ford Lewis Battles [Philadelphia, 1960], I, 80–81) (further citations are to this edition). For Calvin, full assurance is possible—even more so than it was for the Puritans—but only as an individualized form of knowledge. This knowledge could not be communicated to others to supplement revealed knowledge of God.

Cotton Mather, Experience Mayhew, Ebenezer Parkman, and Jonathan Edwards, would persist as long as this particular convergence endured.

As it developed in New England in accordance with a pattern of faith testimony that Winthrop initially described, congregationalism was predicated on knowledge constituted by the privileged body of the elect. The elect were expected to communicate this knowledge, despite the congregational commitment to Calvin's sense of assurance as an inwardly directed, individuated experience that could not be communicated to others with any certainty. At the level of the *congregation*—a body in congregationalism deemed independent from the oversight of any governing body or other congregation—individuated experience was somewhat recuperated. Sectarian disagreements over particular doctrinal and theological interpretations produced variations in the forms of religious practice throughout New England churches. Such interpretive differences largely stemmed from the subtly disparate frames that ministers erected for negotiating the consequences of the Fall. In congregationalism, Reformation theology registered particularly acutely the double bind of knowledge: the social and political desire to know was always coupled with a deep awareness of the theological impossibility of knowing. Yet this double bind was also a reality for seventeenth-century natural philosophy, deeply aware that the myth of the Fall obscured human understanding by destroying the human capacity to reason and accurately intuit certain forms of knowledge.

The Fall marked a crisis that pervaded much of early modern thought as it was received and interpreted by localized congregational practice in New England. Particularly troubling for ministers and the lay religious in the colonies was how to extend the Calvinist limits of assurance and to create a sustainable rationale for the experimental method advanced in their own backyard as well as by the Royal Society, whose own fellows shared the same concerns about the limits of knowledge and the spiritual authority of stepping beyond it. The confession of faith was a historically unprecedented form of spiritual declaration. The practice had clear antecedents in confessional practices in the early Christian church and in a number of Reformed sects that splintered off from Calvin's and Luther's teachings over the sixteenth century. Yet in seventeenth-century Anglo America, the practice changed in an unprecedented way: confession became not only a means of recounting the cycles of sin and redemption but also a fissure in the opaque peripheral wall separating the visible and invisible worlds. Through this fissure, the religious peered into the divine realm in search of uncharted knowledge of God.

Primitive Origins of Soul Science

In the early Christian Church, the public confession of sin performed two crucial functions that led the Puritans to revive the practice in the seventeenth century. In the early church confession tested the sincerity of the postulant's conversion and helped regulate who had authority to interpret divine knowledge. The first formally recorded mention of confession occurs in Origen, the Greek theologian and father of the early church. He delineated a confessional morphology predicated on seven modes of pardon and admonished the clergy that the "wholesome practice" of confession should be public, before the congregation. Despite such advocacy, confession remained voluntary and unevenly practiced among early congregations. Origen recommended that individuals select a competent adviser, either a layman or a cleric, who would facilitate confession as a mechanism for curing spiritual disease or for edifying faith. In the primitive Christian church, the power of the keys to divine knowledge was divinely conferred through ordination. This became the basis for regulating knowledge. With this power, the clergy were responsible for transmitting knowledge to the laity. Throughout ninth-century Europe public confession gained ground as a normative practice in the church.[8]

Throughout the Middle Ages, the practice of confession evolved from its more public configuration in the early church to the annual auricular confession required by the Lateran Council of 1215. With this mandate came a parallel adjustment in the configuration of power, knowledge, and spiritual access embodied in the notion of the keys. The medieval church saw auricular confession as a way of regulating the "power of the keys," an expression derived from Matthew 16:19, when Christ confers authority over the church on Peter. The clerics of the medieval Catholic Church were well aware of the dilemma: to insist too forcefully on confession would negate the sincerity of the practice, requiring postulants to perform the formal structure of the confessional mode without firm commitment to the authenticity of the account. Authorized through the Lateran Council, clerics developed techniques of solicitation to an extreme that Reformation theologians saw as naked coercion.

8. For an overview of the history of confession within the Christian Church, see Henry Charles Lea, *The History of Auricular Confession and Indulgences in the Latin Church* (1896; New York, 1968). Numerous histories written by seventeenth-century Catholics and Protestants debate the use of confession. For example, [John Goodman], *A Discourse concerning Auricular Confession, as It Is Prescribed by the Council of Trent, and Practiced in the Church of Rome . . .* (London, 1684).

Central to the Protestant attack on the Catholic Church was the notion that this hierarchy interrupted faith in a way that prevented revealed truths from becoming manifest on earth to the repentant. Reformed theologians reoriented an epistemology of faith based on the assumption that all humans receive the key of knowledge but that those who are not divinely trained remain ignorant of how to use it. Thus primitive and later Christian and Catholic confessional practices always had an epistemological structure, situated at the nexus between divine and human knowledge and as the intermediary between lay and clerical perceptions of how this knowledge was communicated. The dual functions of confession in early and medieval Christianity, as a means of ascertaining sincerity and of regulating power over the key to divine knowledge, were both integral to the confession of faith within the Puritan church, even though the Puritans would so radically alter the form as to constitute a new genre.[9]

This notion of the keys as the divinely conferred symbol of secret knowledge of God would persist through the Reformation debates of the sixteenth and seventeenth centuries. Protestants argued for a substantial revision of this medieval notion of divine conferment of ecclesiastical authority. Central to their critique was an understanding of the Catholic confessional practice as perpetuating theological error. In a Protestant text written in French and translated into English in 1620, Pierre Du Moulin's *Buckler of the Faith* defends the "confession of faith" of the Reformed churches of France against the objections raised by a Jesuit priest. Here, confession of faith indicated the general theological platform of Protestantism, including sola scriptura and religious knowledge invested in all men. The phrase "confession of faith" thus designates a theological position, generally applicable to Protestantism, while also explicitly referencing the genre of auricular confession that Reformed theologians so strongly opposed.

According to Du Moulin, the Catholic practice of permitting "people [to] make profession to follow without knowledge" is the great "maladie of this age." It means that the Catholic "faith is grounded onely upon the authoritie of men; and by consequence, that [their] religion is humane, and not divine." Moreover, the "humane testimonies" upon which the Catholic

9. In addition to Lea's *Auricular Confession,* an overview of Protestant testimonies of faith can be found in John Bossy, "The Counter-Reformation and the People of Catholic Europe," and Wolfgang Reinhard, "Reformation, Counter-Reformation and the Early Modern State: A Reassessment," in David M. Luebke, ed., *The Counter-Reformation: The Essential Readings* (London, 1999), 85–104, 105–128.

faith is grounded are "the worst and most uncertaine" because they perpetuate and proliferate a religion that is veering further and further away from divine authority and truth. Du Moulin advocates recentering the authority of divine knowledge by investing it within each individual convert such that the truth of God will "alwayes hold" in the "souls" of the converts. *Buckler of the Faith* thus realigns authority by placing the power to read revealed truth in the individual convert rather than the clergy. Du Moulin promises a "faith [that] consisteth not in ignorance, but in knowledge: which equally administreth holy things as well to the rich as to the poor." Thus Du Moulin presents a millennial image in which the Protestant faith "cannot be repressed." As it is set down in writing, it "makes [such] a strong impression in the spirits and hearts of the auditors" that the "truth" of divine "evidence" encoded therein cannot be denied. Ultimately, Protestantism

> is a religion, which in stead of framing and fashioning stones like unto the image of man, seeketh by all meanes possible to reforme man according to the image of God: which in stead of worshipping a crosse of wood, adoreth Christ crucified, trusteth in his passion and glorieth in his ignominy.

Within this treatise, Du Moulin presents several aspects of the Protestant faith that respond directly and explicitly to perceived errors of Catholic theology, particularly with regard to auricular confession. The call to dismantle the hierarchy of the Catholic Church in order to create a new basis of spiritual authority is apparent. Du Moulin proposes a faith that does this by vesting scriptural knowledge within the souls of each individual, rich or poor.[10]

John Cotton's treatise, *The Keyes of the Kingdom of Heaven* (1644), argues against the Scottish Presbyterian insistence that the power of the keys was assigned to the elders rather than the laity. In making his case, Cotton illustrates the extent to which the Protestant theological critique of the Catholic auricular confession centered on redefining the scriptural notion of the power of the keys. Cotton concedes that the passage in Matthew is "Allegoricall," "somewhat obscure," and "therefore controversall." He acknowledges the controversial use of this passage for determining how individuals were admitted into the church or accorded access to divine truth. Part of

10. Peter Du Moulin, *The Buckler of the Faith; or, A Defence of the Confession of Faith of the Reformed Churches in France . . .*, 3d ed. (London, 1631), preface, A3, B.

the contention stems from the tremendous importance assigned to this passage, for recipients of the keys have the power to open or close the gates to the church and, thus, to heaven. In contrast to the Catholic notion of the power of the keys conferred upon the clergy during ordination, Cotton claims that faith and knowledge are intertwined: "The *key of Faith*, is the same which the Lord Jesus calleth the *key of knowledge*." Cotton then expands the power of the keys to "Brethren" who have the "power *to judge*" the authenticity of faith supplied through the testimony and even to the "Indian" and "women." Those who receive faith also receive some knowledge of God, as faith and knowledge coexist for Cotton in a one-to-one correspondence. By adjusting the interpretive frame of power / knowledge access conferred through the keys, Cotton establishes a new semiotic relationship between the natural and heavenly worlds. Each convert is a "visible saint," conferring not only faith but also knowledge on the other godly members of the community, equally vested with the "power *to judge*."[11]

Early Christian confessional practices were integral to their seventeenth-century revivals both as a means of debating the power of the keys and for the purpose of purifying ritual. Seventeenth-century translations accordingly adapted Augustine's *Confessions* to different theological positions. The design of a 1631 edition printed in London with "marginall notes" displayed the errors of a previous "Popish Translation." The frontispiece accompanying the edition reveals the translator's aim of making the *Confessions* accessible to lay readers (Figure 5). The woodcut comes from the Latin edition. The message emanating from the stream of light that connects the kneeling Augustine's mind to divine illumination is taken verbatim from the Latin, "Tolle, lege" ("Take up and reade"). The caption, of course, comes from the famous scene of Augustine's conversion, circa 386 in his mother's garden in Milan. Hearing "Tolle, lege" spoken in a child's voice, Augustine opens the Bible before him to what initially purports to be a random passage. It is, in fact, the scriptural passage that initiates Augustine's conversion.

The illuminated message in the frontispiece evokes one of the core tenets of Protestantism, sola scriptura. By translating the Latin phrase into English, this edition reinforces the connection between Augustine's conversion revelation and the Protestant emphasis on biblical literacy for all Christians and thus the Bible's translation and wide availability in the vernacular. Through this reinforced connection, this seventeenth-century translator of Augustine recuperates Augustine's *Confessions* from its link to

11. John Cotton, *The Keyes of the Kingdom of Heaven* . . . (London, 1644), 1, 11, 14, 44.

Catholicism, suggesting additionally, through the image of the conversion revelation, that Augustine's great work transcended centuries of papal corruption. As William James points out in the eighth lecture of *The Varieties of Religious Experience* (1902), this scene is the quintessential example of how conversion resolves the problem of the "divided self." Augustine's conversion has thus come to embody what Perry Miller defined as Augustinian piety, the "inner core of Puritan sensibility apart from the dialectic and the doctrine."[12]

In a direct response to the perceived Roman Catholic practice of promoting "ignorance" through confession, the Protestants countered with a confessional practice meant to encapsulate this Augustinian ideal. Within this edition of *The Confessions,* a revisionist translation mediates doctrinal and theological differences. Augustine describes what God is and how he is known, the kinds of religious truths that are revealed versus those that remain "secret," the transformation of the sense through faith, and the way that this transformation can ameliorate the intellectual capacities of human reason. Augustine concludes by locating all of these components of divine knowledge within the dangerous rubric of the "curiosity," what he labeled the desire for illicit forms of knowledge. According to the logic of *The Confessions,* desire for knowledge of God must continually be checked by an awareness of the dangers of the human "disease of curiosity." "The lust of the eyes," Augustine warns the lay reader, "usurpe unto themselves, when-so-ever they make search after any knowledge." The good Christian must continually balance the inevitable question of what God is, and how known, with the inextricable danger of aspiring to know too much, a compulsion so deeply rooted in original sin.[13]

This condition of knowledge, the continual interplay between an engagement with what is scripturally defined as knowable and an awareness that to seek knowledge of God is to teeter on the brink of heresy, is the point of contention between this translator and the former "Popish translation" to which he is responding. In a passage central to the epistemologies of faith operative within a Protestant versus Catholic confessional context,

12. Perry Miller, *The New England Mind: The Seventeenth Century* (Cambridge, Mass., 1982), 8. Miller's discussion of this form of piety dovetails nicely with James's, for he also contends that the "piety flows from man's desire to transcend his imperfect self," and it finds all misery ultimately related to the problem of "the relation of the individual to the One."

13. Augustine, *Confessions,* trans. William Watts (London, 1631), 687, 689.

FIGURE 5 • *Frontispiece, Augustine,* Confessions. *Huntington Library, San Marino, Calif.*

the translator juxtaposes Augustine's notion of the human soul with a Catholic misperception:

> For even that naturall darknesse is much to be lamented, wherein the knowledge of mine owne abilities so farre lies concealed; as that when my soule makes enquiry into her selfe concerning her own *powers,* it conceyves it not safe, too light to give credit unto it selfe; because that what is already in it, lies many times so closely muffled up, as nothing but experience can reveale it; nor ought any man to bee secure in this life, (which may well bee called one continued temptation). . . . Our only hope, our onely confidence, the only assured promise that we have, is thy mercy.[14]

The soul described in this passage is much like that Reformed theologians from Perkins to Cotton to Baxter used to describe the soul's function as a repository for grace. In contrast to the natural darkness that speaks to the obscurity of limited comprehension within which most humans dwell, the soul contains certain "powers" or forms of knowledge that may be "muffled" but that may, nonetheless, become apparent through the experience of revelation. As the soul makes proper "enquiry" into these "powers," it communicates knowledge of assurance to the convert with a kind of qualified certainty. Like the Reformed theologians who would implement this conceptualization of the soul, Augustine warns that this knowledge does not offer security, but still allows the continual and sinful temptation facing the convert. Yet, the "powers" of the soul do offer a small piece of evidence that may be used to counteract the lamentable natural darkness. This awakening to the powers of the soul and experience-based evidence is precisely what the "Popish translation" has missed when it reportedly "observes that no man can be sure of his salvation." This translator takes the doctrine of salvation by faith alone—one of the key tenets of Protestantism—directly from this reading of Augustine. Experiential evidence of this doctrine becomes apparent in Augustine's text through the spiritualization that the senses undergo in the moment of conversion.[15]

In a configuration of experiential knowledge that resonates with if it does not quite anticipate Bacon, Augustine describes how the senses convey information to the soul in order to reveal divine truth. Once spiri-

14. Ibid., 671–672.
15. Ibid., 672.

tualized, sensory data lead to reliable forms of "intelligence" concerning the "species" of the soul. The sensory path to divine truth, according to this translator, is another aspect of what has been entirely missed by the Popish translator, who is here deemed to have a limited understanding of philosophy as well as of faith. On the one hand, Augustine seems to oppose Bacon's privileging of the eyes as the sensory vehicle for accurate knowledge of the external world through the caution that the "lust of the eyes . . . make search after any knowledge." Yet, he also describes a formula whereby the senses, redeemed through grace, become the privileged avenue to divine truth. Augustine splits the senses into two categories: the five outward senses that represent sight, sound, smell, touch, and taste and the three inward senses that constitute the "inner part" where divine light shines into the soul. Once aligned through redemption, the outward senses begin to convey information to the soul and lead to divine truth. Experiential knowledge in Augustine is thus homologous with the Baconian turn, but with the important caveat that the outward and inward senses must be aligned in order for divine truth to be communicated clearly without the distracting and erroneous information that can be conveyed through the "lust of the eyes."[16]

As the testimony of faith developed as Puritan practice, it continued to follow an Augustinian model of revelation through experience, extending it into a formal practice for collecting the information conveyed upon the souls of others. The genre incorporated an Augustinian and empirical epistemology to renegotiate the boundaries between internality and externality, spiritual equality and social hierarchy. Whereas in the Perkinsian science of the self, the senses functioned as an avenue to an inwardly directed truth of the divine encounter, the testimony of faith required that the soul convey information to the external senses such that they could accurately display the contents of the soul before an audience of lay and ministerial experts. This constituted one means of negotiating the splintering of spiritual authority that occurred through the sola scriptura and sola fides ideals that came out of the Reformation. In place of the mediating force of the clergy to direct the postulant on the correct path to salvation, the testimony of faith implemented a technique of regulating the *ordo salutis* (way of salvation) that occurred after the experiential fact.

16. Ibid., 585–586, 686–687.

As the testimony of faith developed, it came twinned with new worries. No longer might the individual deceive him- or herself alone, but whole congregations, ministers and elders alike, might deceive themselves. Old concerns emerged anew: How could one could know the status of one's soul? What constituted an authentic sign of grace? Could the divines discern a pattern for such revelations that might lead to greater accuracy across a range of congregations? Finally, To what end might this knowledge be put? The structure encoded within confessional practices and most radically altered in the Puritan testimony of faith explains the centrality of the keys as a doctrinal concept that pervaded debates about this practice while also spanning a range of theological and philosophical discourse from the church fathers and the ecumenical councils of the Roman Catholic Church to the likes of Cotton and Comenius. The keys, of course, generated controversy about the power and authority necessary to gain access to divine and, thus, forbidden knowledge. Implicit within the metaphor of the keys is a tacit understanding of the knowledge lost in the Fall. Even though divine comprehension eluded humanity's rational and logical apprehension, divine knowledge was thought at least partially recoverable through a combination of revealed truths, granted solely at God's dispensational discretion, and the powers of discernment recuperated through divinely authorized training. If sola scriptura and sola fides were the tools ordinary individuals legitimately used to discern personal knowledge of God, the confession of faith represented an attempt to create a corresponding mechanism by which to extract personal knowledge for communal consumption.

The confessional practice was not, however, simply a means of dispersing spiritual authority among the laity, but rather a confrontation with the theological inevitability of recognizing that sola scriptura invested the power of the keys at least partially in the people. The position exemplified by Cotton was not simply a democratizing move, for the practice of testimony in all of the various scenes discussed in this study facilitated the development of a certain ministerial expertise. Taking Cotton's conflation of faith and knowledge as the starting point, ministers listened carefully to what their parishioners had to say and then transcribed and recorded it as a means of accruing new knowledge of God. This process inverted the power hierarchy operative within the Catholic history of auricular confession as a means of regulating the power of the keys, but it did not necessarily dilute it. Rather than receive exclusive discerning authority from God, Protestant ministers gradually accumulated this discerning authority by systematizing and studying the sensory evidence supplied by lay converts. The confession of faith became a disciplinary mechanism for regulating this power over spiritual authority at the same time that it recognized this power.

The Geography of Faith

The Protestant critique of the Catholic practice of auricular confession was a universal feature of post-Reformation theological debates. Across sectarian divides and Reformed communities from New England to Dublin to Wales, Protestants viewed Catholic auricular confession as counter to the individual goal of salvation, the collective goal of millennial fulfillment, and the concomitant arrival at divine truth. But what Protestants across different sects and geographical locations intended to do in response is a more specific phenomenon. Congregationalists decided that a voluntary and public confession of faith as criterion for church membership would address the theological need to redistribute spiritual knowledge within each individual soul while also addressing how to establish a basis for an ecclesiastical community. Presbyterians disagreed vehemently over this arrangement for church government and ecclesiastical practice. More radical sects, such as the Baptists and Separatists, had, in many ways, generated practices that anticipated the testimony of faith through experience-based performances about the individual encounter with God in the moment of conversion. The congregationalists departed from this precedent by engaging a deeper theological caution about the elusive nature of this individual encounter and seeing the need to establish a more formal structure for disciplining and mitigating its potential abuses.[17]

Shortly after the testimony of faith became a mandate for membership, New England Puritans received harsh critiques from across the Atlantic for the orthodox practices of the "congregational way." A flurry of pamphlets reasoned the terms of this debate, as theologians disagreed about the practice's theological and biblical legitimacy. Emerging in an already contentious climate, exemplified by Cotton's discussion of the relationship between knowledge and power, this debate over confession as a threshold culminated in a formal defense of the New England Way in the Synod of 1648. The synod published its conclusions on both sides of the Atlantic in an official *Platform of Church Discipline*. The preface sets the tone, stat-

17. Margo Todd explains that the kirks in Scotland carried the confession of sin over from the Catholic tradition as a means of ritual community building. However, this practice served to enumerate sins rather than test election. See *The Culture of Protestantism in Early Modern Scotland* (New Haven, Conn., 2002).

ing that the "publick Confession of the Faith" is supposed to contribute to "pu[b]lick edification."[18]

The purpose of the synod was both to defend the "lawfulness" of the practice of declaring "a personal and public *confession* and declaring of Gods manner of working upon the soul" and to justify this practice for building a body politic. To make a case for the lawfulness of the practice, the synod cites scriptural precedent. The committee interpreted Acts 2:37–41 as depicting a primitive Christian precedent for the witnessing of faith, whereby "thousands . . . Before they were admitted by the Apostles, did manifest that they were pricked in their hearts at Peters sermon." This is a rather liberal interpretation, to fit the biblical passages to the claim for a scripturally lawful testimony of faith. The idea that the "thousands" "did manifest" that their hearts were pricked is entirely the ministers' own coinage. Those present at the synod interpreted Acts in this manner in order to offer precedent and justification for the most contentious and unprecedented component of testimonial practice: the idea that testifiers could manifest, before a public audience, visible evidence of God's impress on their souls, a feeling over which they had neither agency to initiate nor the rational capacity to accurately discern beyond doubt.[19]

The synod also hoped to legitimate the public dimension of the testimonial practice—developed in explicit response to the Catholic auricular tradition—through scriptural justification. The *Platform* cites Psalms 40:10:

18. *Platform of Church Discipline,* 1. In England, these pamphlets included William Eyre, *Justification without Conditions; or, The Free Justification of a Sinner* . . . (London, 1654); John Graile, *Conditions in the Covenant of Grace* . . . (London, 1655); Benjamin Woodbridge, *The Method of Grace in the Justification of Sinners* (London, 1656), all of which supported the theological legitimacy of understanding grace through a system of visible signs. Contesting this perspective was Thomas Edwards, *Gangraena; or, A Catalogue and Discovery of Many of the Errours* . . . (London, 1646); and John Eaton, *The Honey-Combe of Free Justification by Christ Alone* (London, 1642). This debate has been discussed by Ernest Frederick Kevan in *The Grace of Law: A Study in Puritan Theology* (London, 1964) and Ann Hughes in *"Gangraena" and the Struggle for the English Revolution* (London, 2004). On the New England side, John Cotton made a well-known intervention in this debate in *The Way of the Churches of Christ in New-England* . . . (London, 1645). The Zachary Crofton and John Rogers controversy in Dublin is a parallel debate over similar theological issues.

19. *Platform of Church Discipline,* 17. As reflected in the *Platform* published following the Cambridge Synod, there was a long history of attempts to claim scriptural precedent for this practice. There were London and Cambridge, Massachusetts, editions of the *Platform* shortly after the synod, and the text then went into multiple reprintings throughout the late-seventeenth and early eighteenth centuries, indicating its continued status as an authoritative defense of the congregational way. Two years before the 1749 edition, Isaac Watts published his *Rational Foundation of a Christian Church* . . . (London, 1647), which established scriptural precedent for the testimony of faith (see Chapter 6).

"I have not hidden thy righteousness from the great congregation." This verse expands the logic of Cotton's revisionist argument about the *Keyes* as according the brethren as well as the elders the discerning capacity to judge; it also issues a new mechanism of surveillance. The church community joins the clergy in vigilance against sin and religious hypocrites. Correspondingly, the faithful are not to hide their grace from a collective sense of God's continued favor. In theological and ecclesiastical terms, a primary impetus for the testimony of faith was to create bonds within the church community, to link the individual covenant of grace to the federal covenant, and to ensure that Winthrop's individual "ligaments" of Christ's love thoroughly bound the whole body of Christ, individuals and the communal, church and state.[20]

Many scholars have written on early American women's relationship to the corporate voice, though female testimony has gone remarkably underexplored in the literature interested in this formulation. Instead, the discussion largely focuses on the captivity narrative and the genre's most famous exemplar, Mary Rowlandson. All of the complexities of what constitutes gender difference and female authorship in seventeenth-century New England are condensed within her narrative: How much of the account did she write versus how much did her editor, Per Amicus, or Increase Mather? What is the relationship between the event itself and the narrative about the event? What retrospective literary or theological conventions adjudicate the more ethnographic elements of Rowlandson's encounters? Are there moments within the text that mark a disjuncture between Rowlandson's own sense of self or a particularly female faith and piety, and the Puritan patriarchy? Does Rowlandson's account accord a more nuanced relationship between Puritans and Native Americans than traditionally presumed within the context of Puritan providential design, largely framed after King Philip's War within the jeremiad's apocalyptic representations of the "howling wilderness"? Embedded within each of these concerns is a fundamental question whether Rowlandson, or rather Rowlandson as narrator and heroine, ultimately recapitulates or resists the theological, patriarchal, and racial conventions of her time.[21]

20. *Platform of Church Discipline*, 17.

21. A notable exception is Elizabeth Reis, who in *Damned Women: Sinners and Witches in Puritan New England* (Ithaca, N.Y., 1997) makes a claim for a gendered discrepancy in the testimonies of faith. In *God's Caress: The Psychology of Puritan Religious Experience* (New York, 1986), Charles Lloyd Cohen argues that there is no difference between men's and women's testimonies.

The testimonies of faith, the foundational genre of the conversion narrative, shed important light on these questions. Within the testimonies and the debates surrounding their appropriateness as a theological as well as social genre, the relationship between gendered experience, social hierarchy, and modes of political agency made available through a Calvinist model become apparent. An incipient form of religious experience, testimony, sets the terms for such patterns following the first generation. Gendered patterns were not preset or universal tendencies of Protestantism, but rather emerged from the provisional connection between theology and the body politic that the testimony of faith helped to cultivate. Those involved in the Synod of 1648 sought to secure and formalize the relationship between individual grace and communal covenantal structures through the incorporation of state power to adjudicate church affairs, namely the outward behavior of visible saints, including their political rights. "*Saints by Calling*, must have a Visible-Political-Union amongst themselves, or else they are not yet a particular church." In its ideal form, this "Visible-Political-Union" was designed to be an external manifestation of a perfect godly community. This had been John Winthrop's ideal in 1630, but his famous sermon preached even before the community's departure form the Old World also, famously, could not take into account the vast discrepancies between a vision of a voluntary, godly community of like-minded people and the social and political reality of a wilderness without any English institutions or social structures.[22]

The splintering of authority in the direction of experientially based faith meant that Congregationalists would insist upon the voluntary nature of this testimonial practice, though, given the political stakes of visible sainthood, voluntary is not quite accurate, for not to participate meant exclusion from the visible church community, from the invisible body of Christ, and, at least for property-owning males of the first generation, exclusion from the body politic. To choose not to testify was to relinquish one's political rights. Through a correspondent mechanism of making ecclesiastical structure more democratic, women were included within the population of testifiers deemed appropriate for public reception. Praying Indians would soon be included as well, though their public testimonies took place through a homologous but separate structure of the praying town. But to invite women's experiential testimonies into the congregational setting was

22. " The Cambridge Platform, 1648," in Williston Walker, ed., *The Creeds and Platforms of Congregationalism* (Boston, 1960), 207.

not simply a democratizing move. This was seen as an exclusive context for the female voice, one that took place under special conditions, strictly adhering to theological and ecclesiastical policy. The splintering of authority because of the theological transformations of the Reformation led to improvisational techniques for regulating the new spiritual authority accorded to different groups. The formal rendering of a spiritual experience that displays itself differently in men's and women's testimonies reflects one aspect of this hierarchical and regulatory realignment, prompted explicitly from a practice that embodied the very democratization of spiritual authority encouraged by Reformation theologians. The testimonies collected and recorded by Thomas Shepard and John Fiske sought to develop techniques to assuage concerns about the political authority that women might attain as a consequence of public testimony.

Women's Speaking Justified

The Reformation doctrine of salvation by faith realigned gendered structures of authority around new questions of knowledge. Sola scriptura and sola fides led to a dilemma about what to do with individuals deemed to have insufficient reasoning capacities. How could those with a rational capacity diminished even in relation to that bequeathed to humanity by Adam after the Fall discern the subtlety with which God made his presence known on the human soul? The Reformation both endowed the individual with a new form of potential power and concluded that individuals could not be entrusted with their own discerning capacities. Paul's epistles promised a fundamental equality of human souls but also established a hierarchy between the "weak" and the "strong," which suggested God put the weak before the strong.[23]

Aware of the resulting dual challenge to social and spiritual authority, debates about the spiritual knowledge believed to be embodied by "weaker" human populations intensified through the seventeenth century. Thomas Hooker, John Cotton, John Rogers, Thomas Shepard, and John Fiske were just a few of the theological illuminati who tried to address this dilemma in terms of whether women should be allowed to speak in church, specifically, to practice the testimonial form of speech that would supply ministers and church members with the knowledge of their souls. Women

23. 1 Cor. 1:27 states: "God hath chosen the foolish things of the world to confound the wise, and God hath chosen the weak things of the world to confound things that are mighty."

also took up this question, though, regardless of how measured and deliberate their theological intentions, such discussions were inevitably seen as unruly, if not heretical, by the leading members of the Puritan patriarchy.

In her tract *Womens Speaking Justified,* the Quaker Margaret Fell used scriptural precedent to formulate an argument for women's religious speech. The title page alone cites Acts 2:27, Joel 2:28, Isaiah 54:13, and Jeremiah 31:34 in order to claim that, since biblical times, women have been preaching on the Resurrection and that they were sent by Jesus himself to deliver a prophetic message. The tract opens with the story of the Fall, recounting the serpent's temptation of Eve and the subsequent transgression. Fell uses this story to "stop the mouths of all that oppose Womens Speaking," because, Fell argues, Eve speaks the truth immediately following the Fall. God asks, "What is this that thou has done?" to which Eve replies, "The Serpent beguiled me, and I did eat." Fell then explains that God "put enmity between the Woman and the Serpent." Based upon this enmity, either the "Seed of the Woman" speaks or the "Seed of the Serpent speaks." Fell uses this scriptural precedent to make a more radical claim for women's speech. Preaching rather than testifying argues for a genre that was strictly prohibited for Puritan women because it infused the speaker with too much spiritual authority, believed to come directly from God. Nonetheless, this notion of the "seed" of female speech that ensued from the Fall represents a parallel example of a privileged epistemological potential ascribed to women's speech.[24]

As a member of a more radical sect of Reformation theology, Fell makes this claim based on two principles. Referencing the account of Creation in Genesis, she explains that both men and women were created by God *"in his owne Image*. . . . God . . . makes no such distinctions and differences as men do." In a more modern parlance, Fell essentially views gender difference as a function of social construction, rather than an ontological difference divinely articulated. Even while downplaying the role of gender identity in God's plan, Fell argues that certain women have a divinely ordained prophetic authority. Since biblical times, God has occasionally chosen women to be his earthly representatives, to deliver divine messages. In response to 1 Timothy 2:11, "Let Women learn in silence," Fell explains that this injunction applies only to women "forbidden to speak" because they have "not come to Christ, nor to the Spirit of Prophesie." Fell's justification of

24. [Margaret Askew Fell], *Womens Speaking Justified* . . . (London, 1667), 4.

women's speaking is not so much an argument for gendered spiritual equality as it is a justification for the pursuit of divine truth. To deny women's right to speak when they have the gift of prophecy is to emulate centuries of papal abuse and to espouse a false rather than a true church of God.[25]

Anne Hutchinson marshaled a version of this argument. Not daring to make a claim for her right to public preaching, she made a case for scriptural authorization to hold religious meetings in her home through "a clear rule in Titus, that the elder women should instruct the younger." The magistrates who were instrumental in her prosecution saw this as a clear violation of social and gender hierarchy. But even more threatening to the Puritan elders and integral to Hutchinson's banishment from the colony on charges of heresy was, not the violation of social order, but rather a strictly theological claim to a kind of prophetic knowledge—a claim that seemed to align her more closely with someone like Margaret Fell. What is often seen as the linchpin of the prosecution's case in the court trial is an exchange that focuses on spiritual knowledge and authority. After questioning Hutchinson unsuccessfully for days, the minister, Increase Nowell, in response to Hutchinson's insistence that she has had divine guidance, asks her, "How do you know that that was the spirit?" Hutchinson cannily replies with a question: "How did Abraham know that it was God that bid him offer his son, being a breach of the sixth commandment?" The deputy governor replies, "By an immediate voice," to which Hutchinson famously answers, "So to me by an immediate revelation." When the deputy governor says, "How! an immediate revelation," Hutchinson, without her usual caution, reports, "By the voice of his own spirit to my soul."[26]

Knowing the parameters of Puritan orthodoxy, it is hard to read these lines without a feeling of dismay—a combined sense of awe at Hutchinson's precocity and a sense of dread that she was, at this moment, sealing her fate. Winthrop felt he had no choice but to banish her, to tell of her monstrous births—one for each sin against the state, and to align her with the banished heretic Mary Dyer, for similar crimes of slanderous speech, similarly punished.

25. Ibid., 3, 13.

26. Hutchinson's "rule" is in Titus 2:3–5; David D. Hall, ed., *The Antinomian Controversy, 1636–1638: A Documentary History* (Durham, N.C., 1990), 312, 315, 337. In the opening remarks of the trial, Governor Winthrop explains that the charges have been brought against someone who has "maintained a meeting and an assembly in [her] house that hath been condemned by the general assembly as a thing not tolerable nor comely in the sight of God nor fitting for your sex."

Retrospectively, Winthrop connects the mouth to the womb, making each woman's crime inseparable from her gender, entwining the intricate relationship between theology, Puritan patriarchy, and female subversion that has been an iconographic touchstone for feminist interpretations. Yet part of what Winthrop did historically was to collapse unruly mouths with unruly wombs in order to paper over what he would have perceived as a much more substantial threat. Disorderly women were of little significance for the state or the church, as both institutions had mechanisms in place to deal with them quickly and expeditiously, usually by rendering their actions or speech simply heretical.[27]

If we recall that the question posed to Hutchinson at the turning moment in her trial is fundamentally epistemological—"How do you know that that was the spirit?"—the danger that she poses to civil and ecclesiastical institutions becomes clear. She "knows" "by the voice of his own spirit to [her] soul." By claiming direct access to God, Hutchinson opens the way for the individual to have unequivocal knowledge, to circumvent the learned divines. Undoubtedly some of the anxiety produced by Hutchinson's assertion results from the way that her claim places her in the role of the Protestant to the Puritan leader's role of the Catholic clerisy. After all, the Reformed church premised biblical literacy largely on the theological-political principle of "liberty of conscience." Hutchinson types herself as one of Fell's female prophets; her knowledge combined with her chosen status is much more relevant than her gender. Like some New Testament women, Hutchinson has been elected to carry the "Spirit of Prophesie"; to deny her word is to act against God and to espouse a false church.[28]

From this perspective, the outcome of the Antinomian Controversy seems inevitable and even logical from the Puritan perspective. Winthrop had two options: he could allow Hutchinson to continue, and acknowledge that her spiritual authority was one path to true divinity, or he could banish her and publicly reassert that colonial Puritan doctrine was theologically correct. But behind the threat Hutchinson and her following posed lay the inescapable question about the potential for pious women to wield new and particular forms of religious truth. Hutchinson represented the capacity for any individual to be a vessel of the spirit of prophecy. Moreover, there seemed to be no logic or pattern to which souls God communicated with.

27. For example of feminist interpretation, see Eve LaPlante, *American Jezebel: The Uncommon Life of Anne Hutchinson, the Woman Who Defied the Puritans* (San Francisco, 2004).

28. [Fell], *Womens Speaking Justified*, 13.

To open the door for Hutchinson's way meant opening the door for any other way scripturally justified. While the Puritan leadership denied both Hutchinson's authority and the channel of its reception, it worried that a possibility existed that divine truth might be more effectively conveyed to the community through experiences like those Hutchinson so compellingly described.

During and after the trials, debates about whether women should publicly testify to their conviction of their saving faith initiated what would become a long history of the struggle to discern what constituted both women's spiritual role in the church and the relevance of their spiritual knowledge to the larger community. On his arrival in the Massachusetts Bay in 1633, John Cotton became one of the principal ministers responsible for instituting the testimony of faith. Winthrop explains in his *History* that Cotton immediately joined the Boston church, "signifyied his desire and readiness to make his confession according to order." He also "desired" that "his wife might also be admitted a member, and gave a modest testimony of her, but withal requested, that she might not be put to make open confession, etc. which he said was against the apostle's rule, and not fit for women's modesty; but that the elders might examine her in private."[29]

The scriptural passage to which Cotton refers is Paul's letter to the Corinthians: "Let your women keep silence in the churches: for it is not permitted unto them to speak; but they are commanded to be under obedience as also saith the law" (1 Cor. 14:34). This verse seems clearly and unequivocally to make the case against prospective female church members' offering public testimonies of faith, but other ministers felt so strongly that they should do so that they argued against it. Although from the outset Cotton voiced his opinion that women should be excluded from publicly confessing, in many ways he was less orthodox in his view of women generally. Indeed, his authority early on was such that, had the Hutchinson controversy not emerged so quickly—and had she not been one of his congregants in England—one might have imagined Cotton's coming to advocate more broadly for a limited spiritual role for women, as he did, in fact, for Hutchinson in the initial months of the controversy.

Debates about the extent and context of Paul's proscription against women's speaking emerged quickly, displaying a deep tension between Paul's specific admonition and the larger ideals of sola scriptura and sola fides that seemed to legitimate female testimony. The ministers originally

29. Winthrop, *History*, ed. Savage, 111.

responsible for implementing the testimony of faith confronted the prospect that this practice would realign spiritual authority completely, perhaps even granting that women could achieve a higher spiritual status within the church community. This was an unsettling prospect to most, and in fact not least of all to the female testifiers.

Consequently, the Puritan divines sought to structure a program to manage knowledge gleaned from testimony to replace the dismantled power-knowledge axis that had structured the ordo salutis in the Catholic Church. Testifiers articulated experience in a highly disciplined setting, such that the testimonial form itself was at least in part produced by the witnessing audience. An audience comprising exclusively God's elect determined whether the evidence was sufficient to qualify the applicant for church membership and whether the experience was a reliable source of knowledge of the invisible world. (Women were a part of the witnessing audience but could not vote or hold church office.) What we see in the early testimonies, then, is an unsuccessful attempt to enfold the epistemological potential of the female convert within the incipient empirical project of culling evidence of human souls without according women too much spiritual authority and yielding little if any political or social authority.

Thomas Shepard, John Fiske, and John Rogers disagreed with their ministerial brethren about women's spiritual role in church, as each one's church records indicate by the abundance of female testimony. Rogers and Fiske addressed the putative understanding of Paul's proscription against women's speaking in church. In response to the "agitation" over "women making their relations in public," Fiske, for instance, explains that testimony is a unique kind of speech that doesn't fall within the rubric of what should be kept silent. Testimony is an outwardly directed proclamation, or, "This kind of speaking is by submission where others are to judge." Fiske's construction of the membership as a kind of jury speaks to one of the defining characteristics of the testimony of faith: in contrast to other genres of religious conversion such as the spiritual autobiography or the diary, this first-person account was designed to supply evidence of conversion to others and, significantly, to display that evidence for the witnessing community's approval and put its judgment on record.[30]

Fiske compares 1 Corinthians 14:34 to 1 Timothy 3. He admits that "speaking argues power" but that this power can be restricted in the case of certain

30. Robert G. Pope, ed., *The Notebook of the Reverend John Fiske, 1644–1675*, Colonial Society of Massachusetts, *Collections*, LXVII (Boston, 1974), 4.

forms of public speech, emphasizing the submissive quality of testimonial speech. Whereas voicing an opinion in spiritual meetings or even asking a question signified forms of speech that "argued power," the testimony was the denial of the speaker's authority, as it was delivered in a spirit of submission to the discerning authority of the elect and ministerial judges. As Fiske's *Notebook* makes clear at the outset, the form of public speech generated through the testimony of faith was justifiable because the power of the speaker was transferred to the membership as agents of Christ's kingdom on earth. Ultimately, testimonial speech redounded to communal knowledge and God's glory. It policed the boundaries and protected the purity of the visible church.[31]

In his lengthy introduction to *A Tabernacle for the Sun,* John Rogers, a Puritan minister in Dublin, takes up the question of female testimonies of faith. Rogers not only agrees with Fiske and Shepard that women should be allowed to testify but also makes a claim for the particular value of the spiritual evidence supplied in their accounts. While acknowledging the objection based on Paul's letter to the Corinthians, he, like Fiske, contends that testimonials generate an acceptable form of speech because of the speakers' particular frame of authority:

> *Women* are forbid to speak by way of *Teaching,* or *Ruling* in the *Church,* but they are not forbid to speak, when it is in obedience, and subjection to the *Church* (for this *suits* their *sexes*) as in this case to give *account of faith,* or the like, to *answer* to any *questions* that the *Church* asks, or the like; But I shall answer to this at large afterward, because it is so much opposed.

Rogers confirms that teaching and ruling would not be permissible forms of public speech but argues that testimony is permissible, because it subjects women to obedience while offering crucial information about their experience of faith that, when reviewed by the body of visible saints, assists in protecting the church's spiritual integrity and adds to the fund of that community's spiritual knowledge.[32]

31. Ibid. David D. Hall and Charles Cohen interpret the permission to testify publicly as "empowering" for women as well as men. Hall writes that "lay men and women possessed the confidence to speak for themselves about the ways in which they had experienced the workings of the law and grace." Cohen claims that "no one distinguished between the Spirit's operations in one sex or the other." Cohen, *God's Caress,* 222–223; Hall, *Worlds of Wonder, Days of Judgment: Popular Religious Belief in Early New England* (Cambridge, Mass., 1989), 119.

32. Rogers, *Tabernacle for the Sun,* 294.

Not surprisingly, women were often reluctant to confess publicly under the conditions of submissive speaking, church obedience, and communal judgment. Rogers acknowledged the pattern and proposed a solution:

> . . . if any is to be admitted that is very unable to speake in *publike* (I mean) in the *Church,* as some Maids, and others that are *bashful,* (or the like.) Then the *Church* chooses out some whom she sees fit, against the *next Assembly* to take *in private* the *account of Faith,* the *evidences of Gods worke of grace* upon his or her *heart,* which they either take in *writing,* and bring in into the *Church,* or else (which is most approved) when that *person* is to be admitted, they doe *declare by word of mouth,* whilst some easie questions are (notwithstanding) asked of him (or her) for the *Churches satisfaction,* and for the *confirmation* of what was before *delivered* in *private* to the *brethren.*[33]

The necessary knowledge would be extracted, Rogers assured his readers, and this rather than the public act of submission was what was crucial for the testimonies of faith. Regardless of whether performed publicly or given through private inscription of the *"evidences of Gods worke of grace* upon his or her *heart,"* the experiential knowledge supplied to the individual lay convert in the moment of conversion had to be rendered in the form of communal knowledge. It had to be displayed for those deemed the elect to judge and filtered through a process for systematizing the senses of gracious manifestations such that divine knowledge might be enlarged.

As if these stipulations did not put enough pressure on the genre, regeneration was itself uncertain. Knowledge of regeneration hinged on the individual's capacity to translate often vague and contradictory sense experience into personal feelings that might bridge the gap of understanding engendered by the Fall. Such testimony thus perilously skirted the limits of proscribed knowledge, threatening at any point to transgress the divine penalty for hereditary sin. Testimonies navigated this uncertainty through the characteristic oscillation between anxiety and assurance that became the mark of authenticity. Women experienced and then performed an additional condition of uncertainty. They were being asked to speak through a very precarious set of circumstances. Under most circumstances, to speak in church was to defy what Paul explains is the proper place for women. To speak in ways that evince power, assert agency, or express opinions publicly was to violate the very codes that constituted female piety. Testimony was

33. Ibid., 293.

permissible only because it was meant to reveal the contents of the soul, not the subjectivity of the speaker. To testify publicly as a Puritan woman was literally to bare the contents of the soul.

Conversion Morphologies and Sermonic Form

Early Puritan ministers who implemented the testimony of faith sought to cull the evidence promised through the lay convert's soul but also implement a mechanism of social regulation by developing a morphology of conversion. This morphology had to negotiate a difficult boundary: testimonies of faith had to be unique yet generally intelligible, commensurate with previous experiences of grace while offering a new phenomenal account of the effects of God as evidence of election. The sermon functioned generically to negotiate beyond these theological limits on revealed knowledge, encouraging converted saints to extract evidence of God from their souls in these particular ways. The plain style's structure was intended to inculcate in the auditor the tools to perform the inwardly directed self-searching proposed by William Perkins (described in the previous chapter as a science of the self).

Gradually, over the seventeenth century, the sermonic form evolved to elicit these techniques of inward soul-searching through a highly repetitive structure. In his *Call to the Unconverted,* Richard Baxter writes, "I beseech you presently make enquiry into your hearts, and give them no rest, till you find out your condition." The structure of the sermon urges the repentant to "turn" to Christ, establishing a rhetorical frame that acts as an invitation to the kinds of empirical searching for gracious affections described by Perkins. Yet the sermonic form also incorporates regulatory mechanisms. Baxter explains that the congregation acts as witness and "your own consciences are witness."[34]

As the testimony of faith became an established genre in first-generation New England, Thomas Shepard developed the most thoroughgoing morphology of conversion through the Adamic trope woven into much of his sermon series on conversion. If Cotton was instrumental in instituting the confessional practice on his arrival in 1633, Shepard, a sharp critic of Cotton during the Hutchinson trials, was primarily responsible for the development of a morphology that adhered to the practice's most central tenets while innovating some of the finer aspects of the practice. He instituted the

34. Richard Baxter, *A Call to the Unconverted . . .* (London, 1663), 56, 124.

Adamic form as a way of mitigating the threat of too great a dispersal of spiritual authority among the female congregants from whom Shepard, as opposed to Cotton, demanded a confession of faith. The Adamic trope becomes a way of attaching form to essence as later theorized by John Norton in the sermons surrounding the Cambridge Platform of 1648.

Shepard develops the science of signs around a morphology of conversion that imagines recuperating some of Adam's powers lost in the Fall in sermon series such as *Sincere Convert* (1641) and *Sound Beleever* (1649). Collectively, these sermons foreground the trope of Adam as central to a new ordo salutis that sought, as a vital, redeemed community, to contain the knowledge that results from the regenerated powers of the convert as a new Adam. These sermons establish the basis of the formal structure inhering in each of the testimonies of faith that Shepard recorded. *Sincere Convert* is designed to prove the existence of God through evidence of his grace recorded on regenerated souls as the "regenerated" powers granted Adam at Creation. The sermon explains that the original Adam was "a perfect man." "God made all mankind at first in *Adam,* in a most glorious, happy, and righteous estate." Although humanity resembles the original cast of Adam, this frame was corrupted by Adam's fall. The Fall destroyed Adam's perfection but preserved in his progeny the remnants of his original form as well as the hope that actual perfection might be partially recuperated. Shepard's conversion morphology thus begins with the initial recognition of an inherently sinful self through identification with Adam's sin and then progresses to the phase known as humiliation, in which the convert comes to recognize his own innately sinful self as integrally linked to this original Adamic form. Through humiliation, the convert thus takes on both the debt of original sin and the redemptive possibility of regeneration. True humility comes when the saint accepts that this debt cannot be paid in wages or good works. Recognizing the condition of complete debility before God is the individual's first step toward salvation.[35]

35. Tho[mas] Shepard, *The Sincere Convert: Discovering the Paucity of True Beleevers; and the Great Difficulty of Saving Conversion* (London, 1641), 25, 27; Shepard, *The Works of Thomas Shepard, First Pastor of the First Church of Cambridge Mass.* (Boston, 1853), I, 2. The first part of *Sincere Convert* declares its purpose of proving that "there is a *God:*" "I will beginne with the first part, and prove (omitting many philosophicall arguments) that there is a God, *a true God:* for every nation almost in the world, untill Christs coming had a severall God" (1–2). Setting aside any philosophical arguments that might be made about God's existence, Shepard clears the ground for proving divine existence through the discerning capacities and revealed knowledge transmitted to the descendants of Adam.

Following humiliation, conversion becomes an emptying out of all private feelings and ideas about individual autonomy. In his sermon, Shepard describes it through the metaphor of melting down the tarnished inner self, using the metaphor of a material entity such as gold or silver. The three phases described above—conviction of sin, humiliation, and then this self-emptying—prepare the Puritan for communion with Christ. After preparation is complete, man sees himself as a hollow cast of Adam. It is in this hollow space that the pleasurable, reassuring moment of Puritan conversion occurs. Shepard emphasizes the euphoria of this important moment through an explicitly sexual metaphor: once the saint has emptied out his inner thoughts, feelings, and emotions, he becomes Christ's bride, drawing an analogy between this blissful spiritual moment and an idealized secular marriage.

Encased within the trope of Adamic transmission was the same spiritual jeopardy described earlier as Calvinist optimism's simultaneous promise of increasing divine knowledge coupled with shadows of doubt that recur. As a structuring force that compelled individuals to convert by reminding them of their inherently sinful self, the Adamic trope exacerbated the crisis of knowledge in a fallen state. To dwell in fallenness was to remain "guilty of Adam's sin," perpetuated through the darkness of the world, described in Shepard's *Theses Sabbaticae* (1649). No new knowledge could come to Adam in his sinful state. Only by recognizing the depths of sinfulness and beginning to repay this debt could individual converts start to recuperate some of the knowledge lost in the Fall. But, in emphasizing this condition of epistemological limitation, Shepard also inculcated within his listeners a desire to exceed these limitations of knowledge, to "examine" the Adamic frame and "find" Christ's redemptive promise even though true Calvinists had no agency to do so. Through this encoding of the condition of Reformation optimism within the practice of Puritan conversion, Shepard demonstrated that to intuit signs of election with some qualified certainty was to regain lost knowledge. Shepard establishes the ordo salutis as central to this knowledge acquisition.

The optimistic strand of Reformation theology—that knowledge of God might be partially reclaimed through experience—is thus subtly conveyed to the individual through conversion. Establishing an inductive pattern for the study of this experience, Shepard's sermon resists setting formal criteria for conversion while still attempting to enumerate experiential categories that function as an evidentiary index for each individual saint. Humiliation prepares the heart for melting down the tarnished inner self to create a space for Christ's seed to flow into the hollow cast of the sinner as Adam. This melting can begin through an individual spiritual encounter, mediated through the reading of scripture, or it can come about through listening to a sermon. Humiliation, the recognition of the inherently sinful self, precipitates a second phase, which specifically cannot be conveyed in intellectual terms but, rather, depends completely upon an inwardly directed feeling over which the convert has no control. This inward feeling is the initial form of evidence that the testifier can hope to supply to his or her audience. It begins through an entirely sensory encounter of hearing and feeling. Upon experiencing the heart as tender and melting, Shepard encourages his convert to "cry out, Lord, now strike, now imprint thine image upon me!"[36]

This image is the image of Adam. Shepard's plea is for God to imprint Adam's image on the soul of the convert. It is at this moment that form meets essence in Shepard's morphology. The essence of salvation, which functions not only as qualified proof of election but also as proof of divinity as a philosophical supplement, comes in the form of a feeling and melting heart. Adam functions as a formal channel, a way of understanding the uncontainability of grace in an immediately intelligible form. As the heart melts, Christ pricks the soul and restores Adam's original innocence as the newly sanctified visible saint. Significantly, the visible saint does not become an entirely "new man" through conversion, because the Holy Spirit does not remain "indwelling," but, rather, does its work of renewal and then leaves man with new perceptive capabilities that are only a foreshadowing of the world to come. Preached to the future ministers of New England at Harvard College, *Theses Sabbaticae* explains that these new perceptive capabilities are "nothing else but the Law of nature revived, or a second edition and impression of that primitive and perfect law of nature, which in the state of innocency was engraven upon mans heart." The convert has not acquired a new ontological status; he has merely gained a restored capacity to see the law of nature.[37]

Shepard's use of the image of Adam powerfully conjoined the indwelling fruits of the spirit with the idea of paradise regained, the typological culmination of the Puritan mission. Self-identification of these fruits with

36. Shepard, *Works*, I, 23.

37. Shepard quoted in Jesper Rosenmeier, "New England's Perfection: The Image of Adam and the Image of Christ in the Antinomian Crisis, 1634 to 1638," *WMQ*, 3d Ser., XXVII (1970), 441. My reading of the trope of Adam in Shepard's sermons has benefited greatly from this nuanced article.

Adam permitted the individual saint to see the relationship between his or her own conversion and the larger communal and providential script. While perhaps not explicitly indebted to the writings of Francis Bacon or contemporary philosophical writings that were redirecting the acquisition of knowledge in a generally empirical direction, Shepard's sermon nonetheless manages to harness the crux of experimental religion to the more general program for increasing knowledge of God. Contemporary with Shepard's conversion sermons, these movements also focused on the epistemological possibilities of a redeemed Adamic state. Shepard's sermon makes conversion fundamentally about knowledge not only of one's soul but also of the existence of God and of proof of the validity of the New World mission.

Shepard thus located the kernel of Reformation optimism within the soul of each individual convert and then asked all converts within his congregation to supply the evidence of their experience as new and unprecedented evidence of God's existence. Adamic form directed this process, channeling inductive and experiential knowledge into a productive frame. Redeemed saints would know what Adam knew, but they would also use this knowledge toward the general social and political directions that would benefit the Puritan community. Performed individual knowledge of God was permissible so long as the participants understood that "Adam was the head of mankind, and all mankind naturally are members of that head." The social order must stay intact despite the Protestant remapping of spiritual hierarchies. To disrupt this order was to risk corrupting the very knowledge set forth: If "the head practice treason against the king or state, the whole body is found guilty, and the whole body must needs suffer. Adam was the poisoned root and cistern of all mankind." Knowledge conveyed to Adam must be regulated not only through the body politic but also through the church body, described most famously by Winthrop as the interconnected ligaments of love.[38]

Visible Sainthood and the Body Politic

Women were excluded from these new powers of spiritual discernment by virtue of the congregational practice of testifying to the redeemed Adamic state, implemented largely through Shepard's morphology of conversion as a mechanism of social regulation. The public reclaiming of Adam's perfec-

38. Shepard, *Works*, I, 25.

tion and the knowledge of visible and invisible phenomena that comes with this process of reclamation pedagogically reinforced a masculine model as the ideal knower. Shepard implements this process of gendered social control by regulating spiritual experience through a practice of communal verification through witnessing and transcribing the oral account. Whereas a purer version of Calvinism, such as that represented by the more radical antinomians engaging in the pamphlet war, proclaimed that the feeling of religious experience was the first and only authentic evidence of grace, Shepard perpetuated a supplemental form of verification. Experience must be accompanied by evidence that is visible to the human eye: "You can hardly perceive in the seal what is engraven there, but set it on wax, you may see it evidently."[39]

Rendering such evidence metaphorically visible requires translating an inwardly directed phenomenal feeling into communal knowledge. This translation begins in the reciprocal exchange between testifier and audience, as the testifier offers an oral account of the grace recorded on his or her heart for the discerning judgment of the elect. Additionally, Shepard inaugurates the practice of recording such evidence as an additional form of verification. His sermonic metaphor of "engraving" the "seal" of grace as if it were set with "wax" partakes in the more general Protestant idea that evidence of conversion could be more fully verified by setting it down in writing.

The practice of recording the testimonies of faith precisely to guarantee their authenticity was one of Shepard's primary contributions to the development of the genre. Following 1636–1638, years ripe with concern over the problem of hypocrisy, delusions, and deceits from the antinomians and Familists, the accuracy of the testimony of faith became of utmost importance. Shepard even discarded an entire group of oral testimonies of faith intended for the formation of a second church at Dorchester in 1636 on the grounds that they did not supply adequate evidence. In a letter to the Dorchester pastor, Richard Mather, Shepard explained that the confessions were indeed "orthodox and satisfactory." Formalism was not an issue; rather, the experience supplied did not strike Shepard as satisfactory enough to constitute a stable church foundation. Shepard urged Mather not to be "slight" in laying this foundation, for a true church community depends on the strength of its "godly men." "Many godly men are weak, and simple, and unable to discern, and so may easily receive in such as

39. Ibid., I, 19, II, 80.

may afterward ruin them, hence unfit to lay a foundation." A godly church depends on the autonomous self-scrutinizing powers of such godly men who can "search themselves more narrowly" in order to "cast away all their blurred evidences." Clarity is the foundation of a godly church and the goal of the testimony of faith, for, Shepard explained, "we came not here to find gracious hearts, but to see them too." It was not enough to see godly evidence within one's self, for "'tis not faith, but a visible faith, that must make a visible church, and be the foundation of visible communion." Visibility brings truth to such evidences of grace, reducing the blurry nature of grace into a clear and systematized order. For Shepard, this was the mark of a strong church foundation.[40]

Transcription augments verification beyond the initial witnessing approval by members of the elect. Henry Dunster presents his testimony of faith on uncertain grounds: "For assurance of faith I can't or dare not say but I hope I have closed with the Lord Jesus as mine have the condition of the presence wrought in me." Dunster does not know whether he has assurance, but he senses it enough to hope that the "presence wrought" in him will become visible through testimony. While the inward experience of assurance may leave only a faint outline of God's grace, this faint outline can be transferred to "wax," an outward impression of experience through testimony that makes the effects of grace visibly evident for the individual convert as well as the witnessing community. The perceptive capacities that the convert acquires by being renewed to Adam's image are thus applied to humans in order to discern the forms of grace upon the soul as divine law appears in material form.[41]

As Shepard extends the observational capacities of the converted saint to make the inward experience of conversion visible, the waxen image becomes a way of mitigating the problem of hypocrisy. In an attempt to dispel the specter of hypocrisy, "*the Formalist,* who contents himself with some holinesse, as much as will credit him," Shepard turns the perceptive capacity of the elect into a pedagogy of transparency. He warns his congregants, "Feare to sinne therefore in secret, unlesse thou canst find out some darke hole where the eye of God cannot discerne thee." The practice of testimony introduced to Shepard's Cambridge congregation in 1638 demanded a "wax-like impression" that would cast inward experience as a visible tem-

40. Thomas Shepard to Richard Mather, Apr. 2, 1636, in John A. Albro, ed., *The Life of Thomas Shepard* (Boston, 1847), 211, 214–216.

41. Morgan, ed., *Diary of Michael Wigglesworth,* 113.

plate before the members of the elect who had acquired Adam's "new eyes" to see it. Testimony served as a partial remedy for the human proclivity toward misreading, as witnesses "skild in the deceits of mens hearts" could identify those who had been so deceived. Through the waxen image and the Adamic trope that supplied testifying saints with a formal structure through which to experience and narrate the feeling of grace, Shepard sought to extract the kernel of Reformation optimism from its recurrent and ever-present shadow of doubt.[42]

Exhibiting the transparency that Shepard imagined in his waxlike impression, male testimonies performed and recorded in Cambridge narrated a science of self-scrutiny that externalized grace through a self-authorizing capacity to see Adam's sin and discover the prepared heart as an empty frame. The inductive method of discerning grace enabled men to act as their own witnessing subjects, recounting the inward encounter through "observed particulars" from scripture that matched the numinous ethereality of one's own soul to the divine.[43]

Exemplifying this process, Mr. Haynes states, "I examined had I that thirsting frame, which I found I had in some measure." The Reverend John Fiske asks Norton to consider "the frame of his own heart." Other men relate their experience of preparatory discovery through the use of specific active verbs: Captain Gookin, "saw I had no power to believe." Comfort Starr, "saw I knew not God but was ignorant of him." John Shepard, "saw my inability to prepare." Robert Browne testifies, "I saw myself such a one, for I was ignorant if Christ knew me." Mr. Collins recalls, "It pleased God to let me see I was one needed preparation." "I saw," "I found," and "I examined" are often, almost exclusively, used by men to convey a self-authorizing testimony of empirical discovery. The empiricism of the self embodied in these verbs and the identification of the "I" with Adam's sin

42. Shepard, *Sincere Convert*, 20–21. The phrase "skild in the deceits of mens hearts" appears in W. Greenhill's introduction to the author of the sermon in "To the Christian Reader," [xi]; Shepard, *Sincere Convert*, 34. Shepard balances the Calvinistic suspicion of the "self's own instrumentality" with a technique for tracing visible forms of divine election in newly transparent spiritual subjects (65).

43. I borrow the phrase "observed particulars" from Mary Poovey, *A History of the Modern Fact: Problems of Knowledge in the Sciences of Wealth and Society* (Chicago, 1998). Poovey examines the relationship between Puritan plain style and the Erasmian ideal of "brevitas," or fitness between words and meanings, in order to argue that the material practices that ensued from this nexus of cultural influences contributed to the rise of the Baconian fact. The practices of transcribing experimental data expressed in Puritan conversion narratives would be an example of this, although I am proposing a more-prolonged and bidirectional period of influence between Puritanism and empiricism.

translates experience into visible evidence for a witnessing audience that trusts the speaker's capacity to be a reliable witness to his own soul. "I" plus the active verb renders the reliability of the speaker visible for the witnessing audience, which then accompanies him on an inward journey into the soul. The testimony serves as a wax template of the conversion experience through the testifier's authorizing vision. Spiritual truth comes through the discerning capacities of the individual. In New England congregations, a process of self-identification with an externally visible Adamic frame enabled a newly transparent male subject to look toward a new model of masculinity.[44]

John Stansby, a farmer and clothier who joined the church in 1641, describes the inward gaze of a self-authenticating masculine convert. He begins his experience as many others begin, with a recollection of a sinful childhood. As he grows, he tempts others into sin; his life follows a path of drinking and adultery until the day that he recognizes the vile, "rottenness" of his heart and begins to turn to Christ:

> I saw how the evil of sin how it separated me and God, greatest God, and that nothing provoked the Lord nor grieved Him more than sin . . . by seeing my vileness, I was drawn to hunger and thirst after Christ and made me feel my need of Christ.[45]

Stansby's autonomous self-presentation as an "I" both separate from and connected to God constructs the boundaries of his spiritual self as one that can reveal and also withhold in the moment of testimonial performance. What we know of Stansby's spiritual journey comes purely through what he chooses to tell us. His sin separates him from God until he practices a form of pious self-scrutiny that allows him to see his own "vileness," which then produces the pivotal narrative turn toward Christ. Stansby recounts this as an allegorical turn from darkness to light: "I saw my hellish, devilish nature opposite to God and goodness, between light and darkness." This form of inwardly directed seeing initiates a spiritual transformation

44. George Selement and Bruce C. Woolley, eds., *Thomas Shepard's "Confessions,"* Colonial Society of Massachusetts, *Collections,* LVIII (Boston, 1972), 170; McCarl, "Thomas Shepard's Record of Relations," *WMQ,* 3d Ser., XLVIII (1991), 444, 451, 456, 462; Morgan, ed., *Diary of Michael Wigglesworth,* 107. When these verbs of discovery appear in women's testimonies, the minister often records them in the third person, employing an editing technique that implicates the witness-transcriber in the process of identifying inward signs of grace.

45. Selement and Woolley, eds., *Thomas Shepard's "Confessions,"* Colonial Society of Massachusetts, *Collections,* LVIII, 86, 87.

that causes Stansby to feel his "heart" break as he realizes his own implication in dragging "Christ to the Cross." Through his own experiential journey, Stansby sees his own Adamic sin, establishing a frame through which the testimony turns further inward and progresses smoothly to receiving divine grace. Those witnessing the testimony in Shepard's congregation observe this reception of divine grace through the "evidence" as it is recognized and supplied by Stansby. He proclaims: "I have an evidence my nature is changed because when my sin ariseth I go to the fount opened." Stansby's narrative continues to offer such proof of his changed nature, experienced and expressed, not in good works, but rather through the feeling that his once broken "heart hath been straightened for God." He exclaims, "Love came to me in Cambridge."[46]

Stansby's conversion works by recognizing an inherently sinful self that separates him from God. This recognition changes the course of his life and behavior as he begins to "thirst" after Christ until he experiences a changed inward nature that unifies rather than severs him from God. As this divine unification begins, Stansby experiences evidence of grace that he can then recount to the community as a sign that his nature has changed. Finally, correspondent to this changed nature is a feeling of integration within the Cambridge community; love comes to Stansby in Cambridge, as he can live there while supplying evidence of Christ with qualified certainty.

Stansby's individual experience of conversion connects him to a community of love where he derives strength that stands in contrast to Shepard's own repeated sense of inadequacy and weakness to perform his public ministerial duties. In a journal entry following a 1641 sermon, Shepard narrates a self-scrutinizing "I" that traces inwardly discernible evidence that stands in marked contrast to the male confessors exemplified by Stansby:

> I saw my own weakness (1) of body to speak, (2) of light and affection within and enlargement there, and that my weak mind, heart, and tongue moved without God's special help. (3) I saw my weakness to bless what I did. Hereupon I questioned whether the Lord would ever bless one so impotent that did my work without his power and sinned so much with such dead, heartless, blind works.

Shepard describes himself as "weak" and "impotent" because he does not feel the strength of assurance that can come only from the saving power

46. Ibid.

of Christ. The Puritan theological problem of assurance was caught up in constructions of gendered identity to such a degree that, as Jane Kamensky has argued, anxiety over "public speaking" "reflected" and to some extent "constituted Puritan masculinity" for precisely this reason. This passage in Shepard's *Journal* reveals not only cultural anxieties about salvation but also a specific uncertainty about his own manhood and ability to exercise political authority.[47]

Because he cannot imagine his own private body as "spotless," that is, emptied of original sin and prepared as an empty Adamic frame within this moment, Shepard feels his own inward inadequacy to perform his outward duties as a minister. Shepard's exacerbated sense of his own depleted capacity was compounded through the experience of the Antinomian Controversy, where he not only played an integral role but also took to heart the sheer difficulty of maintaining a transparent and pure congregation. Alongside his own inability to imagine a "spotless" interior self, Shepard vows that his church would remain "spotless from the contagion of the opinions" of the Antinomians and protected from the invasion of the "Indians."[48]

Following the failure of the testimonies at the Second Church at Dorchester to supply clear evidence of salvation, Shepard insists upon surveying the private spiritual experience of each member of his congregation through a rigid pedagogical structure so as to eliminate all "blurred evidences." His goal is to establish a secure foundation for the congregation both within its original pillar members and as the subsequent generations continued to produce proof of their own election. The male testifiers recorded by Shepard in his minister's notebook represent his ideal sense of this foundation, expressed to Mather through an explicit desire for strong "godly men" who in contrast to the "weak and simple" can "discern" evidence properly and build a solid foundation. Shepard depends upon the autonomous self-scrutinizing capacities of certain members of the elect because of his own admission to Mather that it was his own "weakness" that prevented him from "seeing" faith "visibly" among the first professors in Dorchester. Consequently, the masculine models of piety recorded in Shepard's Cambridge represent narratives of the fulfillment of a certain

47. Michael McGiffert, ed., *God's Plot: Puritan Spirituality in Thomas Shepard's Cambridge,* rev. ed. (Amherst, Mass., 1994), 98; Jane Kamensky, "Talk like a Man: Speech, Power, and Masculinity in Early New England," *Gender and History,* VIII (1996), 23.

48. McGiffert, ed., *God's Plot,* 70.

kind of empirical desire for clear evidence that stands in contrast to Shepard's own deeply personal and unfulfilled spiritual desire.[49]

Divine grace becomes the nourishing energy source, the content that fills the empty Adamic form and, in Bacon's words, "quickneth the spirit of Man." The quickening of the spirit through assurance constructs a homology between the male convert's newly redeemed status and the Puritan typology of the New World mission. The masculine form instituted through the practice of visible testimony enabled men to serve as reliable witnesses to the legitimacy of the unfolding providential design. Shepard's idealized bridegroom of Christ, whose image was perfected in the *Parable of Ten Virgins* preached as a sermon series from 1636 to 1641, is a fallen Adam who can be saved only if he becomes passive, humble, and submissive—that is, feminized—before God. Therefore the "prick" into the hardened heart makes the heart soft so that it can receive God's seed. Shepard's bridegroom feels the pleasurable, assuring moment when God's seed is implanted into his heart, and then looks forward to the next phase, in which he will embody Christ's goodness and be able to act accordingly. The masculine external body in Thomas Shepard's "Confessions" is an empty shell, capable of fulfilling secular requirements once the empty internal soul desires and experiences the penetration of Christ's love.[50]

The inner experience of the soul and the outward experience of living in a godly community coalesce in the masculine testimonial form to create an allegorical link between the physical and spiritual journeys to the New World. Roger Clap describes this idealized function of testimony when a stranger walks into the Dorchester congregation. Although he "had no love to his Person," upon hearing the "Report" of his fear of God and conversion, Clap's "Heart was knit unto him, altho' I never spake with him that I know of." John Stansby's proclamation that "love came to me in Cambridge" echoes this sentiment. William Andrews revisits his initial impression of New England after his conversion experience: "When I saw the

49. Thomas Shepard to Richard Mather, Apr. 2, 1636, in Albro, *The Life of Thomas Shepard,* 214, 215, 216.

50. Francis Bacon, *The Confession of Faith* ([London], 1641) 9. My reading of the experience of Shepard's bridegroom builds upon what Ivy Schweitzer calls "redeemed subjectivity" in her reading of the lyric poetry of Edward Taylor. Schweitzer argues that the contradiction between sociopolitical demands and pious Puritan humility becomes resolved through the "difference" between visible and invisible realms. This difference enables the translation of passive spiritual humility into masculine public agency. Ivy Schweitzer, *The Work of Self-Representation: Lyric Poetry in Colonial New England* (Chapel Hill, N.C., 1991).

people, my heart was knit to them much and thought I should be happy if I should be joined and united to them." The Adamic form prevalent throughout male testimonies through the trope of the "thirsting frame" that, as we see in Stansby's confession, generates and directs desire for Christ also conjoins these various elements of Puritan conversion. Adam is the link to the evidence that Stansby sees upon his own soul as an indication of his own redeemed status as a new creature; Adam also indexes the expansion of this redeemed status toward a New World of dawning light where Adam's posterity can reclaim its original powers of discernment in order to see the natural world in new light.[51]

Invisible Testimony

In contrast to the transparency of male testifiers, first-generation women show a reluctance to engage in experiential religious empiricism, to claim the signs of grace as their own autonomous discovery, and to reliably witness to the evidence of their own soul. This gendered discrepancy hinges on the problem of assurance in relation to testimony as a visible sign of grace. Standing before an audience of congregants in 1649, Goodwife Jackson states her knowledge that "the heart is deceitful above all things," a standard theological premise of Puritanism, based upon Jeremiah 17:9, which nonetheless translates in women's oral testimonies into an inability to perform qualified evidence before a witnessing audience. The fear of exposing a deceitful rather than a pious heart before a community of judges supplements the doctrinal insecurity of one's unworthiness for election with a social anxiety about publicly proclaiming signs of grace. Imagine the scene of Goodwife Jackson's testimony, tentatively delivered with the hope that her experience might provide enough evidence for her admittance into the church body of elect saints in Cambridge. She captures the hesitant and cautious poignancy of this scene through the metaphor of

> the opening of the prison to them that are bound. For I have sometime looked on myself as a prisoner shut up in sin and corruption. Yet that place being once opened, I was much helped by it, and though I do not feel help from it daily, yet sometime I do.[52]

51. Roger Clap, *Memoirs of Capt. Roger Clap* . . . (Boston, 1731), 7; McGiffert, ed., *God's Plot,* 180, 191.

52. McCarl, "Thomas Shepard's Record of Relations," *WMQ,* 3d Ser., LVIII (1991), 446, 449.

That the testimony supplies a small opening to Goodwife Jackson provides some relief. Yet testifying publicly also produces an uncomfortable sensation of self-exposure, which intensifies as she proclaims feeling, as she feared, "the deceit of the heart being opened." "I did not think there had been so much deceit in my heart till then that the minister did discover the same." In explicit contrast to the repeated self-discovering claims of her male brethren—like Jonathan Mitchell, who confesses, "I found my heart unwilling to stick to Christ," or John Shepard, who claims, "I saw my inability to prepare"—Goodwife Jackson is unable to know the deceit of her own heart. A ministerial guide must "open" her heart to discover the evidences recorded therein. Moreover and more devastatingly, the act of publicly recounting this opening finds more "deceit in [her] heart" than she originally thought.[53]

Joanna Sill hears repeatedly of the New World promise of spiritual fulfillment: "Diverse ministers came to apply [the] promise but she could apply none." Finally, while living under the ministry of Jose Glover, the rector at Sutton in England, Joanna becomes convinced of her purpose to go to New England: "Oft troubled since she came hither, her heart went after the world and vanities and the Lord absented Himself from her so that she thought God had brought her hither on purpose to discover her." Sill presents herself as an object rather than subject of discovery. Her "inward terrors or miseries" supply the evidence sought by the witnessing community rather than a self-discovering "I." The narrative gaze—conveyed in Jackson's case through passive voice and in Joanna Sill's case through a confession recorded almost entirely in the third person—is outwardly directed, belonging to the eyes of God that, as Shepard's sermons explain, see all and to "man's sight," intent upon seeing what God sees. The hidden heart "inclosed" upon itself is deceitful, yet the hearts of Shepard's female congregants have been "opened" before him and others who have the discerning capacity to see the displayed contents. This does not lead to the same form of spiritual self-fulfillment promised by "diverse ministers" and convincingly by Mr. Glover, for, upon arriving in the New World, Sill discovers that "though she did not neglect duties, yet she found no presence of God there as at other times."[54]

Displaying a similar experience of objectified empirical discovery and exclusion, Elizabeth Olbon explains that, even though she has the empiri-

53. Ibid., 446.

54. Selement and Woolley, eds., *Thomas Shepard's "Confessions,"* CSM, *Collections,* LVIII, 50, 51.

cal desire for the soul's evidence that results from repeated sermonic instruction, her desire does not lead her down a path of salvation. Her narrative lists a series of frustrated attempts and failures to move toward the path of self-knowledge that will lead to salvation:

> Yet burdened, and speaking with her, he pressed her rather to be fitted for comfort than to seek for comfort. . . . And then he showed what that hiding place was. But seeing her evil, she saw she had no right to it. . . . But she felt so much evil in her own heart, she thought it impossible so poor a creature should be saved . . . she saw she must come to a naked Christ, and that she found the hardest thing in the world to do.[55]

Rather than a desire to convert to a community of saints, Shepard's instruction seems to have produced a sense of doubt and despair that mirrors the personal despair he records in his "Journal." Elizabeth Olbon is so afraid of her depravity that, unlike the male confessors, she cannot let her inner self melt away. In his "Journal," Shepard confesses a similar inability when he repeatedly refers to his inner body as "weak" and "vile" in contrast to "Christ's righteousness." During her confession, Olbon falls down before the entire congregation, publicly performing Shepard's deeply personal feelings of impotency and weakness as a public speaker. Whereas the male confessors fulfill Shepard's desires for divinely sanctioned public agency, the female confessors often mirror this anxious inner self. Olbon's list of attempts and failures signals her own inability to claim the Adamic form that will enable the transforming inward experience. Her relation rests in the troughs of anxiety more consistently than do her fellow masculine congregants', who readily find their "hiding place" through their gendered "right to it." While Olbon's testimony concludes with a hint of assurance—"[She] hath since witnessed the Lord's love to her"—it is an assurance that appears without any logical narrative progression and lingers at the end as an ethereal experience detached from its grounding in the material form of the Adamic frame. Without this frame, the spiritual domain remains elusive and full of mystery. A consequence of this mystery is that female testimonies reflect a reluctance to perform the inward gaze and a deferral of empirical autonomy in the moment of public expression.[56]

To compensate for this failure, women's narratives depend upon min-

55. McGiffert, ed., *God's Plot,* 152–153.
56. Ibid., 114–115, 117, 152–154.

isterial guidance leading up to the moment of imputation, the stage of the conversion morphology that offers evidence of the presence of divine grace. Elizabeth Oakes states: "[Shepard] showed [me] that the soul could do nothing without Christ." Goodwife Stevenson hears from "Mr. S[hepard] that Christ would come in flaming fire, etc. and hence desired the Lord." In contrast to her husband's discovery of his own "filthy nature and actual sins," Mistress Gookin claims that "the Lord first began to work on my heart by Mr. Tompson," a minister in Billerica. Mary Angier Sparrowhawk testifies that "living under a powerful ministry of Mr. Rogers of Dedham, she was convinced that her estate was miserable." Such testimonies cite the evidence of humiliation produced by ministers in place of the discovering "I" that sees and finds the form of Adam. After examining and finding his thirsting frame, Mr. Haynes states: "I considered what need I had of Christ. And the Lord witnessed I did thirst and so the Lord did draw my heart to Himself and then I had manifest light of my estate." He refers to Matthew 5:6, a passage often cited by men as scriptural evidence that their own "hungering" and "thirsting" frame will soon be "filled" with the assuring infusion of divine grace.[57]

Differences in the scriptural metaphors used by men and women stem from contrasting abilities to imagine the self within the trope of the Adamic frame. Testifying on September 11, 1648, Captain Gookin writes that, even though he "had not then any clear evidence that [he] had closed with Christ[,] Yet the Lord made him appear so beautiful to my heart, and the more desirable in respect of my misery, that I saw he was the only physician fit to apply a plaster." Christ is the only fit "physician": whatever transformation takes place and enables Gookin to see this evidence more clearly will be instigated directly by God. Plaster figures within the Bible as a metaphor for the reworking of a falsely laid foundation or remolding of man within God's image. Captain Gookin imagines his own intact frame prior to imagining this plastering. In contrast, Mary Angier Sparrowhawk learns from a neighbor that "she was God's clay," a metaphor from Isaiah 64:8 that reaffirms a sense of her own malleability. Isabell Jackson finds comfort in a "godly minister's instruction" that teaches her that she is the "bruised reed"

57. McCarl, "Thomas Shepard's Record of Relations," *WMQ*, 3d Ser., LVIII (1991), 442, 443, 458–459; McGiffert, ed., *God's Plot*, 168–169; Selement and Woolley, eds., *Thomas Shepard's "Confessions,"* CSM, *Collections*, LVIII, 170. In the Geneva Bible, Matt. 5:6 reads: "Blessed are they which honger and thirst for righteousness: for they shal be filled." When women reference this passage, they express a different relationship to it. Olbon follows her reference to Matt. 5:6 by stating, "She saw she must come to a naked Christ, and that she found the hardest thing in the world to do."

from Isaiah 42:3. Goodwife Gookin cites this image too, explaining that, upon realizing that "the Lord still affected my heart with deadness," she "read Isaiah 42, A bruised reed he will not break, etc." The bruised reed image is a comforting one, as it is for Goodwife Gookin, signaling resilience in times of spiritual trial, yet it is also an image of malleability that stands in contrast to the secure Adamic foundation traceable across male testimonies of faith. The bruised reed and clay-potter images constitute a discernible pattern in women's testimonies in which the interior remains "bruised" throughout conversion. Without the self-emptying described in *Sincere Convert,* the soul lacks the frame to support the infusion of Christ's seed.[58]

In John Fiske's Chelmsford congregation, many of these gendered patterns recur. A woman there expressed knowledge of humiliation and "a willingness and desire wrought in her to turn from her sins, but mixed with great fear and doubting considering *her own ability* to turn [i.e., convert]." In Fiske's *Notebook,* one consequence of the absence of a female frame to support the infusion of divine grace is a pattern of questions and answers between minister and testifier that is most common to female saints. John Fiske asks of Ann, the wife of Phineas Fiske: "What evidence was there of God all this while for that time hitherto?" "Hath the Lord helped you to see any such failings as whereby you justly may be hindered?" "How she made out the love of God to her sake from that scripture?" In each case, the questions are designed to supplement Ann's own self-discovery, to guide her on a path toward the recuperation of the particular forms of evidence within her soul. She answers that "she [has] examined herself and found that it rested with God." This statement repeats verbs commonly found within such testimonies of faith; Ann examines her soul to find the evidence requested of her through the minister's question. But, in contrast to the male testimonies, Fiske records Anne's discovery in the third person, deflecting the ascribed legitimacy to the self-discovering "I" that we saw in John Stansby. Ann further distances her experiential relation from her own

58. McGiffert, ed., *God's Plot,* 152–153; McCarl, "Thomas Shepard's Record of Relations," *WMQ,* 3d Ser., LVIII (1991), 446, 454, 459. The biblical passages cited in relation to Gookin: Mark 2:17 and Luke 5:13; the direct use of plaster seems also to draw from Ezek. 13:14: "So will I break down the wall that ye have daubed with untempered morter, and bring it down to the ground, so that the foundation thereof shall be discovered, and it shall fall, and ye shall be consumed in the midst thereof: and ye shall know that I am the lord."

Jackson: Isa. 42:3: "A bruised reed shall he not break, and the smoking flax shall he not quench: he shall bring forth judgment unto truth." McGiffert, ed., *God's Plot,* 170, 213.

self-scrutiny by stating that her discovery is that all such evidence rests with God. As such, the testimony augments the Calvinist deferral of human volition to the divine.[59]

Although almost all of the testimonies recorded by Shepard and Fiske produced adequate evidence for church membership, women's testimonies were occasionally recorded as empirical failures. When Brother William Fiske's wife desired to have her conversion considered for church membership, the witnessing audience objected to her "reservedness" in delivering her oral testimony. But, then considering that she was "usually observed to be silent from speaking of heavenly matters and spiritual matters," they decided to excuse her failure to adequately produce a "particular" of her conversion. Elizabeth Olbon's collapse before the entire congregation publicly performs the formal inadequacy of the testimony to capture her experience of grace. This pattern of empirical failure publicly externalizes the private confessions of deep anxiety and doubt that Shepard records in his "Journal," suggesting a connection between the inner recesses of the anxious Puritan self and the public performance of female piety. The dark impenetrability of spiritual interiority becomes reconfigured as a feminized space. Shepard's sermons warn of a "dark hole" in which one can hide from the sight of God, drawing a contrast to the light and clarity that he wishes to bring to more certain knowledge of the soul's evidence and figuring this dichotomy in explicitly gendered imagery. Those with visible hearts keep their "lamps trimmed, [their] lamps burning, [their] wedding garments on to meet the bridegroom." Those who seek refuge within the dark hole that obscures this clarity threaten Shepard's system for communal surveillance.[60]

In the spirit of the aftermath of the Antinomian Controversy, which constructed a heretical woman as a scapegoat for larger theological and epistemological concerns, Shepard explains that the human heart's tendency to become this dark hole makes it like "the womb that contains, breeds, brings forth, suckles all the litter, all the troop of sins that are in thy life; and therefore, giving life and being to all other, it is the greatest sin." Sin issues forth from the dark shadow of a feminized space. Slanderous speech that would obscure the clarity of the soul's evidence is once again linked to the womb. In contrast to the Adamic form, which can be emptied

59. Pope, ed., *Notebook of Fiske*, CSM, *Collections*, XLVII, 30 (my emphasis), 44.
60. Ibid., 4; Albro, ed., *Life of Thomas Shepard*, 214, 217.

of its contents, such imagery suggests that the "blinder strata" of religious experience are a mysterious, feminized domain that cannot be so transparently displayed.[61]

But, far from merely a patriarchal means of social regulation, this conflation of the inner recesses of mysterious spirituality and gendered imagery transforms the epistemological blindness of a particular strand of Calvinist thought into social anxiety. Female testifiers incorporate into their public performances the dark shadow of doubt that is intrinsically linked to the kernel of Reformation optimism. They absorb this condition of incertitude such that the male testifiers might perform its counterpart: the desire to "cast away" "blurred evidences" and the hope that the knowledge deemed unattainable since the Fall might be reclaimed.

The deferral of self-discovery in women's testimonies registers self-discovery's inextricable antithesis: evidence of the soul cannot be known with certainty, and it certainly cannot be displayed clearly to others. Women testifiers embody the condition of unknowable knowledge in order to permit the community to pursue this knowledge in spite of its possible heretical consequences. The result is a disjuncture between their personal assurance and their experience of divine love in New England. Olbon states that, "since she came hither, she hath found her heart more dead and dull, etc., and being in much sickness when she came first into the land, she saw how vain a thing it was to put confidence in any creature." Barbary Cutter recalls that, although "she embraced the motion to New England . . . after [she] came hither, [she] saw [her] condition more miserable than ever . . . and spoke to none as knowing none like me." Brother Crackbone's wife discovers upon coming to New England that she has a "new house" yet "no new heart." Jane Palfrey Willows describes desiring to go to New England to "know more of [her] own heart," but, without the supporting form for this self-discovery, she "wondered [that the] earth swallowed [her] not up." Mistress Joseph Cooke explains, "When I came here I found my heart altogether dead and unprofitable under means." Each of these statements marks a sense of spiritual depletion upon arrival in New England. The inability to find a self-discovering "I" within the formal register of the testimony doubles through an inability to imagine this

61. Shepard, *Works of Thomas Shepard*, I, 29–30; quote from William James, speaking of "the recesses of feeling" as "the darker, blinder strata of [human] character," in Clifford Geertz, *Available Light: Anthropological Reflections on Philosophical Topics* (Princeton, N.J., 2000), 167.

"I" within a social order that invests the New World mission with meaning and order. In the absence of an inwardly gazing self and a form that could bear witness to imputation of grace, women struggle to see their own experience of assurance and fail to connect this experience to Shepard's maxim, "'Tis not faith, but a visible faith."[62]

This pattern lasted until 1660, when Charles II required the New England colonies to expand the franchise of church government. The crown's modification coincided with the Halfway Covenant, a theological agreement designed specifically to expand the franchise of the federal covenant through an ecclesiastical rather than a religio-political hierarchy, creating two tiers of church members. The purpose of the Halfway Covenant has long been read as a response to the perceived problem of declension. This was in part the case, but the Halfway Covenant also functioned as an explicit response to Charles II's mandate, which disrupted the empirical inquiry encoded within the sociopolitical structure of the first generation's use of the testimony of faith. The Halfway Covenant permitted the social hierarchy to persist as a structure partly mitigated by ecclesiastical order and partly by the state while also shifting the epistemic gaze such that data of the soul could be culled from alternate sources. Divine election to the visible body of Christ, homologous and homogenously linked to the body politic, was no longer the sole criterion by which a human soul could be deemed an appropriate empirical site for increasing knowledge of God.

Also, by 1662, certain clever ministers interpreted this restructuring of congregational hierarchy as affording them more freedom in religious matters. One of the key ways in which they exercised this freedom was focusing ever more intently on the testimonial evidence that could be culled from an increasingly diverse population of women, children, and Praying Indians. Ministers were freer in "matters of religion" to look to human souls, seated in individuals that no longer needed to be enfolded into an external form of political enfranchisement. And these individuals were increasingly asked to supply evidence of grace that did not fit a prescribed mold because it did not need to be fully integrated into a cohesive body of Christ and collective body politic. The decoupling of the church from the state, the visible saint from the active political member, supplied a developing ministerial expertise with greater potential to find anomalous patterns

62. McGiffert, ed., *God's Plot*, 153, 183, 194, 199–201; McCarl, "Thomas Shepard's Record of Relations," *WMQ*, 3d Ser., XLVIII (1991), 460; Albro, ed., *Life of Thomas Shepard*, 214.

of grace across a diverse array of lay souls. Correspondingly, lay populations soon begin to exhibit greater freedom of religious expression; the patterns of silent reservation that characterize Shepard's women are soon replaced by increasingly emphatic testimonial voices, but this freedom of expression was much less tangible in any realizable political way. Moreover, it was a freedom increasingly mitigated by the discerning capacity of the ministry, who were regaining the authority that had splintered through the aftermath of the Reformation by accumulating divine knowledge of God through an empirical study of human souls.

Toward a Natural Philosophy of Grace

The conversion publications coming out of London in 1653 and 1654 made the science of the soul available to a natural philosophical community that was increasingly interested in empirical methods of increasing divine knowledge. Samuel Petto's *Roses from Sharon* (1654), Vavasor Powel's *Spirituall Experiences of Sundry Beleevers* (1653), John Rogers's *Tabernacle for the Sun* (1653), and John Eliot and Thomas Mayhew's *Tears of Repentance* (1653) each contain exemplary testimonies that record the experiential process of grace in an attempt to catalog the evidence of the soul. While the conversion morphologies represented in these tracts were not uniform, the published collections put into circulation new models for more radical forms of spiritual autobiography and religious experience.[63]

Rogers announces this science of signs in the pages leading up to his catalog of testimonies: "*Assurance* is a *reflecting act* of the *soule* by which a *Saint* sees clearly that he is in the *state* of *grace*." "Marks and evidences" contained in testimonies are "infallible," he tells the reader; "experience teaches more and better than all," an important and surprising declaration in a religious climate consumed with the danger of hypocrisy. Such a declaration privileges the stability of human experience as the most secure form of data collecting. As in New England's congregations, the evidence, taken from the mouth of the testifiers, qualifies each convert for membership within the church because it stands as adequate proof that he or she is among the elect.

63. Crawford Gribben traces the theological distinctions across these collections of conversion testimony. He also points out that the questions posed on the title pages highlight the epistemological stakes of Puritan conversion. *God's Irishmen: Theological Debates in Cromwellian Ireland* (New York, 2007), 69.

But Rogers's prefatory remarks to Oliver Cromwell in *The Heavenly Nymph* also propose a supplemental use of such evidence, expressing an interest in how *"theology"* might supplement *"philosophy"* in seeking *"Interest* of *Christ Jesus."*[64]

A portrait frontispiece figures the ministerial divine as a kind of philosophy of the soul: it fashions an image of the divine as the philosopher, reasserting a continuity between the two forms of expertise after Calvin had severed the "rude and untutored crowd" of "philosophers who have tried with reason and learning to penetrate into heaven." In a poetic epigraph to his treatise, Rogers explains that the universe remains mysterious behind the "Veile *of* Clay," covering all of the inhabitants throughout the "milkie Way." The only means of transcending this mystery is through a practice of gazing and attuning human perceptive capacities increasingly to the "Shadow" and "Substance" of "Christ's motion's *Swift*." Sight is integral to the discernment of spiritual truth for Rogers: "The truth is, not only mine eyes are *dim,* and *dull-sighted,* but so much *dirt* hath been lately *dashed* into them, that I am forced to sit still to *picke* and *cleanse* them, before I can come *forth* againe to this *golden worke* in the *Mines*." This is not the plain vision of observational empiricism, but rather the particular variation of visual techniques supplemented by gracious affections and religious truths. Rogers imagines that spiritual sight is the key to spiritual truth as the mysteries of the universe and the Milky Way unfold upon the souls of each individual convert within his Dublin congregation. His goal is to supply evidence of an unfolding millennium, but his method of arriving there is deeply empirical.[65]

Rogers describes the testimonies collected in the remaining pages of his treatise as "Examples of Experience," or "Discourses and Discoveries" that come through "hearing the voice of the Son of God." The techniques developed by Rogers establish criteria for discerning marks of grace on human souls according to the methods that would soon be codified by the Royal Society. Since the introduction of the practice in New England in the 1630s, the testimony of faith reflected an application of experimental religion that drew inductive knowledge from the senses. By the 1640s the genre coalesced and transformed through the heightened convergence of

64. John Rogers, *Tabernacle for the Sun,* 293, 376, *Challah: The Heavenly Nymph; or, The Bride . . .* (London, 1653), 16.

65. Calvin, *Institutes,* I, 65; Rogers, *The Heavenly Nymph,* 45.

FIGURE 6 • *Frontispiece, John Rogers,* Ohel; or, Beth-shemesh: A Tabernacle for the Sun *(London, 1653). Huntington Library, San Marino, Calif.*

Comenian linguistic and natural philosophical and theological debates (as discussed in the previous chapter). Consequently, Rogers, Petto, Powel, and Eliot, the authors and compilers of the 1653 and 1654 publications, augment the experimental religious tradition with new methods of observation from the natural sciences.[66]

Members of Rogers's church in Dublin provided "a clear *account* . . . of the *work of Grace* upon their *hearts*." Some were converted through ordinary means and some through extraordinary means; in each case, Rogers tracks the means through a sequence of highly regulated questions: "When? where? and how? with the effects." The questions establish a sense of uniformity with regard to the technique employed for study of experiential evidence of grace. Each testimony supplies such evidence according to a morphological structure that tracks each phase of conversion according to the repetition of such questions in the margins. Rogers emphasizes the authenticity of each account by announcing that each experiential relation "*was taken out of his own* mouth *in* Dublin." In Rogers's collection, this reiteration of the place and direct method of transcription encases each testimony with a discourse of verbatim legitimacy.[67]

The questions elicit experiential knowledge of the work of God's grace upon the human soul. Answers are performed orally by a group of converts who believe themselves capable of supplying such experiential evidence and then are recorded as "evidences of the work of grace" in the "wilderness." The wilderness stands metaphorically in this case as a landscape capable of wielding new truths of divinity to those with the discerning capacity to listen and then transform the oral evidence supplied by the testimony into a taxonomy of visible signs. Dublin, Wales, Suffolk, and Natick are the sites for enacting these practices of reading "EXPERIMENTAL *Evidences of the* work *of* GRACE *upon his* SOUL." This geographic diversity speaks to the widespread effects of a remarkable early modern convergence between theology and natural philosophy that taught Baconians and Calvinists alike to seek new locales and new experiences to expand the scope of experimental discovery. This sense of a diaspora of Puritan practitioners fueled the millennial eschatology woven into each of these collections. Through witnessing such testimonies, Rogers writes: "We may say, that have well *observed*, how vigorously and quickly *Christ* hath mastered *England! Ireland!*

66. Rogers, *Tabernacle for the Sun*, 392.

67. Ibid., 392, 396.

and Scotland! over and over! O *Lord,* go on! these cannot *satisfie!* go on to *France! Holland! Italy!* and *all* about, till all the *tribes* of *Israel* be *gathered* into *Hebron!* and *crown thee* their *King!*"[68]

Yet the collections also reflect the particular way with which the techniques for observing experiential evidence were implemented across these various locales. For example, Rogers, more freely than Shepard, privileges the experiential knowledge contained in women's testimonies. He goes beyond Shepard's and Fiske's positions on female testimony, claiming that, once members, women as well as men have an equal stake in church affairs: "*In the* Church *all the* Members, *even* Sisters *as well as* Brothers, *have a* right *to all* Church-affairs; *and may not* onely implicitely *but* explicitely *vote and offer* or object." Consequently, his female testifiers do not display parallel performative or rhetorical forms of shyness, reluctance, and reserve, but, rather, speak freely within a theological frame that recognizes them as equal contributors to the church community. They, along with the men, convey "the *best* and *choicest*" examples of experience that should be brought to a "*publique light*" as evidence of an unfolding millennium. Recounting an experience "*as it came out of her own mouth in the Church at* Dublin," Ruth Emerson declares that she "came to see that [she] was *lost,* and left by *Adam* in a woful condition." Anne Hewson confesses that she "came to see no *happiness* but in *Christ* alone." Anne Bishop recounts, "I Have tasted much of *God* upon my *spirit.*" Mary Barker learns from the preaching of Mr. Rogers how she "might *know Christ is in us of a truth.*" And Sarah Barnwell declares that she was "brought to this assurance which I have."[69]

These female testifiers represent some of Rogers's "*best* and *choicest*" examples, and so confident is he in their words that he chooses to bring them to "public light" such that they might reveal unfolding evidence of the millennium. Their testimonies are commensurate with the male testimonies, displaying parallel self-identification with Adam and the narrative rendering of a self-discovering "I" that wields the discerning expertise to correctly identify evidence displayed upon their own souls. Rogers is more interested in the epistemological potential encoded in their speech than in its regulation. These subtle differences illustrate spiritual knowledge as a spatially constituted phenomenon within the early modern world, and the widespread appearance of such testimonies in diverse locales reflects a clear

68. Rogers, *The Heavenly Nymph,* 21, *Tabernacle for the Sun,* 354.
69. Rogers, *Tabernacle for the Sun,* 411, 412, 414, 415, 450, [463].

and consistent convergence across diverse systems of knowledge in the late-seventeenth century. As this portrait of John Rogers suggests, natural philosopher and minister alike were gaining expertise by borrowing from each other. Both used the methods of empiricism for "keeping the soul under a continual sense of God" while simultaneously searching the soul for new domains of knowledge and new *"hidden Treasure"* of wisdom.[70]

70. Jos[eph] Glanvill, *Philosophia Pia; or, A Discourse of the Religious Temper, and Tendencies of the Experimental Philosophy*... (London, 1671), 13, 15.

{3} Praying Towns
Conversion, Empirical Desire, and the Indian Soul

A testimony proposed or offered . . . is not effectual unlesse received.—SAMUEL PETTO, *The Voice of the Spirit* (1654)

In the winter of 1652, ten Indian proselytes assembled in an English-style meetinghouse in Natick, Massachusetts, to orally recount what their conversion to Christianity felt like. Puritan missionaries John Eliot and Thomas Mayhew's *Tears of Repentance* published the record of these testimonies. In 1653, *Tears* appeared in London in multiple editions alongside John Rogers's *Tabernacle for the Sun,* a similar collection of Irish conversion testimonies. Their publication corresponds to the year of Oliver Cromwell's installation as the lord protector of the Commonwealth of England, Scotland, and Ireland. *Tears* and *Tabernacle* appeal directly to this context: epistolary dedications to Cromwell frame the conversion testimonies gathered, translated in Eliot's case, transcribed, and collected therein as "tokens of more Grace in store to be bestowed on Indian" and Irish souls, respectively.[1]

These tokens honor Cromwell's reign with exemplary proof of an imminent millennium that accorded with the goals of Cromwell's troops as they advanced through the British Isles in a concerted effort to reinforce English authority. Eliot and his collaborator, Thomas Mayhew, hoped to position their missionary endeavors as pivotal to the Cromwellian vision for the spread of Protestantism throughout the Anglo world. They described their work in Natick accordingly, through a dramatic apocalyptic portrait of "poor captivated men (bondslaves to sin and Satan)" who have been brought forth by the hundreds "to renounce their false gods, Devils, and Pawwaws" in designated "meetings" before many "witnesses." By the

1. Thomas Mayhew, letter, in [John] Eliot and Mayhew, *Tears of Repentance; or, A Further Narrative of the Progress of the Gospel amongst the Indians in New-England* . . . (London, 1653), B2. Eliot expresses similar sentiments in his prefatory letter.

hundreds, Eliot and Mayhew emphasize, these "poor Indians" have come forth to "def[y] their tyrannical Destroyer the Devil."[2]

Written and published after Charles I's beheading in 1649, it is hard to read this image apart from its clear allegorical reference to the defeat of a tyrannical king who was wholly unsympathetic to the Puritan cause. Eliot and Mayhew align themselves with Cromwell's troops in their epistolary dedications, performing the lord protector's will in the far reaches of the Anglo world to ensure that Protestantism takes root so as not to be so easily threatened again. This was an opportune time for Eliot to make a case for his mission's centrality to the Protestant cause. Parliament had just formed the Society for the Propagation of the Gospel in 1649 and bestowed praise on Eliot as exhibiting "the pious care and pains of some godly English of this Nation, who preach the gospel to [the Indians] in their own Language." Through this praise and through the formation of the society, Parliament recognized the importance of missionary work to a British Empire thriving under the renewed canopy of Protestantism.[3]

If the confession of faith developed in the 1630s on the eastern seaboard of North America so as to be safe from the persecution of Archbishop William Laud, 1653 was not a political time to keep quiet the evidence of grace listened to, collected, and recorded from human souls. Rogers and Eliot quite boldly and directly laid out a religious practice that up to this historical moment was the exclusive purview of the Puritan elect, an enclosed community wherein the evidence of grace was tacitly understood as fleeting and as yet inconsequential to the world beyond the congregation. In contrast to the data quietly recorded by Thomas Shepard, John Fiske, and Jonathan Mitchell and stored in their private ministerial notebooks for centuries, Mayhew, Eliot, and Rogers recognized that, in a time of militant millennial belief and practice, the evidence of God recorded on human souls could speak unequivocally as empirical verification that God's "promise to his plantation" was finally bearing its fruits. The audience witness-

2. Ibid.

3. Thomas Weld, "An Act for Promoting and Propagating the Gospel of Jesus Christ among the Indians and Others in New England," MS Rawlinson, Bodleian Library, Oxford University. A minister in Dublin, John Rogers was also one of the Fifth Monarchy men, a Puritan sect of extreme millennialist believers who supported Cromwell at first but then turned against him during the establishment of the Protectorate. Many were arrested for violent agitation, including Rogers.

Kristina Bross makes a compelling case for the publication period for the Eliot tracts (1643–1671) as coinciding with the English Civil War. *Dry Bones and Indian Sermons: Praying Indians in Colonial America* (Ithaca, N.Y., 2004).

ing these signs of grace transformed at once from a self-selecting group of communal judges to a diverse, transatlantic population of those for (and against) the Puritan cause. Perhaps in part for this reason New Englanders chose to send their "poor Indians" as exemplary souls. There was much less at stake in rendering the work of grace upon an Indian soul up for scrutiny by all. But, if accepted as authentic, the work of God among this population could also serve as greater proof that the gospel was in fact taking hold in the wilderness, for the natives did not begin their spiritual journey as Puritans, but, rather, under the more ominous directive of a devilish and tyrannical ruler. Their salvation, therefore, carried more evidentiary weight as proof of an impending millennium.[4]

Set in the broader philosophical, religious, and political context of 1653, *Tears of Repentance* indexes the coalescing threads within the science of the soul. Evidence of grace reaches a broader audience than previously imagined, one that at least potentially included members of Parliament and natural philosophers as well as divines. Eliot seeks to buttress his missionary work against theological and ecclesiastical critique while also promoting the revelatory power of an unprecedented form of testifying to the experience of religious conversion. The "pious care" with which Eliot learns Algonquian and receives praise from Parliament facilitates his capacity to elicit conversion while also aligning him with the group of language philosophers most commonly associated with Samuel Hartlib's circle. For Eliot as well as for Rhode Island's Roger Williams and Chappaquiddick's John Cotton, Jr., the study of Algonquian at least aspired to be a scholarly endeavor, extending from beliefs in language's capacity to unlock the mysteries of the divine to a universal language movement.

With the help of his printer, Marmaduke Johnson, and his native translator, James the Printer, Eliot translated several Christian texts into Algonquian to be used as pedagogical tools in missionary schools where young Massachusett Indians were taught to read in their native tongue. Such holy texts as Eliot's *Indian Primer* (1669), Richard Baxter's *Call to the Unconverted* (1663), and the Bible comprised Eliot's Indian Library, or the collection of Algonquian texts used in missionary schools. Eliot not only preached in Algonquian; he required his converts to testify in Algonquian. Consequently, the oral narratives recorded in *Tears of Repentance* couple the experiential

4. I am quoting the title of John Cotton's sermon, *Gods Promise to His Plantation* . . . (London, 1630), which was delivered with the goal of encouraging New World migration and reprinted several times, in 1630 and 1634.

data of God's manifestations upon the souls of America's "heathens" with the descriptive capacity of a native tongue that was believed, in its redeemed form, to encapsulate the sacred essence of an ancient biblical people, scattered and lost through the dispersion of Babel. Indian testimony presented a new use of language as a medium of communication between the natural world and the mysteries of the invisible realm. As extensions of the natural landscape—"poor souls" who are still in their natural state—Indians, speaking in their native tongue about God's manifest presence on their souls, presented a wealth of empirical possibility.

Following the epistemological structure described earlier, the testimonies of faith spoken in the praying towns conjoined the enigma of grace and Baconian procedures of natural science such that a holy empiricism of sorts became a hallmark of Puritan practices of faith. Self- and communal scrutiny surrounding the testimony of faith proceeded with the same care that the new categorizing energy around natural phenomena demanded. Yet the publication of the Praying Indian testimonies of faith marked the science of the soul as colonial in new ways, setting the practice apart from its Anglo congregational precedent. Framed as proof of an impending millennium in 1653, the Natick testimonies mark the implementation of the new science in an internally contradictory practice by presupposing the outcome. These are the experiences of "poor captivated men," Mayhew announces to Cromwell. The possibility of an objective, inductive, empirical approach to the work of grace upon their souls ceases in this initial generic description of a "Pawwaw" whose "Imps" continue to torment his "flesh"; even after his public renunciation of his "Devillish craft," he "could never be at rest, either sleeping or waking." Generic description preempts an empirical account, transforming the testimony into an artifact of communal knowledge for an audience of theologians, natural philosophers, ministers, and lay readers struck by the "curiosity" of the phenomenon of grace among the Indians.[5]

5. Quotations in text come from Thomas Mayhew's letter to Cromwell that begins *Tears of Repentance.* Based on their belief that complete cultural transformation necessarily accompanied Indian conversion, English missionaries set up a system of "Praying Towns," or missionary communities set up to impose English standards of landownership and social order as a prerequisite for membership within Indian congregations. Jean M. O'Brien, *Dispossession by Degrees: Indian Land and Identity in Natick, Massachusetts, 1650–1790* (Lincoln, Nebr., 1997), 27, 29.

The curiosity always signified licit and illicit knowledge of nature, the extraordinary, and the marvelous. P. Fontes da Costa, "The Culture of Curiosity at the Royal Society in the First Half of the Eighteenth Century," *Notes and Records of the Royal Society of London,* LVI (2002), 147–166;

Yet the testimonies recorded in the *Tears of Repentance* never deliver the curiosity promised to the reader in this initial generic description. Instead, the tract records a sequence of frustrated attempts and failures to observe and bear witness to anomalous forms of grace. The Native Americans, presumably doing just as they were taught, recite accounts of conversion that accord with what one might expect to find in an Anglo congregation but with the important exception that they express assurance much less emphatically. In a clerical decision reminiscent of Shepard's rejection of the testimonies at Dorchester, Eliot registers his frustration by refusing to approve Natick as an independent congregation. He attempts to rectify the problem by using other means of extracting the evidence, such as the question-and-answer method employed in a subsequent tract, *A Late and Further Manifestation of the Progress of the Gospel amongst the Indians in New-England* (1655). While still somewhat unsatisfactory to Eliot, his work in this tract refines his empirical and ethnographic techniques toward more effective means of soliciting the responses he desires in native testimonies.[6]

Both *Tears of Repentance* and *A Late and Further Manifestation* are part of the eleven well-known pamphlets referred to as the "Eliot tracts." All were printed in London between 1643 and 1671 as evidence of the "light of grace" among the Pequot, Wampanoag, Mashpee, and Massachusett Indians of southern New England, Massachusetts, Nantucket, and Martha's Vineyard. From the rise of natural theology in such texts as Walter Charleton's *Darknes of Atheism Dispelled by the Light of Nature* (1652) to the Baconian attention to experience and nature to the formation of the Royal Society, the Eliot tracts span the formative decades of England's Scientific Revolution, crystallizing the science that I am describing as an effort to catalog the soul alongside curiosities and natural taxonomies. The Eliot tracts reflect a concerted effort to mobilize the potential for new spiritual knowledge encoded within the Praying Indian testimony. Ministers who recorded these testimonies produced new evidence of grace by focusing on the scene of witnessing as well as on contemporary theories of the Al-

Susan Scott Parrish, "Women's Nature: Curiosity, Pastoral, and the New Science in British America," *Early American Literature*, XXXVII (2002), 195–245; Parrish, *American Curiosity: Cultures of Natural History in the Colonial British Atlantic World* (Chapel Hill, N.C., 2007); Lorraine Daston and Katharine Park, *Wonders and the Order of Nature, 1150–1750* (New York, 1998).

6. Thomas Shepard gives his reasons for rejecting the Dorchester testimonies in Shepard to Richard Mather, in John A. Albro ed., *The Life of Thomas Shepard* (Boston, 1847), 214–217.

gonquian language's sacred potential. A key audience and recipient for this evidence was Robert Boyle, who would become a founding member of the future Royal Society in 1660 and governor of the New England Company's mission to the Indians in 1661. The extended relationship between Eliot and Boyle was not merely a philanthropic arrangement. Boyle's notes and essays, Eliot's published tracts, the correspondence between the two, and the larger contextual overlap between the New England Company and the Royal Society demonstrate an emergent theory of Indian testimony as an object of ethnographic inquiry and a resource for knowledge about the divine as well as the natural world.[7]

Henry Whitfield (*Strength out of Weakness,* 1652), John Eliot and Thomas Mayhew (*Tears of Repentance,* 1653), and Eliot (*A Late and Further Manifestation,* 1655) imagine a London audience of natural philosophers. By the 1660s, when Eliot and Boyle formalize their relationship, Indian testimony encapsulates a conversion experience that is novel, anomalous, and marked by an emergent natural philosophical discourse of primitive populations. An ethnographic record of the specific ways that grace affects a primitive and natural soul never appears in the published Eliot tracts, which instead record the frustrating silences of empirical failure. But, ultimately, Eliot fulfills his empirical desire in a manuscript collection of seven conversion testimonies collected from the Mashepog Indians and recorded on the Sandwich Islands in 1666. By this time, the combined theological frame of North American natives as more deeply fallen sons of Adam and an emergent natural philosophical proclivity toward taxonomizing humans permitted Eliot to assign a racialized logic to the Indian soul. In so doing, his ethnographic and empirical eye bore witness to a Mashepog conversion experience that exceeded the conventions of Protestant theology.

The seven Praying Indians whose testimonies Eliot recorded in this manuscript display with surprising conviction and certainty that they have "found the presence of God on their souls." They define faith explicitly as

7. The Society for the Propagation of the Gospel in New England (or the New England Company) was formed through an act of Parliament in 1649; when Charles II came into power in 1662, the charter was reissued. "Copy of the Charter for the Propagation of the Gospel in New England," MS Papers of the New England Company, 7909, Guildhall Library, London. According to George Parker Winship, there were two periods of evangelical effort, instigated by financial groups in London: the Society for the Propagation of the Gospel in New England, in 1649, and the New England Company, in 1662. *The New England Company of 1649 and John Eliot* (Boston, 1920).

"a light let into my heart," and they repeatedly proclaim not only an awareness of the knowledge lost in the Fall but also emphatic declarations that their "faith" has restored much of this knowledge generating new "understanding," "will," and "memory." Struck by the power of these claims and confident in the confinement of their potentially heretical implications to a group of Indians on the Sandwich Islands, Eliot sent them to Robert Boyle.[8]

Boyle never published this document, though it is possible that Eliot wanted him to, given the formality with which he composed it. Like the curious but cautious ministers in the Massachusetts Bay, Boyle carefully and discreetly preserved this collection of testimonies in the records of the Royal Society. It seems, though, that Boyle read them quite carefully and gleaned what he could from the exciting epistemological implications of the religious affections recorded on native North American souls. It was only a few years later that Boyle would conclude that the mysteries of religion could be discerned as man became more empirically adept. *The Christian Virtuoso,* the magnum opus of Boyle's attempt to employ experimental philosophy in the service of discovering Christian truths, would reference the Indian soul as an empirical site upon which to increase the imaginative intellectual capacity of the Christian philosopher.

Christian Virtuoso reveals the impact that the manuscript collection of Mashepog testimonies sent by Eliot had on Boyle's philosophical imagination. Read alongside each other, Eliot's missionary manuscript and Boyle's philosophical treatise exemplify profound epistemological and political continuities between the New England Company and the Royal Society, between the new science and New World discovery. As if registering this connection, when Charles II chartered the Royal Society in 1662, he outlined his effort to "'extend not only the boundaries of the Empire, but also the very arts and sciences,' especially those philosophical studies 'which by actual experiments attempt either to shape out a new philosophy or to perfect the old.'" While the society's experimental objective created a rapport between philosophers from Old and New England, the praying towns also constituted laboratories of grace, where divine phenomena could be more

8. A fair copy of this manuscript is in the Boyle Papers at the Royal Society of London, entitled "A Brief History of the Mashepog Indians, called by the English the Sandwich Indians, inhabiting within the Colony of Plymouth, who did publicly make their Confession of Jesus Christ, upon the 11th of the 5th month, 1666." These confessions have been edited with an introduction by J. Patrick Cesarini, "John Eliot's 'A Brief History of the Mashepog Indians,' 1666," *William and Mary Quarterly,* 3d Ser., LXV (2008), 101–134.

accurately discerned and where the divine essence encoded in the Algonquian tongue might finally be comprehended and redeemed.[9]

Indian Keys to Ancient Wisdom

Talal Asad's question, "What kind of epistemic structures emerged from the evangelical encounter?" has not been fully addressed for colonial America, because scholarship has directed most of its attention to a debate about the extent to which natives were or were not able to secure avenues of resistance to the colonizing practices of Christianity without due attention to these practices themselves.[10]

Indeed, John Eliot knew the Praying Indians at Natick were ready for conversion when they built "a very sufficient Meeting-House, of fifty foot long, twenty five foot broad." This meetinghouse demonstrated their successful emulation of the civilizing standards of English society. But there was an exception to these seemingly totalizing practices of civilization. Missionaries worked to preserve the language of the proselytes so that, when they stood within this meetinghouse, they narrated their religious

9. The Royal Society formed in 1660 as the Philosophical Society but asked Charles II for an official charter in 1661; he issued it in 1662. Charles II quoted from the "Translation of the First Charter, granted to the President, Council, and Fellows of the Royal Society of London, by King Charles the Second, a.d. 1662," Royal Society, London. Quotation in text appears in a seminal text on this connection, Raymond Phineas Stearns, *Science in the British Colonies of America* (Chicago, 1970), 90.

10. Talal Asad, "Comments on Conversion," in Peter van der Veer, ed., *Conversion to Modernities: The Globalization of Christianity* (New York, 1996), 264; James P. Ronda, "Generations of Faith: The Christian Indians of Martha's Vineyard," *WMQ*, 3d Ser., XXXVIII (1981), 369–394; Neal Salisbury, "Red Puritans: The 'Praying Indians' of Massachusetts Bay and John Eliot," *WMQ*, 3d Ser., XXXI (1974), 27–54. See also Robert James Naeher, "Dialogue in the Wilderness: John Eliot and the Indian Exploration of Puritanism as a Source of Meaning, Comfort, and Ethnic Survival," *New England Quarterly*, LXII (1989), 346–368; James Axtell, "Were Indian Conversions *Bona Fide?*" in Axtell, *After Columbus: Essays in the Ethnohistory of Colonial North America* (New York, 1988), 100–121; Harold W. Van Lonkhuyzen, "A Reappraisal of the Praying Indians: Acculturation, Conversion, and Identity at Natick, Massachusetts, 1646–1730," *New England Quarterly*, LXIII (1990), 396–428; Kristina Bross, "Dying Saints, Vanishing Savages: 'Dying Indian Speeches' in Colonial New England Literature," *Early American Literature*, XXXVI (2001), 325–352; James Axtell, *The Invasion Within: The Contest of Cultures in Colonial North America* (New York, 1985).

Hillary Wyss, Kathleen J. Bradgon, and Ives Goddard are exceptions to this general consensus. Where their work looks at the structures of cultural retention afforded "writing Indians," this chapter will explore the missionary apparatus motivating this project. Hilary E. Wyss, *Writing Indians: Literacy, Christianity, and Native Community in Early America* (Amherst, Mass., 2000), Ives Goddard and Kathleen J. Bragdon, *Native Writings in Massachusett* (Philadelphia, 1988).

experience in Algonquian. The Praying Indians spoke in a Massachusett tongue that they had relearned from the missionaries who had transformed it into a written language by translating key catechistical literature and treatises on conversion. The field of missionary linguistics thus involved a deliberate act of preservation that was not simply a way of facilitating conversion or augmenting civilizing practices.[11]

While promoting a diversity of tongues and voices, missionaries espoused a belief in a universal Christianity that regarded Indian languages as derived from the events of Babel. To collect them was to move toward recapturing the moment described in Genesis 11:1 when "the whole earth was of one language and of one speech." Philosophers since Augustine have expressed an interest in this endeavor, but there was also something unique and historically specific about the intersection between universal language theory and missionary linguistics in the Anglo-American context: In the mid- to late-seventeenth century, a group of Royal Society philosophers tried to develop and implement a universal language that would be common to all humanity. Their hope was that the universal language movement would unlock the secrets of God by recuperating semiotic perfection, where words would be linked to nature and their correspondent referent in the invisible world. From Roger Williams's *Key into the Language of America* (1643) to Eliot's *Indian Grammar Begun* (1666), missionary linguistics bore an intrinsic relationship to the universal language movment as it developed through Jan Comenius, Samuel Hartlib, Cave Beck, and John Wilkins.[12]

This Royal Society program originated in the teachings of Czech philosopher Jan Comenius, who propounded a theory that language could be used as a vehicle for unlocking divine essence. Comenius taught that the key to revelation was a linguistic system, or a "universal language," that followed a one-to-one correspondence between language and nature. His

11. Eliot and Mayhew, *Tears of Repentance,* 2. For readings of the implications of missionary linguistics, see Laura J. Murray, "Vocabularies of Native American Languages: A Literary and Historical Approach to an Elusive Genre," *American Quarterly,* LIII (2001), 590–623; Jill Lepore, *The Name of War: King Philip's War and the Origins of American Identity* (New York, 1998); Edward G. Gray, *New World Babel: Languages and Nations in Early America* (Princeton, N.J., 1999). Murray looks at the genre of the Indian vocabulary itself and reads documents from John Eliot to Emma Gardiner's "Vocabulary of the Language of Penobscot Indians" (1821).

12. Rhodri Lewis and Hannah Dawson show that universal language theory had a much wider reach within seventeenth-century philosophy than scholars have previously imagined. Lewis, *Language, Mind, and Nature: Artificial Languages from Bacon to Locke* (Cambridge, 2007); Dawson, *Locke, Language, and Early-Modern Philosophy* (Cambridge, 2007).

theory of linguistic isomorphism suggested that revelation was the result of the proper use of language to unlock the mysteries of the divine. Ultimately, Comenius hoped to promote religious harmony through a universal character that was "both an attempt to renew contact with divine harmony in the universe and a crucial effort to bring about a reconciliation between men that would lay the foundations for an enduring, religious peace." The method for achieving religious peace consisted of a redemptive language that was mathematically and empirically precise. Mathematics compensated for human fallibility by providing a "demonstration [of] everything not only in nature, but also in morals and politics, according to principles that are universal, fixed and subject to no exception."[13]

This reliance on mathematics compensated for the problem of knowledge lost in the Fall, a problem that constituted a central site of epistemological dilemma in the early modern world. Since the destruction of the tower of Babel and subsequent centuries of misuse, natural languages were corrupt and consequently full of irregularities. The universal character that would emerge out of this system was necessarily artificial rather than natural because it was impossible to eliminate the irregular and corrupt elements of natural languages. Comenius himself gained popularity in England in the 1630s and 1640s through his own travels as well as through collaboration with English linguist Samuel Hartlib. Hartlib promoted Comenius by publishing some of his tracts in English; additional versions of his theory were available in such texts as Ezekias Woodward's *Light to Grammar* (1641) and John Saltmarsh's *Dawnings of Light* (1644). Comenius's utopian vision for establishing religious harmony through the stabilizing framework of mathematical science influenced the Invisible College that began meeting at Cambridge and Oxford in the 1640s. In addition to Hartlib, members of the Invisible College included John Wilkins, Seth Ward, Robert Boyle, Sir William Petty, Sir Christopher Wren, and Thomas Sprat. Comenian philosophy thus had a significant impact on the group that would form the Royal Society. Through the course of their discussions, Wilkins and Ward began to formulate their own plan for the discovery of a universal character. This program served practical political as well as epistemological functions, for they believed that language could facilitate trade

13. James Knowlson, *Universal Language Schemes in England and France, 1600–1800* (Toronto, 1975), 14–15; G. H. Turnbull, *Hartlib, Dury, and Comenius: Gleanings from Hartlib's Papers* (London, 1947), 344.

between nations and ultimately expand English national territory, justified through the millennial vision of a global Christianity encoded within.[14]

Comenian theory first appears in Roger Williams's *Key into the Language of America*. It begins with the following declaration: "This *Key,* respects the *Native Language* of it, and happily may unlocke some *Rarities* concerning the *Natives* themselves, not yet discovered." Williams then draws upon Comenius's lexical system for intuiting God in nature. Williams applies Comenian linguistics to the concept of the rarity. The "rarity" was a defining category of empirical discovery in the scientific revolution from Bacon forward. Yet Bacon also warned that rarities could be falsely replicated as resemblances appearing to the human mind. In the *Key,* the native speaker attains the status of a natural object of inquiry through which observers might discover unexplored dimensions of the invisible domain. By conjoining the Baconian term "rarity" and the Comenian metaphor of unlocking, the *Key* promotes the study of Indian language as the point of intersection between empiricism and the discovery of divine essence. As the place for this discovery, Williams identifies New England as the *"Land* of *Canaan."* Consisting of the geographical as well as national space of Israel's typological descendants, it must be "attended with *extraordinary, supernatural,* and *miraculous* Considerations."[15]

This association of New England with the Land of Canaan reflects the typological tradition in New England. But Williams does not simply narrate a story of spiritual fulfillment upon New World arrival. Rather, he explains that the land itself is infused with an indigenous sacred power that is worth observing in its natural state. Through his unique typological association of New England as an "extraordinary and supernatural" land of Canaan, Williams inhabits a unique position as a kind of spiritual eth-

14. G. H. Turnbull, "Samuel Hartlib's Influence on the Early History of the Royal Society," *Notes and Records of the Royal Society,* X (1953), 101–130.

15. Roger Williams, *A Key into the Language of America,* ed. J. Hammond Trumbull, in *The Complete Writings of Roger Williams,* I (New York, 1963), 19 (Comenius's *Janua linguarum reserata* translates as "the gate of languages unlocked"); Williams, *The Examiner Defended, in a Fair and Sober Answer to the Two and Twenty Questions Which Largely Examined the Author of "Zeal Examined,"* in *The Complete Writings of Roger Williams,* VII, ed. Perry Miller (New York, 1963), 251.

Bacon constructs his warning about the mind's ability to falsely construct resemblances from the Platonic school of knowledge: "The sense of man carrieth a resemblance with the sun, which (as we see) openeth and revealeth all the terrestrial globe; but then again it obscureth and concealeth the stars and celestial globe: so doth the sense discover natural things, but it darkeneth and shutteth up divine." *The Advancement of Learning,* in Brian Vickers, ed., *Francis Bacon: A Critical Edition of the Major Works* (New York, 1996), 125.

nographer. He presents the natives as exhibiting a special connection to the land that makes them key experimental subjects for ascertaining the sacred essence encoded in primitive languages. Through the metaphor of the key and the typological association of New England as the land where such evidence can be revealed, Williams's text represents a transformation in the history of missionary linguistics. Specifically, he adapts Comenian theory to the particular case of America in order to make a claim for the sacred and secret contents of American languages through the idea that we need a key to gain access to them. As such, Williams's *Key* sets the stage for a history of evangelical encounters by claiming that both the land and its inhabitants contain the potential for recapturing something rare, wonderful, and original in native languages.[16]

Eliot began to study Algonquian languages in 1643, the year of the publication of Williams's *Key*. His tutor was a Long Island Indian captive from the Pequot War (1636–1638) and servant of a colonist in Dorchester. In 1644, the General Court ordered Indians in several counties to be "instructed in the knowledge and worship of God," and in 1646 Eliot went to Nonantum, where he preached his first Algonquian sermon in Waban's wigwam. The insistence that the Native Americans retain their own language in pious practice reflected a belief in the missionaries' power to actualize Comenian potential: the possibility of a universal language that would make divine phenomena visible in their pure, spontaneous, and unadulterated forms. Boyle recognizes this potential in a 1663 letter to Eliot: "We desire care may be taken that [the Indians] retain their own native Language." Missionary records from the 1660s emphasize that such care has been taken. Thomas Shepard's 1673 letter to a Scottish Presbyterian minister in Edinburgh describes several Indians reading "in the Indian Language" from the Bible or the *Indian Primer*. In his diary, John Cotton, Jr., recounts preaching to the Indians on Chappaquiddick in "their own language" and then listening to their testimonial replies. Eliot's "Brief History of the Mashepog Indians" begins with a description of "prayers powered out before the Lord in both languages."[17]

16. Missionary linguistics had a well-established colonial presence in New Spain. Starting in the mid-sixteenth century, a printing press in Mexico City put a number of Indian grammars into circulation. Pedro de Gante's *Doctrina cristiana en lengua mexicana* (Mexico City, 1547), for example, compiles a Nahuatl grammar from twenty-nine manuscript leaves in different hands, each drafting a preliminary Indian vocabulary of words learned in the missionary field.

17. Notes from Company Records, New England Company, MS, box 1, fol. 1, American Antiquarian Society, Worcester, Mass.; Robert Boyle to John Eliot, Sept. 18, 1663, MS, Ecclesiastical

Missionaries made such declarations while their contemporaries devoted much time to the study of sacred languages. Samuel Hartlib and the members of the Royal Society developed a series of biblical translation projects, directed toward the translation of scripture into ancient languages such as Arabic, Hebrew, and Latin as well as languages where the gospel was spreading to: Gaelic, Turkish, and Algonquian. Hebrew held privileged sway as the original sacred language, because it was believed to most closely approximate the language spoken in Genesis before the confusion of Babel. Universal language theorists believed Hebrew to contain the mystical qualities as well as fewer radicals (underived words) than any other language. This grammatical feature appealed to the development of a universal character, where the basic technique was to minimize the number of radicals in order to simplify and expand intelligibility. Hebrew also appealed to John Eliot, who, partaking in what one scholar has called "an indigenous American Hebraic occultism," wrote in his preface to Thomas Thorowgood's *Jews in America:* "It seemeth to me, by that little insight I have, that the grammatical frame of our *Indian* language cometh *nearer to the Hebrew,* than the *Latine,* or *Greek* do." According to this logic, Anglo missionaries tailored Indian grammars according to two inextricably linked goals. The first involved proving narratives of native origins, and the second sought to unlock the divine essence encoded in their languages. The theory of the Algonquian Indians as one of the ten lost tribes enfolded them within the Anglo plot to recuperate the language of Genesis.[18]

While Eliot might have been the most public and prominent advocate of this perspective, he certainly was not the only one. Lesser-known ministers attended to missionary linguistics with pious care. A page from the diary of Cotton, Jr., reflects that he was also invested in capturing the sacred essence

---

Papers, I, doc. 7, Connecticut State Archives, Hartford; Thomas Shepard, Jr., to———, Sept. 9, 1673, MS, Qu. CV xi (fol. 105), Wodrow Collection, National Library of Scotland, Edinburgh; John Cotton, Jr., Diary and Indian Vocabulary, 1666–1678, MS, Massachusetts Historical Society, Boston; Cotton, "Brief History of the Mashepog Indians," Royal Society, London.

18. Noel Malcolm, "Comenius, Boyle, Oldenburg, and the Translation of the Bible into Turkish," *Church History and Religious Culture,* LXXXVII (2007), 327–362; "3 Memorandum of Agreement between Boyle and Robert Everingham for printing of Irish New Testament" MS, fol. iv, Royal Society.

The first quote is from Shalom Goldman, *Hebrew and the Bible in America: The First Two Centuries* (Hanover, N.H., 1993), xviii. Eliot's quote comes from his "Preface" to Thomas Thorowgood, *Jews in America; or, Probabilities, That Those Indians Are Judaical, Made More Probable by Some Additionals to the Former Conjectures* (London, 1660), 19.

of redeemed Algonquian. "Witness" is a key term within the empirical endeavor of studying human souls for evidence of divine grace. Tracts from the widely printed and circulated *Tears of Repentance* to these lesser-known manuscript records verify the authenticity of Indian conversion through a language of witnessing that includes the number and identity of those ministers present as well as testimony by multiple parties that the utmost care has been taken to properly translate and record the oral performances. That the younger Cotton takes the time to list this word with all of its possessive pronominal variants in a relatively limited vocabulary list indicates some awareness of his own practice of experimental philosophy, implemented in the small, contained tiny island off the coast of Martha's Vineyard.

The standard interpretation of these grammars by scholars like David Murray is that early missionaries simply inserted what they perceived as flawed indigenous languages into a Latin grammatical frame, but this is not entirely the case. To test his theory of Algonquian and Hebraic grammar, the elder Eliot compiled and published *The Indian Grammar Begun* (1666). Dedicating it to Boyle, Eliot states that, because Boyle was "pleased" with the "Testimonies" of "the effectual Progress of this great Work of the Lord Jesus among the Inhabitants of these Ends of the Earth," he would also like to supply "a *Grammar* of this Language, for the help of others." Like Williams, Eliot sought the linguistic rarity in order "to satisfie the prudent Enquirer [perhaps Boyle to whom the work is addressed] how I found out these new wayes of *Grammar,* which no other Learned Language (so farre as I know) useth." Eliot intended a record of linguistic anomalies that might yield a grammatical structure and reveal Algonquian's proximity to Hebrew. While early missionaries were not adept at thinking about grammatical structures outside a European context, Eliot's goal in pointing to the *"Rarities"* was, not to point to deficiencies (as Murray proposes), but rather to gain access to the epistemological potential encoded therein.[19]

Eliot presented his *Indian Grammar* to Robert Boyle as his contribution to a philosophy of language acquisition advocated by Comenius, Cave Beck, and John Wilkins. Each philosopher believed that compiling the new languages discovered by missionaries around the world and comparing these

19. John Eliot, *The Indian Grammar Begun* . . . (Cambridge, Mass., 1666), A2, [66]. According to David Murray, an Enlightenment universalist assumption informed the construction of these texts, leading to "the search for common structures of mind revealed in a common grammar." *Forked Tongues: Speech, Writing, and Representation in North American Indian Texts* (Bloomington, Ind., 1991), 15.

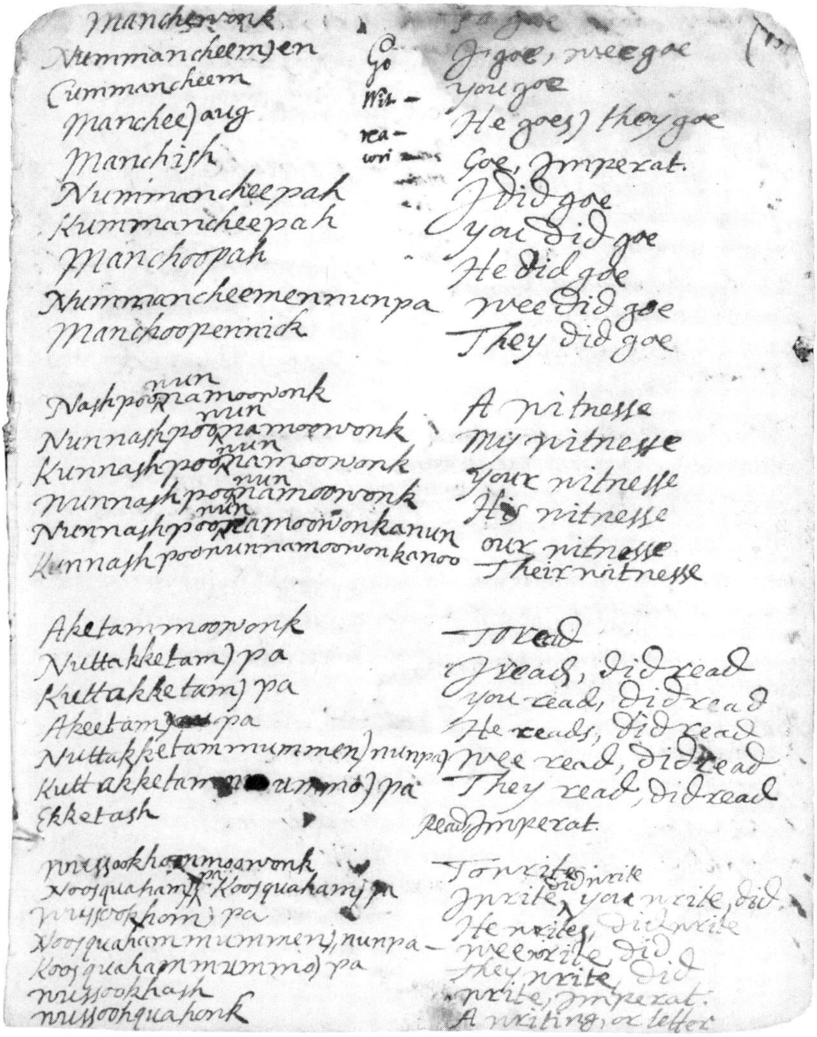

FIGURE 7 • *John Cotton, Jr., Diary and Indian Vocabulary, 1666–1678. Massachusetts Historical Society, Boston*

languages to those of the ancients, including Egyptian, Hebrew, and Chinese, would advance the discovery of the universal character. Cave Beck's frontispiece to *The Universal Character* shows universal language theory as an outgrowth of colonial history. It displays a European consigning Beck's project to a Hindu, an African, and an American Indian. The American Indian remains standing and expresses himself with a gesture of his hand. Europe's elected leader seems to be in a position of power that immediately evokes the idea of *translatio imperii,* or the westward course of empire. The scroll he holds echoes the scroll above; both uncurl slowly to reveal the unfolding potential of the universal character. But what is the source of this universal character? The Hindu and African gaze toward the European, suggesting that he wields power over the unfolding scroll. But the European gazes at the American Indian, whose eyes, in turn, are raised toward heaven. A poem opposite the title page refers to his gesture as "dumb Signes," which in the seventeenth century would most likely have meant that he was deaf. (The deaf were believed to have certain divinely inspired powers.) A deaf American Indian seems to occupy some kind of knowledge position within this scene. Beck is invested in the use of the universal character for free trade between nations and frames his treatise clearly as a tool for English expansion. But what we see in this image is a striking disruption of the translatio imperii narrative. The frontispiece masks—in order to powerfully reveal—Beck's deep awareness that the deaf but heaven-gazing Indian supplies the ancient wisdom necessary for this endeavor.[20]

20. Cave Beck, *The Universal Character by Which All the Nations in the World May Understand One Anothers Conceptions, Reading out of One Common Writing Their Own Mother Tongues* (London, 1657), B2. Entitled "The Mind of the Frontispiece," the poem announces its explicatory capacity:

> Men to this lower world fou[r] parts assign,
> Since Neptune's Trident, purchas'd a fourth Tin[e].
> All by their Representatives here meet,
> And by dumb Signes, each other kindly greet.
> Europe Elected Speaker, takes the Chair,
> Laws of one common Voice enacted are.
> May your next Vote be, Let us Sion build,
> Sion, that now lies like a plowed field.
> For when you Babels Tower reare in vaine,
> God builds a Babel in your lips again.

In the seventeenth century, the deaf were thought to be close to prelapsarian blessedness. Douglas C. Baynton, *Forbidden Signs: American Culture and the Campaign against Sign Language* (Chicago, 1996).

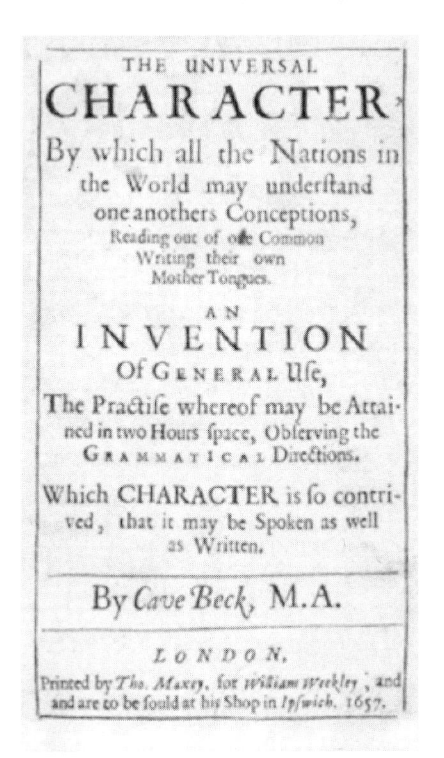

THE UNIVERSAL
CHARACTER,
By which all the Nations in the World may underſtand one anothers Conceptions,
Reading out of one Common Writing their own Mother Tongues.

AN
INVENTION
Of GENERAL Uſe,
The Practiſe whereof may be Attained in two Hours ſpace, Obſerving the GRAMMATICAL Directions.

Which CHARACTER is ſo contrived, that it may be Spoken as well as Written.

By *Cave Beck*, M.A.

LONDON,
Printed by *Tho. Maxey*, for *William Weekley*; and are to be ſould at his Shop in *Ipſwich*. 1657.

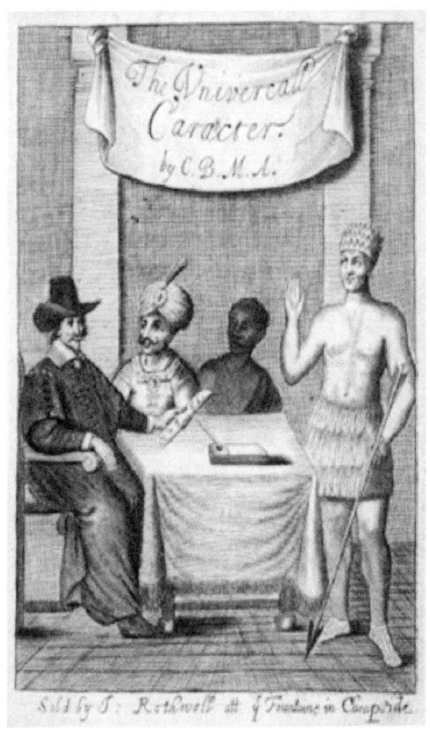

FIGURE 8 • *Cave Beck,* The Universal Character. *Huntington Library, San Marino, Calif.*

Grammars and Laboratories of Grace

The American Indian's gesture organizes the allegory of empire depicted in Beck's scene. Without the context of the scroll unfolding through the authorizing capacity of the European, the gesture remains a dumb sign in the more familiar sense: a signifier severed from its clear and revelatory referent. The epistemological potential of the wisdom-purveying American Indian remains untapped and inaccessible without the accompanying presence of his European scribe, who wields both the power of the pen and the capacity for semiotic decipherment. The gesture thus refers to a complex process of locating the power of Christian scriptural authority in the heathen other and then reclaiming this authority as an ancient, sacred essence unfolding within the present. A similar dynamic structures the relationship between missionary linguistics and the practice of recording Indian testimony in ministerially managed experimental sites, organized according to a colonial logic. Grammarians began their Indian translation work by searching indigenous languages for anomalies. The anomaly becomes a way both of marking the American Indian as different and of framing the utterance in such a way as to suggest the recapturing of something that has been lost through the Fall. While the *Indian Grammar Begun* provided an occasion to study a "primitive language," the testimonies contained the promise that such language would be the key to the divine mystery. The scene of witnessing Indian testimony, much like the scene in Beck's frontispiece, encapsulates the purported wisdom of an ancient and original other in order to rescript it within a presentist eschatological frame.[21]

The Eliot tracts rhetorically construct such an experimental scene of deciphering the wisdom encoded in the dumb signs of Massachusetts natives. In *A Late and Further Manifestation of the Progress of the Gospel amongst the Indians* (1655), William Walton, one of Eliot's coauthors, accordingly testifies to the authenticity of the recorded account:

> The Elders saw good to call upon the Interpreters to give a publick testimony to the truth of Mr. *Eliots* Interpretations of the *Indians* Answers, which Mr. Mahu and the two Interpreters by him did . . . *the Interpretations which Mr.* Eliot *gave of their Answers was, for the substance the same which the* Indians *answered, many times the very words which they spake, and alwayes the sense.*[22]

The disciplined discernment of sensual data within this scene authenticates the Indian experience of Christianity in its aural and visual expressive form, according to what Boyle would later establish as the proper criteria for witnessing testimony in experimental philosophy. In *Christian Virtuoso* (1690), Boyle writes, "The Outward Senses are but the Instruments of the soul, which Hears by the intervention of the Ear, and the Eye . . . the Sense does but Perceive Objects, not Judge of them." Applying the same sensual techniques to Indian testimony, a group of native and English interpreters hears and remembers the spoken Algonquian while Eliot records the oral account with, presumably, help on the translation. The witnessing audience visually verifies the proper transmission of aural evidence to "Examine, whether the testimony be indeed Divine and whether a Divine Testimony ought to be . . . believ'd." The authors of the tracts then frame the evidence through a technique of ethnographic authorization that assures European

21. Robert Boyle, *The Christian Virtuoso: Shewing That by Being Addicted to Experimental Philosophy, a Man Is Rather Assisted, than Indisposed, to Be a Good Christian* (London, 1690), 50.

22. William Walton, in John Eliot, *A Late and Further Manifestation of the Progress of the Gospel amongst the Indians in New-England* . . . (London, 1655), 20.

readers that they are encountering the *sense,* if not the exact *words,* of a "heathen's" first experience of grace.[23]

Eliot and other evangelical ministers wrote ethnographically but in a way that was markedly different from the nineteenth-century definition of the discipline as "writing about customs . . . or cultures based on first-hand observation and participation in fieldwork." For seventeenth-century evangelicals, grace rather than human culture and custom was the focus of the empirical quest motivating the observational record. The Eliot tracts racialize images of Native Americans, not through a fascination with the "barbaric" other, but rather through an effort to demonstrate empirically and to witness ethnographically a species-specific form of grace.[24]

Michel de Certeau's analysis of Michel de Montaigne's "Of Cannibals" offers a useful elaboration of what we might call ethnographic authorization through his definition of the heterological tradition as that "in which the discourse about the other is a means of constructing a discourse authorized by the other." Specifically, Certeau explains how the ethnological encounter authorizes the text by naming the "other" through discourse and then framing the utterance in such a way as to suggest the recapturing of something that has been lost through language, much like "Europe's Elected Speaker" interpreting the American Indian's dumb sign in Beck's frontispiece. In *Tears of Repentance* this technique of framing represents the sense of the heathen's divine encounter as a "faithful and verifiable speech" that approximates the primitive origins of Christianity. Ethnographic authorization affirms that empirical evidence of divine grace upon a non-English soul has been properly seen, observed, recorded, and ultimately reinscribed within the symbolic order of seventeenth-century natural philosophy and divinity.[25]

23. Boyle, *Christian Virtuoso,* 115, 118.

24. David Levinson and Melvin Ember, eds., *Encyclopedia of Cultural Anthropology,* II (New York, 1996), 416; Ronald L. Meek, *Social Science and the Ignoble Savage* (New York, 1976); Margaret T. Hodgen, *Early Anthropology in the Sixteenth and Seventeenth Centuries* (Philadelphia, 1964); Anthony Pagden, *The Fall of Natural Man: The American Indian and the Origins of Comparative Ethnology* (Cambridge, 1982).

The Puritan's quest for species-specific grace did not disrupt their prevailing view in monogenesis. They understood Native Americans as Adam's descendants. Theories of polygenesis did not emerge until the mid-eighteenth century. For an account of this shift, see Colin Kidd, *The Forging of Races: Race and Scripture in the Protestant Atlantic World, 1600–2000* (Cambridge, 2006).

25. Michel de Certeau, *Heterologies: Discourse on the Other,* trans. Brian Massumi (Minneapolis, Minn., 1986), 68.

This idea of capturing the sense as initially expressed through the "primitive simplicity" of the Massachusett language displays what would become a central tenet of the Royal Society and New England Company's empirical objective. As Thomas Sprat explains in the *History of the Royal Society:* truth needs to be represented "cloth'd with Bodies" in order to meet the scientific goal of bringing "*Knowledge* back again to our very senses." In an attempt to develop a linguistic method that follows this model, Sprat explains that the members of the society have returned to the "primitive purity" of speech, developing "a close, naked, *natural way* of speaking; positive expressions; clear senses; a *native easiness:* bringing all things as near the Mathematical plainness, as they can: and preferring the language of the Artizans, Countrymen, and Merchants, before that, of Wits, or Scholars" (my emphasis). Sprat's method of speaking returns language to the Society's commitment to "nullius in verba," a saying that marks an epistemological turn from reliance on textual language to reliance on a disciplined kind of sensual data discernible through techniques of performed speech and witnessing.[26]

Through their conversion, Indians were represented with the unique ability to conjoin Algonquian with an Augustinian understanding of language as a potential vehicle of truth and divine communication. Compounded with Comenian linguistic theory, conversion testimonies would express a "natural" form of religious primitivism, one that had been lost to "civilized" "wits and scholars," even in their moment of conversion. While Eliot's efforts to translate Algonquian from an oral to a written language marked an attempt to make this "primitive form" a usable tool for English Christians, the translation of testimonies into English presented an idealized vision of a universal language to his London audience. Alongside this universal language, the ministers anticipated a racialized testimonial utterance: through the public performance of Praying Indians, theologians and scientists sought to discern the "attributes of God" in a "Heathen" people,

26. Thomas Sprat, *History of the Royal Society* (1667), ed. Jackson I. Cope and Harold Whitmore Jones (Saint Louis, Mo., 1959), 112, 113.

For modern work on this epistemological turn and its cultural and social implications, see Joyce E. Chaplin, *Subject Matter: Technology, the Body, and Science on the Anglo-American Frontier, 1500–1676* (Cambridge, Mass., 2001); Mary Poovey, *A History of the Modern Fact: Problems of Knowledge in the Sciences of Wealth and Society* (Chicago, 1998); Barbara J. Shapiro, *Probability and Certainty in Seventeenth-Century England: A Study of the Relationships between Natural Science, Religion, History, Law, and Literature* (Princeton, N.J., 1983); Peter Dear, *Discipline and Experience: The Mathematical Way in the Scientific Revolution* (Chicago, 1995).

"deeply engaged in polytheism, idolatry, magical rites and superstitions," who became "worshippers of Christ." This effort would ultimately lead Boyle to conclude that what "could never be learned" in Aristotle or Ptolemy is revealed in the "supernatural testimonies" of America's "Heathen" as they convert to Christianity. Supplied with the "supernatural testimonies" that had been collected by *"Eye-Witnesses and Ministers"* and other *"Travellers* to America," the experimental philosopher could begin to discern "attributes of God" and "knowledge of what is infinite."[27]

Three pamphlets published in the early 1650s collectively establish Indian testimony as both a return to this universal language and a tool for bringing forth the secrets of the divine. Henry Whitfield's *Light Appearing* (1651) and *Strength out of Weaknesse* (1652) set the stage for Eliot and Mayhew's *Tears of Repentance,* the most ambitious attempt to record evidence of grace through individual accounts of conversion. In *Light Appearing,* Whitfield goes with Eliot to Watertown, where he records the questions that follow Eliot's sermon to the Indians. "I . . . have set down some of their Questions, wherby you might perceive how these dry bones begin to gather flesh and sinnews," Whitfield writes, evoking Ezekiel 37 to create a material image through which the reader can bear witness to the visibility of grace. Flesh and sinews signal, quite literally, the corporeal presence of grace within the praying town. On a more symbolic level, this metaphor foreshadows the records of testimony that will make the presence of grace perceptible to the witnessing audience reading the tracts. It is as if the numinous, ethereal quality of grace becomes visible to the discerning eye through the oral relation as experience would seemingly take on an embodied presence. In the same tract, Eliot refers to the soul as an "eye of faith," expanding upon the theological and scientific theory that the ocular sense, transformed through faith, makes the divine manifest within the natural world. The soul is both the place in which the individual discovers his or her spiritual status and the text that others read to discern the work of grace.[28]

27. Boyle, *Christian Virtuoso,* 50, 75–76, 94. Augustine experiences a new relationship to a universal language that corresponds to the invisible realm in book 9 of the *Confessions* when words rise above their fallen status to form the visible signs of the invisible realm.

28. Henry Whitfield, *Light Appearing More and More towards the Perfect Day; or, A Further Discovery of the Present State of the Indians in New-England* (London, 1651), 18. Ezek. 37 was often preached to American Indians and was highly effective in eliciting conversion, quite possibly because the healing powers of Christianity described in the passage spoke directly to the needs of an American

In *Strength out of Weaknesse,* Whitfield and Eliot describe the preparatory stages of conversion. The Indians have "form[ed] themselves into the Government of God" and are now ready for the "Church-estate." Anticipating this estate, Eliot instructs his native congregants "that the Visible Church of Christ is builded upon a lively confession of Christ" and "exhort[s] them to try their hearts by the word of God to finde out what change the Lord hath wrought in their hearts." He then explains to his readership, "This is the present worke we have in hand." The telling pronominal shift from "them," the Indians, to "we," the readers of the tracts, locates authority in the discerning judges who are witnessing the work of grace upon the Praying Indians. This self-examination is preparation for the testimony that will narrate the results of this introspection to a discerning audience of reliable witnesses. Eliot conveys to the Indians the work they must do in anticipation of forming the visible church without telling them what he expects to hear in the oral relation. The reader understands that the Indians have not been coached in what counts as evidence of grace, but only in how to find it. Although the praying town will mirror an English congregation, the authority to judge the evidence produced in the testimonial account does not belong to the Native Americans themselves. The evidence instead becomes the "present work" of theologians and scientists investigating Nonantum, Natick, Martha's Vineyard, Mashpee, and other praying towns as laboratories of grace wherein the complexities of divine mystery might be further understood.[29]

Even though *Strength out of Weaknesse* prepares natives for a science of self-scrutiny, the pronominal shift from "them" to "we" signals that the Praying Indians cannot be entirely trusted in this introspective science and that the translators and transcribers ultimately take responsibility for tracking evidences of the soul. Through this transference of discerning au-

---

Indian community suffering from illness. See Chaplin, *Subject Matter;* Bross, *Dry Bones and Indian Sermons;* Richard W. Cogley, *John Eliot's Mission to the Indians before King Philip's War* (Cambridge, Mass., 1999); O'Brien, *Dispossession by Degrees.*

Eliot's description of the soul in these terms comes from Heb. 11:1: "Faith is the ground of things, which are hoped for, and the evidence of things unseen." Mid-seventeenth century theologians and scientists, from Bacon to Richard Baxter to Boyle, subtly transform this scriptural sense through a new use of faith to tap the willed observational capacities of the elect. Eliot employs this ocular transformation here as he presents the soul as a text whose evidence others can discern.

29. Henry Whitfield, *Strength out of Weaknesse* . . . (London, 1652), 10. The visible church marks the Puritans' attempt, following Augustine, to create a visible approximation of Christ's invisible church on earth. See Chapter 2, above.

thority, the authors anticipate what the experience of grace will look like upon an Indian soul by attributing to the Native American worshippers a peculiar and explicitly non-English physicality in their public performances. Generalized descriptions of the scene of witnessing native conversion in the tracts leading up to *Tears of Repentance* describe this racialized spectacle of religious affect:

> Truly it did give to us who were present a great occasion of praising the Lord, to see those poore naked sones of *Adam,* and slaves to the Devill from their birth to come toward the Lord as they did, with their joynts shaking, and their bowells trembling, their spirits troubled, and their voyces with much fervency, uttering words of sore displeasure against sin and Satan, which they had imbraced from their Childhood with so much delight, accounting it also now their sinne that they had not the knowledge of God.[30]

Through such descriptions of religious affect, the tracts construct a difference between English and native testimonies, registered as variations in religious experience. Mayhew describes an embodied, performative, and emotive experiential testimony. "Shaking," "trembling," and "fervency" depict a mode of religious utterance that was neither practiced nor permitted in English congregations. Ministers labeled this very form of religious expression "enthusiastic" and in fact urged their English congregants to practice restraint and follow decorum in their own relations of faith. The description, "poore naked sones of *Adam,* and slaves to the Devill from their birth," situates the Praying Indian within a biblical typology, both locating Indians within the story of the Fall and its accompanying promise of redemption while also differentiating them from the English according to their natural and more fallen state.

In contrast to English converts, these sons of Adam are not only fallen but also long-forgotten remnants of the Old Testament. The moment of conversion reinstantiates Praying Indians within God's covenant, but only as relics from an ancient past. The ministers hoped that in addition to speaking in an Algonquian tongue, believed to be one of the lost languages, Praying Indians would complement this language by performing a mode of

30. Whitfield, *Strength out of Weaknesse,* 29. Boyle's letters provide the strongest evidence that he read the Eliot tracts. A 1663 letter quotes the line "poore naked sones of *Adam*" verbatim. Boyle to Eliot, Sept. 18, 1663, Notes from Company Records, New England Company, box 1, fol. 3, AAS; Ecclesiastical Papers, I, doc. 7, Connecticut State Archives, Hartford.

primitive affect. This primitive form of conversion fits neatly within the Puritan narrative of the New World as a site of typological fulfillment: in an unfolding, providential design, the conversion of Old Testament Native Americans mirrored New Testament fulfillment through the conversion of English congregants. Collectively, the reinscription of the "Jews in America" and the English in America within God's covenant would signal the imminence of Christ's return to earth, the reform of New England, and the subsequent reform of the world. Encased in Christian primitivism, the figure of the Praying Indian also fitted the burgeoning empiricism of Royal Society natural philosophy. As if to anticipate the Royal Society's desire for a primitive speech "clothed in bodies," Mayhew's description promises atavistic forms of native worship to accompany Algonquian testimonial utterance. Here, a version of "nullius in verba" involves a return to a primitive and pure form of language that would demonstrate a one-to-one correspondence between the sign, the oral sound of the sign, and the body's physiology in relation to speech.[31]

In his letter to Cromwell that begins *Tears of Repentance*, Thomas Mayhew foregrounds Bacon's method of the observed particular as a technique for identifying manifest forms of grace upon non-English souls when he establishes that the text will convey "more particulars" and evidence of the "tokens of more Grace in store to be bestowed on Indian souls." Providing material for the ethnographic imaginary of Boyle's *Christian Virtuoso*, Mayhew describes encountering the Native Americans on Martha's Vineyard as a group of "heathens" who had gone beyond moral and ethical degeneracy into a state of spiritual captivity. The description of the failure of a powwow (medicine man) to terminate his "devilish craft" rhetorically underscores the extent of this captivity: "After he had been brought by the Word of God to hate the Devil, and to renounce his Imps (which he did publickly) . . . his Imps remained still in him for some months tormenting of his flesh, and troubling of his mind, that he could never be at rest, either

31. "The Philological and Miscellaneous Papers . . . ," in John Lowthorp, ed., *The Philosophical Transactions and Collections, to the End of the Year 1700*, III (London, 1716), 373. As a corrective to Max Weber, Theodore Dwight Bozeman explains the primitivist dimension of Puritan typology in which "moving forward also meant retrogression to the primitive origins of Christianity," represented through sermonic "plain style," the institution of the Augustinian church, and the invocation of a "mythic time" of the ancient past. I am adding to Bozeman's account an explanation of the racialized dimensions of Puritan primitivism. Bozeman, *To Live Ancient Lives: The Primitivist Dimension in Puritanism* (Chapel Hill, N.C., 1988), 17. On the theory of the Native Americans as one of the ten lost tribes that was much debated in the seventeenth century, see Cogley, *John Eliot's Mission to the Indians.*

sleeping or waking." Yet, God's "mercy," Mayhew attests, allowed eight powwows to convert to Christianity and "brought two hundred eighty three Indians (not counting yong children in the number) to renounce their false Gods, Devils, and Pawwaws" and to "publickly" defy "their tyrannical Destroyer the Devil . . . in set meetings, before many witnesses." Here we have the description of witnessing Indians' "supernatural testimonies" that Boyle would make central to Christian philosophy: Mayhew describes a scene of witnessing that highlights the contrast between heathen darkness and divine light, dramatizing the ideology of the savage to appeal to religio-scientific desires for evidence of divinity through a form of Christian primitivism. Mayhew's letter promises that the testimonies recorded in *Tears of Repentance* will offer evidence of grace as a natural phenomenon occurring within the new world.[32]

Through the biblical framework of heathen ideology, the tracts construct the Indian as the proper hermeneutic site through which to witness the spread of the gospel. In contrast to "learned Nations," still "loth to yield to Christ,"

> poore Indians they have no principles of their own, nor yet wisdome of their own (I meane as other Nations have) . . . and therefore they do most readily yeeld to any direction from the Lord, so that there will be no such opposition against the rising Kingdome of Jesus Christ among them.[33]

While this paternalistic statement clearly presents the Indian in a state of primitive simplicity, Eliot's remark is not simply a disparaging ethnographic one in the self-critical terms of our modern sense of ethnography. His racial observation, rather, is an attempt to establish the Indian subject as an appropriate text upon which to witness the workings of God in a distilled form. Because the Indians have no history, politics, or "human wisdom," their religious experience offers purer evidence of grace. Eliot's account of native depravity as a particularly conducive site for witnessing manifestations of grace reminds the reader of criticisms of English temporal concerns in English revolutionary tracts such as Samuel Hartlib's *Faithful and Seasonable Advice* (1644). Hartlib attributes the delay in the millennial realization of the national covenant to the absolute "Temporall

32. Mayhew, "To the Much Honored Corporation in London," in Eliot and Mayhew, *Tears of Repentance*, B2.

33. Whitfield, *Light Appearing*, 28.

Monarchy," which claims dominion over men's spiritual "estates and bodies." The Indians embody a promising space upon which to witness, in Old England and New, the progress of the Protestant cause.[34]

For Puritan theologians and natural scientists, the phenomenon of grace among the Praying Indians supplied this politico-religious national agenda with visible evidence of the invisible realm, a space that might ultimately be understood empirically through the careful recording and collecting of such testimonies. *Tears of Repentance* also attempts to translate, record, and circulate such evidence for scientific and religious communities engaged in a shared desire to collect proof of what Richard Baxter and Boyle call "things unseen." Situated between Bacon's concept of the "natural theologian" and Boyle's theory of the "Supra-Intellectual" that can know "things above reason," *Tears of Repentance* participates in a burgeoning scientific method that seeks to discover divine mystery. The task of presenting evidence of grace in the recorded testimonies was crucial not only to the integrity of the visible church at Natick and the legitimacy of the evangelical mission but also to the establishment of Natick and other praying towns as laboratories of grace for natural philosophers.[35]

Tears of Repentance reflects a contradiction between the observed particular, or the idea that the "fruits" of grace might be witnessed empirically, and the desire to hear and present certain forms of grace to London's natural philosophers through the record of testimonial relations. This contradiction stems from the authors' attempt to ensure that the evidence of salvation produced in the oral account met the spectacle of religious affect that they claimed to be witnessing among the Indians. But in place of the "shaking," "trembling," and "fervency," promised in *Strength out of Weaknesse,* the testimonies offer a version of the ambivalent assurance that would count as adequate proof of grace in English congregations. Monequassun "thank[s] Christ for all these good gifts which he hath given me." Totherswamp declares, "This is the love of God to me . . . I trust my soul with him for he is my Redeemer." Ponampam testifies, "My heart is broken, and melteth in me." Owuffumag unequivocally states, "My heart turned to praying unto God, and I did pray . . . then I purposed to pray as long as

34. [Samuel Hartlib], *A Faithful and Seasonable Advice; or, The Necessity of a Correspondencie for the Advancement of the Protestant Cause* (London, 1643), 4. The very notion in the title of a necessary correspondency indicates a political and religious climate in England that would have welcomed the evidence of grace and the advancement of the Protestant cause presented in the tracts.

35. Robert Boyle, *A Discourse of Things above Reason: Inquiring Whether a Philosopher Should Admit There Are Any Such* (London, 1681), 33, 79.

I live." Each of the above statements *could* have been spoken by an English congregant, and this was precisely the problem. The ministers were not simply interested in knowing that the Indians believed themselves to be saved; they also wanted to understand the peculiar way in which grace affects a "heathen" soul. The Indians, in other words, failed to produce the "curiosity" that Mayhew and others claimed to be observing among the native population. In order to become a laboratory of grace for the scientific community and a space for securing the purity of English religious identity, the praying towns had to demonstrate what grace looked like among a population represented as historyless and in a depravedly fallen state. The authors of the tracts did not know exactly how such grace was supposed to look, but they knew that it would reflect a more radical break with Protestant standards of piety.[36]

In response to their failure to record empirically the curiosities of the Indian soul, the ministers employed a surgical method of abstracting desired evidence. Whereas the English congregants delivered a unified, cohesive account of saving grace *once* to a community of elect witnesses, the Indians were often requested to give their account twice or even three times in the hope that they would approximate the account that the ministers wanted to hear. In response to the ministers' prompting for a second confession, Waban stated, "I do not know what grace is in my heart." This clearly frustrated the ministers witnessing and recording the event, who refused the proposed testimony by deeming it "not so satisfactory as was desired." Waban's belief in Christianity was not their concern, for he repeatedly attested to his faith in God. The ministers desired "to see and hear" evidence of his soul rather than a statement of his belief. In place of the evidence that Waban could not (or would not) deliver orally, Eliot attested to the "exemplar[y]" nature of Waban's conversion, explaining to the reader as well as to the witnessing ministers, "His gift is not so much in expressing himself this way, but in other respects useful and eminent."[37]

Waban is not alone in his inability to know the contents of his heart, despite repeated prompting. Eliot makes a similar intercession for Nishohkou: "When he had made this Confession, he was much abashed, for he is a bashful man; many things he spoke that I missed, for want of

36. Eliot and Mayhew, *Tears of Repentance*, 6, 7, 15, 24, 44.

37. Eliot, *A Late and Further Manifestation*, 20; Eliot and Mayhew, *Tears of Repentance*, 8. For examples of this testimonial practice in New England congregations, see Michael McGiffert, ed., *God's Plot: Puritan Spirituality in Thomas Shepard's Cambridge*, rev. ed. (Amherst, Mass., 1996).

through understanding some words and sentences." Blaming the first narrative's failure on a combination of Nishohkou's timidity toward public speech and the difficulties of transcription, Eliot offers a privately related testimony in its stead. Bashfulness, timidity, and an inability to "know" directly contrast the anticipated embodied performance with an insecurity as to the testifier's worthiness for election. While doctrinal insecurity was part of a Calvinistic understanding of grace, the Praying Indian further supplemented theological ambivalence with social forms of hesitation and reluctance to perform before a witnessing audience.[38]

Eliot displays disappointment in his inability to capture "heathen" affect empirically. While determined to prove the effectiveness of the praying town as a laboratory of grace, his subsequent attempts to extract evidence of grace in several of the testimonies not only contradict the methodological goal of the observed particular—applied to the soul in Boyle's sense that the senses "Perceive Objects, not Judge of them"—but also cause the empirical failure of the oral testimonies to produce an account of the laws of grace working upon an Indian subject. Robin Speen's confessions, offered in three sequential versions, reveal the minister's attempt to extract tangible proof of grace from the relation and Speen's inability or unwillingness to produce it. In the middle of his second confession, Speen stops and asks Eliot, "What is Redemption?" This question signals Speen's struggle to match his narrative with the witnessing audience's idea of what an experiential testimony should sound like. By this point, Speen was undoubtedly familiar with several definitions of the term from Eliot's sermons and translation of the catechism into Massachusett. Speen stops in the middle of his own narrative of religious experience in order to verify that it matches the forms of evidence he is expected to produce. Eliot's answer does not help Speen out, for it merely reiterates that redemption is "the price which Christ paid for us, and how it is to be applied to every particular person." Speen continues repeating his desire for redemption and his knowledge of sin, without delivering the "particular" application that Eliot requested. The final iteration of his testimony offers a version of the ambivalent assurance that counts as membership qualification in the visible church: "I give my self unto Christ, that he may save me." Although Speen does not meet Eliot's specific narrative demand, he ultimately gives adequate testimony to his experience of grace.[39]

38. Eliot and Mayhew, *Tears of Repentance*, 33.
39. Ibid., 31, 32; Boyle, *Christian Virtuoso*, 115.

The interplay between testimonial voice and witnessing audience reflects how the ministers attempted to extract empirical evidence of grace from the experiential authority of the Indian's narrative. Although many Indians narrate the transition from Indian religious practice to Puritanism as an allegory for their spiritual movement from darkness to light, their method of testifying itself mimics the decorum modeled for them by their English proselytizers, occasionally supplemented with an exaggerated social insecurity. Ministers presented the towns as spaces for witnessing the forms of divine intervention required to "change the heart" of such "barbaric natives," yet the Indians would not narrate their experience in terms adequately different from the English.[40]

In response to the failure of *Tears of Repentance,* Eliot's *Late and Further Manifestation* attempts to ameliorate the adequacy of the evidence the Indian testimonies produced, through a more direct method of extracting evidence of grace. Reflecting on *Tears of Repentance,* Eliot expresses a subtle disappointment in the first tract of recorded testimonies, noting, "I did much desire to hear what acceptance the Lord gave unto them, in the hearts of his people there." Eliot's desire seems not to have been fulfilled, for the 1655 tract pursues the quest of extracting evidence of grace through a persistent set of questions that demand an account of the particular nature of experiential religion: "What doth he put into your heart, that causeth your heart to break?" "What worke of the Spirit finde you in your heart?" "What change hath God wrought in you of late, which was not in you in former times?" Varying the language and sequence of each question for each confessor, Eliot puts forth such questions as a method of soliciting the exact nature of the divine encounter experienced by each of the six Indians offering testimonies of faith. The questions extract the very evidence that the testimonies failed to produce, but the answers replicate the patterns of misrecognition. The Indians provide answers that describe a conventional pattern of conversion—"I find my heart turned," or, "the Spirit turneth us from our sins"—without the specificity typically requested by the ministers.[41]

40. Whereas Charles L. Cohen has argued that the testimonies recorded in *Tears of Repentance* failed to provide adequate evidence to legitimate the formation of the visible church at Natick, I am suggesting here that the testimonies *did* provide adequate evidence for the praying church but not for the natural taxonomy of grace that the authors of the tracts hoped to develop. Cohen, "Conversion among Puritans and Amerindians: A Theological and Cultural Perspective," in Francis J. Bremer, ed., *Puritanism: Transatlantic Perspectives on a Seventeenth-Century Anglo-American Faith* (Boston, 1993, 233–256.

41. Eliot, *Late and Further Manifestation,* 3, 10–11, 18–19.

The Algonquian Word and the Spirit of Divine Truth

The failure of Natick's converts to publicly perform adequate evidence of grace resulted in an eight-year delay in establishing the praying church at Natick. It was not until 1660 that the elders finally sanctioned full communion for eight Indian confessors. This delay clearly registers Eliot's frustration, but the ultimate establishment of a Massachusett congregation also indicates a renewed optimism toward the missionary project in 1660. This was a time of institutional consolidation. The Royal Society formed; the king, Charles II, ascended to the throne in 1661; Parliament officially founded the New England Company in 1662, and Eliot imported a printer named Marmaduke Johnson from London to publish his Indian library.[42]

Working alongside a Massachusett Indian named James the Printer, Marmaduke Johnson translated key Christian texts into Algonquian. These texts helped to create an aural culture of spoken Algonquian that was believed to be partially redeemed to its original, prelapsarian state by the missionaries. Anglo ministers as well as native converts participated, creating a cross-cultural exchange of speaking and listening to a preserved and sanctified Algonquian language. Translation tracts often contain pronunciation keys for this purpose. Eliot explains, "Such as desire to learn this language, must be attentive to *pronounce right* . . . and accustome their *tongues* to pronounce their *Syllables* and *Words*." His goal was to generate a community of native and Anglo preachers who would speak in an Algonquian language renewed with its original, divine essence. English translations of Algonquian speech struggled to preserve the aural quality by carefully recording, as promised by Eliot himself, *"the very words which they spake and always the sense."* Native Americans were taught to read this Christian library in their own native tongue so that they could recount their conversion experiences in Algonquian. In his conversion narrative, Wuttinnaumatuk mentions reading Baxter's *Call to the Unconverted* and learning about his own sinful situation from it. As his testimony demonstrates, the Praying Indian meetinghouse thus became a site of sensory, signifying practices, an empirical

42. This delay is according to O'Brien's history of Natick in *Dispossession by Degrees*, 51. Compared to the parallel Anglo process of sanctioning congregations, this was a highly anomalous occurrence. In fact, I believe that Thomas Shepard's one-year delay in the formation of a church at Dorchester is the only instance in which full communion for the founding members was not approved right away.

testing ground for discovering how the universal character might be heard as the embodiment of divine essence.[43]

Sent to Boyle in 1666, the "Brief History" presents the "Church of Sandwich" more generally as such a site. Just as *Tears of Repentance* established the Natick Indians' preparedness for conversion through their construction of a "sufficient Meeting-house," the "History" describes the "Indians meeting place" in the colony of Plymouth. The Indian confessors assemble into this meeting place before a group of honored judges. Composed of ministers as well as magistrates, the witnessing audience included the "Government" of the "Colony of Plymouth" along with the "preaching and ruling Elders of the Churches." Five members of the latter group, including Mayhew, Cotton, Richard Bourne, Eliot, and Eliot's son, "would use [their] pens" to render the confessors words as precisely as possible. This opening description clearly frames the meeting place as an experiment and aligns the words of the confessors with the experiential testimonies so central to Boyle's experimental philosophy.[44]

The examination begins with a series of questions designed to test the Indians' knowledge of the Fall: "Q. What is the State of Man by Nature?" The examiners ensure that the confessors know that this is the irreversible "State of Man by Nature," resting with satisfaction upon the answer that "man can not take sin out of his heart, for there is a root which desires to sin." Sin is the state of nature, the condition to which all sons of Adam are bound. The correct response to this question completes the first phase of the Puritan morphology of conversion: humiliation, or identification with one's complicity in the consequences of original sin. It also assures the ministers, magistrates, and natural philosophers to whom this "Brief History"

43. Eliot, *Indian Grammar,* 5; Eliot, *A Late and Further Manifestation,* 20; Eliot, "Brief History of the Mashepog Indians." Eliot's Indian Library included *A Primer or Catechism* (1654), Richard Baxter's *Call to the Unconverted* (1664), the Holy Bible (1685), Lewis Bayly's *Practice of Pietie* (1613), John Eliot's *Indian Primer* (1669), and [Eliot], *A Christian Covenanting Canfession* (1660), among other texts. We do not have a lot of information on the circulation of these texts and the reading practices of Native American communities, but a few scholars, such as Hillary Wyss and Kathleen Bragdon, have carefully sifted through the spotty archive. We learn from Bragdon's archival research that "sermon delivery was a flourishing vernacular genre among the native Christians" and "several natives enjoyed reputations as powerful and accomplished preachers." Bragdon, "Gender as a Social Category in Native Southern New England," *Ethnohistory,* XLIII (1996), 573–592, esp. 581.

44. Quotations come from the prefatory section of the "Brief History of the Mashepog Indians," signed by Eliot.

is directed that the confessing Indians are fully aware of the knowledge lost in the Fall.[45]

Given this awareness, their testimonies are bold, daring, and surprising. Paumpmunot, alias Charles, "finds the presence of God in [his] Soul." Wuttinnaumatuk "finds" his heart "opened to Christ in the promise." Kanoonus proclaims with Lazarus, "I am the resurrection and the Life." Nonqutnumuk explains, "Faith is as it were a light." William Poase exclaims, "I see in my heart, there is some trust in Christ Jesus." These testifying Indians reiterate the self-scrutinizing "I" plus active verb that we saw in the Anglo male testimonies recorded in Shepard's church, yet they dispel the uncertainty surrounding election to an even greater degree. Not bound to direct their spiritual quest toward the same political and communal responsibilities, the Mashepogs of the Sandwich Islands are encouraged to step even further outside the orthodoxy of Protestant decorum.[46]

In his testimony, Paumpmunot supplies a telling example:

> I heard that Scripture opened John 3.19.15.16 as Moses lifted up the Serpent in the wilderness, even so must the Son of man be lifted up, that whosoever believeth in him should not perish but have eternal Life; for God so loved the world that he gave his only begotten son, Christ by this I saw that Christ did bring healing and Life for my Soul. . . . Then my heart said truly this is the true and living God; and as the Israelites did much rejoice when they saw the Arc of God, so did my heart much rejoice when I found the presence of God in my Soul; and according to 2 Cor: 5:17, I see I am become a new creature, having now understanding, now will, now faith, now hope, now joy, now memory, now considerations in the having and inquiring of Jesus Christ for my Savior.[47]

Paumpmunot sees and hears proof of Christ's promise of salvation by equating his own wilderness experience with that of the Israelites. As sight and sound take on an intuitive capacity that exceeds natural sensory limits, Paumpmunot claims empirical knowledge of the "presence of God in [his] Soul." In a declaration that would become a normative feature of Protestant conversion only through the theological transformations of the first Great Awakening, Paumpmunot declares, "I am become a new creature." Conversion indexes for him a sense of differentiation from his state of na-

45. Ibid.
46. Ibid.
47. Ibid.

ture as innately depraved. He collapses the distinction between the event of his conversion experience and his retrospective narrative retelling of this experience through a shift from past to present verb tenses at this moment in the testimony. "I see I am" signals a new ontological status separate from "I found." The sequence of *nows* that proceed from Paumpmunot's self-declaration as a "new creature" reclaim the knowledge lost in the Fall: memory, will, joy, and understanding. This shift from past to present, the proclamation of the self as a "new creature," and the claim to structures of knowing typically unavailable to fallen humans depart radically from the norms of Paumpmunot's contemporary Anglo converts; in fact Paumpmunot's testimony aligns more closely with the conversion narratives generated from the theological transformations of the first Great Awakening. Conversion locates Paumpmunot's life within a temporal frame that permits a radical ontological and epistemological transformation. As a new creature, Paumpmunot can differentiate his Christian self from his previous self to claim new forms of agency and understanding.[48]

Contained within this brief conversion testimony, these statements starkly contrast Waban and Nishohkou's inability to know, producing new evidence of grace soon to be recognized by ministers and natural philosophers as ripe with epistemological potential. It is easy, even from the secular perspective of a twenty-first-century reader, to be seduced by the spiritual power recorded in this manuscript testimony, easy to marvel at the confidence and bold assertions of Paumpmunot's claims, particularly when contrasted to Nishohkou's and Waban's troubling silences recorded only thirteen years before. Is this "Brief History" then an instance of a particular form of spiritual authority or empowerment that the Mashepog Indians claimed through their Christianity? Possibly, but, however imaginative and compelling, such a reading can be made only if we suspend the maxim proposed by Samuel Petto: "A testimony proposed or offered . . . is not effectual unlesse received." And this particular testimony was produced and received by an audience of ministers who had rendered such claims permissible only because of the primitive, simple, and utterly fallen status of the people speaking them. Moreover, they had done so to serve the goals of a natural philosophical community rapidly unfolding the taxonomies of species distinctions that would ensconce a primitivizing biblical logic within the temporal organization of natural history. Paumpmunot could attain this level of spiritual empowerment only from his position as a primitive whose

48. Ibid.

language and customs would soon be rendered ancient remnants of a naturally disappearing past.[49]

The scene of witnessing described by Eliot indexed the descent of a universal religious harmony, also described by Thomas Shepard, Jr., in his 1673 letter to a Scottish Presbyterian minister. The younger Shepard describes this harmony in similar scenes throughout New England: the spread of the gospel among Richard Bourne's Mashpee proselytes, the natives preaching in Algonquian on Martha's Vineyard, and the praying town system established by John Eliot in Massachusetts. Each instances a microcosm of the universalizing Protestant spirit as it descends across sectarian divisions on both sides of the Atlantic. The aural culture of Algonquian communication produced in each setting is evidence of this process. Eliot "begins his prayers in the Indian's language." Then the son of Waban reads the proverbs from Eliot's Indian Bible, "which [according to Shepard, Jr.] has been printed and is in the hands of the Indians." A native named Job prays for half an hour in "the Indian Language" and then preaches from Hebrews 15:1. Several natives stand up and read from the *Indian Primer* or from Eliot's Bible. The younger Shepard emphasizes that the allure of such scenes is in the aural quality of a divinely redeemed Algonquian tongue, serving as proof of God's presence, as the aural sound lifts the spiritual essence from the Algonquian words printed in Eliot's library. Along with his letter, Shepard sent a copy of the *Indian Primer* to the Scottish minister in an attempt to illustrate the precision with which New Englanders pursued this linguistic path to a universal Protestant spirit.[50]

Revived with spiritual, sensory significance, spoken sounds and their carefully corresponding written marks achieve a semiotic unity with the divine. Revealing the divine essence encoded in the Algonquian tongue, New England missionaries presented a landscape through which such grammatical spiritualizing was taking place. Redeemed Algonquian words unlocked a secret and invisible divine essence that became immediately intelligible to those witnessing the scene. This goal of semiotic perfection stemmed in part from Augustine's proposal for a language restored from its fallen state to a pure, direct connection to the divine in the moment of conversion.

If the younger Shepard's letter conveys this ideal as the scriptural words uttered from the lips of Native Americans, Eliot's translation of the en-

49. Samuel Petto, *Voice of the Spirit; or, An Essay towards a Discoverie of the Witnessing of the Spirit* . . . (London, 1654), 8.

50. Thomas Shepard, Jr., to———, Sept. 9, 1763, National Library of Scotland.

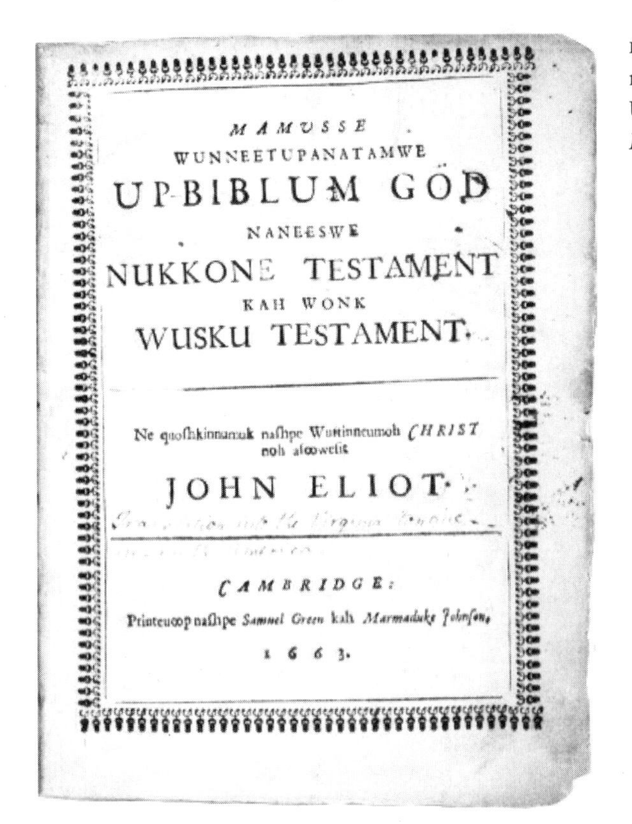
MAMUSSE
WUNNEETUPANATAMWE
UP-BIBLUM GOD
NANEESWE
NUKKONE TESTAMENT
KAH WONK
WUSKU TESTAMENT.

Ne quoshkinnumuk nashpe Wuttinneumoh CHRIST
noh asoowesit

JOHN ELIOT.

CAMBRIDGE:
Printeuoop nashpe Samuel Green kah Marmaduke Johnson.
1663.

FIGURE 9 • *John Eliot,* Mamusse Wunneetupanatamwe Up-Biblium God. *Huntington Library, San Marino, Calif.*

tire King James Bible into Algonquian grounds these performative sounds within the sola scriptura ideal. Eliot's Indian Bible was a novelty. It was the first printed in the colonies, at least one of the first printed in a non-European tongue, and the first printed for which an entire phonetic writing system was devised. Eliot's translation of the King James Bible conjoined the sola scriptura ideal and the universal language project. Reflecting on this accomplishment in a letter to Boyle, Eliot wrote, "My work is translation, which, by the Lord's help, I desire to attend unto." This statement conveys the sense of translation as God's work as much as the business of saving souls. Published in Cambridge in 1663, Eliot's Indian Bible was part of a biblical translation effort connected with the universal language movement as well as a more general Royal Society practice of biblical translation. Society papers contain several examples of "specimens" of scripture designed to decode and illuminate the hidden meaning within the text. This manuscript table gives the elements of the oriental languages and references in Hebrew, Latin, and Arabic. It shows how the ancient languages

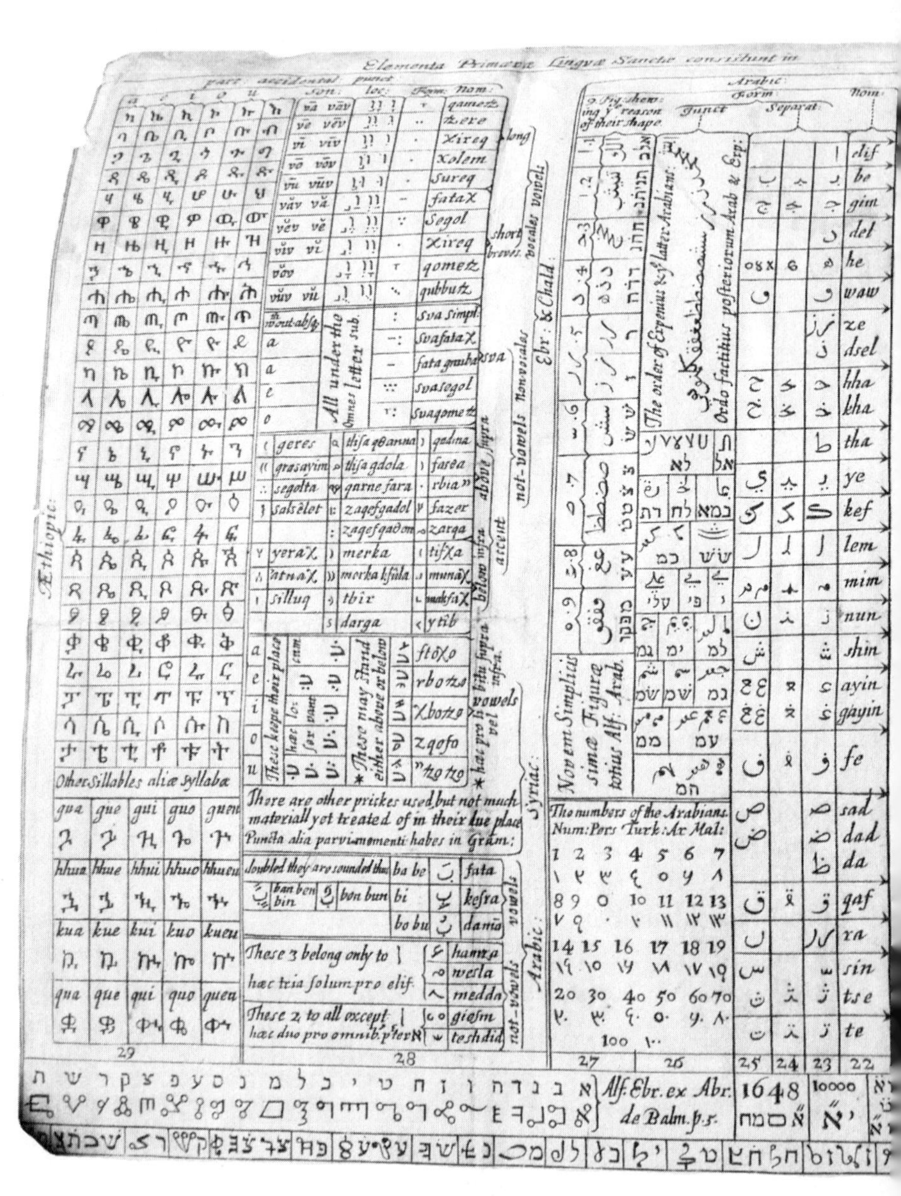

FIGURE 10 • *"Printed Table Giving Elements of the Oriental Languages and Notes by Thomas Hyde." Boyle Collection, 1670s–1680s, Ref RB 1/11/13, Royal Society, London.* ©The Royal Society

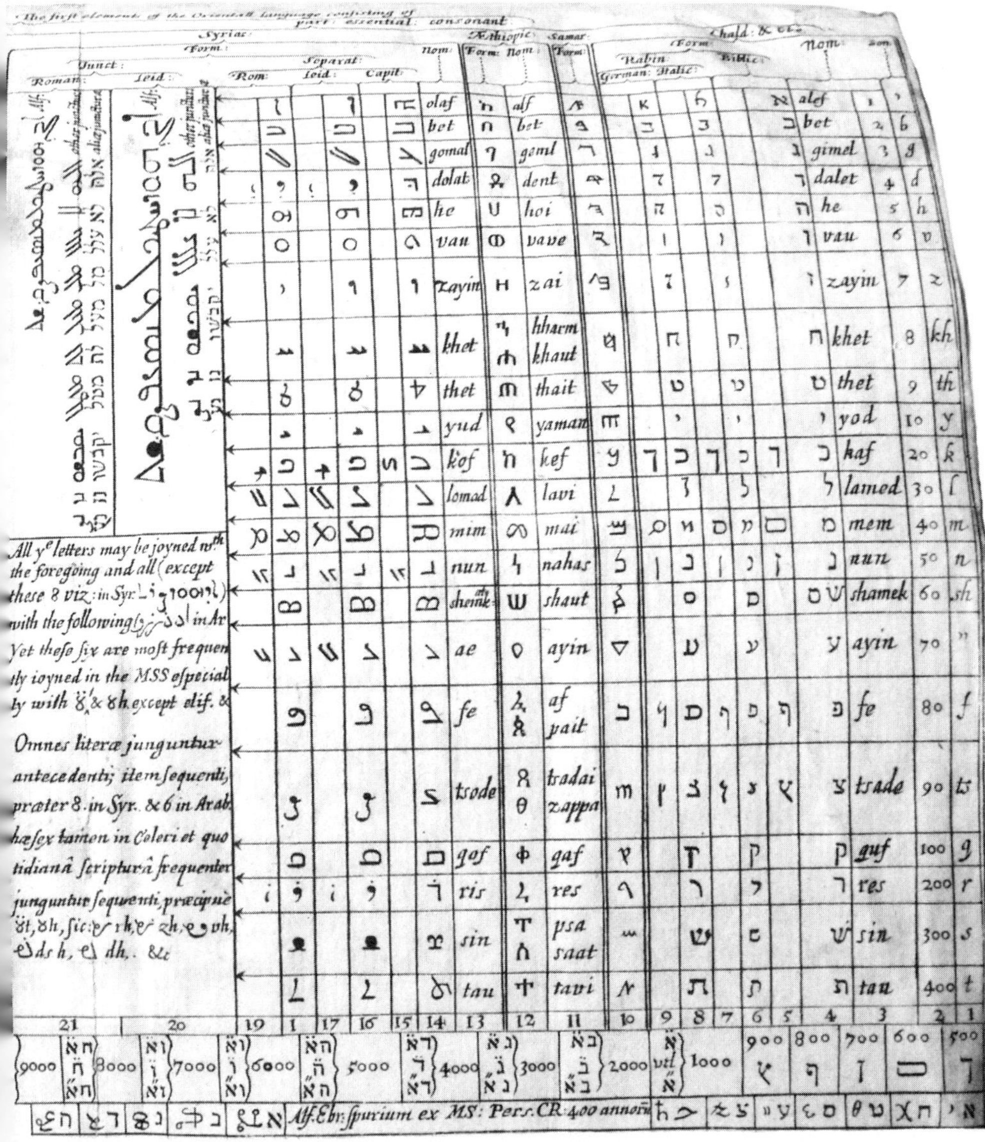
The first element of the Orientall language consisting of part: essential: consonant:
Syriac:
Form:
Junct:
Separat:
Roman:
Leid:
Rom:
Leid:
Capit:
Nom:
Æthiopic:
Form:
Nom:
Samar:
Form:
Chald: & c.
Form
Rabin:
Biblic:
German:
Italic:
Nom:
son
olaf alf alef 1
bet bet bet 2 b
gomal geml gimel 3 g
dolat dent dalet 4 d
he hoi he 5 h
vau vave vau 6 v
zayin zai zayin 7 z
khet hharm khaut khet 8 kh
thet thait thet 9 th
yud yaman yod 10 y
kof kef kaf 20 k
lomad lavi lamod 30 l
mim mai mem 40 m
nun nahas nun 50 n
sheink shaut shamek 60 sh
ae ayin ayin 70
fe af pait fe 80 f
tsode tradai zappa tsade 90 ts
qof qaf quf 100 q
ris res res 200 r
sin psa saat sin 300 s
tau tavi tau 400 t
All ye letters may be joyned wth the foregoing and all (except these 8 viz: in Syr.) with the following in Ar. Yet these six are most frequently ioyned in the MSS especially with & h except elif. &
Omnes literæ junguntur antecedenti; item sequenti, præter 8. in Syr. & 6 in Arab. hæ sex tamen in Celeri et quotidianâ scripturâ frequenter junguntur sequenti, præcipuè t, h, sic: rh, zh, vh, dsh, dh. &c
21 20 19 1 17 16 15 14 13 12 11 10 9 8 7 6 5 4 3 2 1
9000 8000 7000 6000 5000 4000 3000 2,000 vel 1000 900 800 700 600 500
Alf. Ebr. spurium ex MS: Pers. CR. 400 annorum

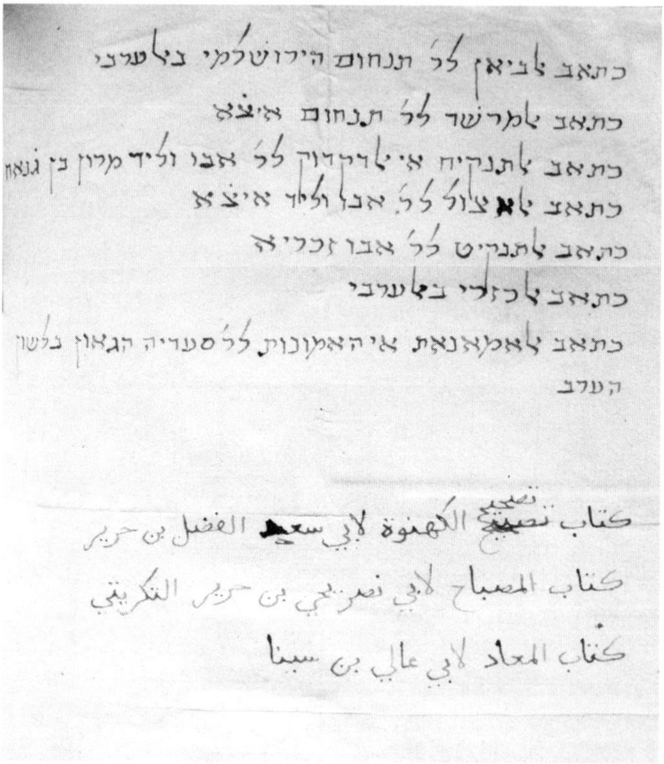

כתאב אלביאן לר' תנחום הירושלמי באלערבי
כתאב אלמרשד לר' תנחום איצא
כתאב אלתנקיח אי אלדקדוק לר' אבו אלוליד מרון בן גנאח
כתאב אלאצול לר' אבו אלוליד איצא
כתאב אלתנקיט לר' אבו זכריא
כתאב אלכזרי באלערבי
כתאב אלאמאנאת אי האמונות לר' סעדיה הגאון בלשון הערב

كتاب تصحيح الكهنوة لابي سعيد الفضل بن جرير
كتاب المصباح لابي نصر يحيى بن جرير التكريتي
كتاب المعاد لابي علي بن سينا

FIGURE 11 • *"Specimens of Hebrew and Arabic by Thomas Hyde." Boyle Collection, 1670s–1680s, Royal Society, London. ©The Royal Society*

were studied by members of the Royal Society interested in this project of biblical translation. Translations of scripture into ancient tongues coincided with translations of the Bible into indigenous tongues where the spread of the gospel was taking hold.[51]

While Eliot prepared the Indian Bible in Massachusetts, Robert Everingham secured funds for three thousand copies of a Gaelic translation of the New Testament, first printed in 1681 and distributed in Ireland and Scotland. The simultaneous efforts to study these biblical specimens and to translate the Bible into Gaelic and Algonquian reflect an interest in native tongues as both ancient and new, conveying anomalous secrets of divinity

51. John Eliot to Robert Boyle, Aug. 26, 1664, in Boyle, *The Works,* ed. Thomas Birch (1772; rpt. Hildesheim, 1966), VI, 510; "Specimen of Hebrew and Arabic by Thomas Hyde," Boyle Papers, XI, F 323–324, Royal Society.

that might be extracted from the utterances of the people from whom they came. This juxtaposition of the ancient and the new, the learned and the primitive, was a defining feature of the universal language theory, which looked to the sacred content of pre-Babel languages and attempted to further unfold their mysteries through comparison with the diversity of tongues spoken throughout the world.[52]

Eliot published his Indian Bible with the hope that it would sanctify Indian languages and contribute to this sensible knowledge. Two supplemental texts bookend the scripture to reveal the Bible's status as a connecting thread between two very different audiences. The Bible concludes with "Rules for Christian Living in Algonquian," reflecting the proselytizing and civilizing function of the text for native people. At the beginning of the book an epistle dedicatory to Charles II establishes the publication as a valuable commodity within the economy of English nationalism. The copy presented to Charles II in 1664 was a symbol and a gesture of English nationalism rather than something he could read and understand. The dedication explains that this is the "First . . . in this Language, or from this *American World,* or from any Parts so Remote from *Europe* as these are, for ought that ever we heard of." Eliot's Indian Bible both conveyed the homogenizing effects of subsuming more of the Earth's scattered and degenerate "Sons" into the canopy of English nationalism and announced itself as a New World curiosity: the first of its kind. The Indian Bible might also have appealed to Charles II as a symbol of his agenda for imperial expansion through scientific discovery.[53]

As quoted above, Charles's 1662 charter expressed a desire to "extend . . . the boundaries of the Empire." Eliot's Indian Bible and the universal language movement that contributed to its production symbolized this vision of the new science as commensurate with an agenda for expanding national territory. Condensed within the quest for a universal character, this vision involved a desire for expansion and progress as well as primitivism and recuperation. Comenius's ideal system of names that would signify

52. The memorandum establishing the Irish Bible project in fact references the "Case of the Corporation for New England," where there "hath been a very considerable progress in having the Gospel preached to the Indians in their own language the new Testament . . . printed in their own language," Boyle Papers, IV, 122, Royal Society.

53. John Eliot, *The Holy Bible: Containing the Old Testament and the New* . . . (Cambridge, Mass., 1663). Bragdon and Goddard *(Native Writing in Massachusett)* believe that these Bibles were widely available in praying towns; one copy at the American Antiquarian Society contains marginalia written in Algonquian by a Native American.

the essence of things promised enduring religious peace and reconciliation among men. Cave Beck expanded this logic to promote a homologous relationship between commerce and communication. The elimination of all "Equivocal words, Anomalous variations, and superfluous Synonomas," Beck explains, "would much advantage mankind in their civil commerce, and be a singular means of propagating all sorts of Learning and true Religion in the world."[54]

The Bible projects and the universal language movement both reached their zenith in the late 1660s with the publication of John Wilkins's *Essay towards a Real Character, and a Philosophical Language*. The *Essay* included a table in which Wilkins sought nothing less than to classify the whole knowable universe in taxonomic form, construct a written character based on the tables, and generate a spoken language that conveyed an isomorphic relationship to the real character. Wilkins's table contains more than 2,030 "primitives," or natural philosophical species. Arranging them taxonomically, he assigns each group a character that is supposed to stand as a sign capable of signifying all of its variations. Wilkins chose his characters arbitrarily, but, once inserted into a sequence, their signifying power was so complete that they were "supposed to mirror the very composition of the denoted thing."[55]

Wilkins constructed his table to uniformly and systematically unite a character to the thing or concept that it named, thereby mathematically achieving a level of semiotic perfection that eliminated linguistic differences and inconsistencies. The graphic symbol assigned to each universal concept formed a universal character that could be intelligible in any language. As the basis for this mathematical compilation, Wilkins proposed that the "*Hebrew* Tongue as consisting of fewest Radicals" of any language, should "be the fittest ground work for such a design." Hebrew would form the basis for a natural grammar that not only would be understood by all men everywhere but would also reclaim the voice of God in spoken language. At the bottom right of the chart, we see the first line of the Lord's Prayer written in the universal character with the pronunciation key below. The ability to translate the first line of the Lord's Prayer into the universal character marks the goal of the universal language movement: to reclaim "the first Language [that] was con-created with our first Parents" who "im-

54. Beck, *Universal Character,* "To the Reader."

55. Umberto Eco, *The Search for the Perfect Language,* trans. James Fentress (Oxford, 2005), 250.

mediately underst[ood] the voice of God speaking to them in the Garden." While Wilkins's chart foreshadows the promise of such reclamation, he cautions elsewhere in the text that the table is not complete. An adequate classification of this original language requires the continued ethnolinguistic efforts of scientists working over vast regions of the globe for a considerable time. Wilkins solicited the collaboration of the Royal Society for this endeavor and observed with praise, "*American* histories relate" evidence of these original languages found in "*Peru,*" "*Florida,*" and "the Northern parts of *America.*"[56]

Knowledge of Attributes Human and Divine

Like Hebrew, Algonquian constituted the mathematical basis of a grammar, one that would help natural philosophers such as Wilkins collect and compile essential linguistic components, match them to their immediate referent in the natural world, and then artificially reconstruct an Edenic language that would facilitate universal intelligibility as well as knowledge of the universe in its human as well as divine dimension. Within the universal language movement, theological constructions of North American natives as forgotten sons of Adam merged with philosophical conceptions of the natural world as divisible into the kinds of taxonomic categories featured in Wilkins's chart. For missionaries, empiricists, theologians, and philosophers of the late-seventeenth and early eighteenth centuries, North American natives, whether converting to Christianity or existing in a natural state, were repositories of an ancient sacred essence and scriptural wisdom, increasingly understood as part of an irretrievable and rapidly decaying past.

In 1670, only four years after Eliot composed his "Brief History" of the radical spiritual declarations of the Mashepog Indians, he wrote to Boyle describing the "rare work of God" that had recently taken place "in Watertown," where Indians learned a particular "root" that allows them to "read" spiritual phenomena not discernible to the English. Eliot asks for money to recompense those Indians who are able to "bring in a desirable experiment" that would explain the work of God in nature. Such experiments could facilitate the Royal Society's mission "to study *Nature* rather

56. John Wilkins, *An Essay towards a Real Character, and a Philosophical Language* (London, 1668), [xi], 2, 3, 5.

A Summary of Direction[s] both for the *CHARACTER* and *LANGUAGE*, Relating to

Integrals; whether considered as

Radicals; either

Direct.

Genus's; being 40. Transcen. Rel.

	Lan.
Genera[l]	Bα
Mixed	Ba
Of Act[ion]	Be
Discou[rse]	Bι
God	Dα
World	Da
Elemen[t]	De
Stone	Dι
Mettal	Do
Plant — Herb Leafe	Gα
Plant — Herb Flower	Ga
Plant — Herb S. Vef.	Ge
Plant — Shrub	Gι
Plant — Tree	Go
Animal — Exanguious	Zα
Animal — Fish	Za
Animal — Bird	Ze
Animal — Beast	Zι
Part — Peculiar	Pα
Part — Generall	Pa
Quantity — Magnitude	Pe
Quantity — Space	Pι
Quantity — Measure	Po
Quality — Na. Power	Tα
Quality — Habit	Ta
Quality — Manners	Te
Quality — Sens. Qual.	Tι
Quality — Sicknefs	To
Action — Spiritual	Cα
Action — Corporeal	Ca
Action — Motion	Ce
Action — Operation	Cι
Relation — Oeconomical	Co
Relation — Possessions	Cy
Relation — Provisions	Sα
Relation — Civil	Sa
Relation — Judicial	Se
Relation — Military	Sι
Relation — Naval	So
Relation — Ecclesiastical	Sy

Differences / *Species*

C. L.	Differences	Species
1	b	α
2	[illegible]	[illegible]
3	g	e
4	p	ι
5	t	o
6	c	8
7	z	y
8	S	yι
9	N	y8

Char. Lang.

Collateral; being pared with othe[rs] by

- *Opposition*; in Monasyllables / Dissyllables by Preposing the Opposite Vowels / Subjoyning the Letter S. (α to ι, a to o, e to 8, or contrariwise.)
- *Affinity*; in Monasyllables / Dissyllables by Preposing the Radical Vowel / Subjoyning the 2d. Consonant

Derivations; relating to such words as [are]

Concrete; whether

		Genus's	Differences	Species
Substantive	Neuter	Bα	Bαd	Bαdo
	Active	Bαl	Bαld	Bαldo
	Passive	Bαm	Bαmd	Bαmdo
Adjective	Neuter	Vα	Vαd	Vαdo
	Active	Vαl	Vαld	Vαldo
	Passive	Vα[m]	Vαmd	Vαmdo
Adverbe	Neuter	Bα[ι]	Bαιd	Bαιdo
	Active	Bαι[l]	Bαιld	Bαιldo
	Passive	Bαι[m]	Bαιmd	Bαιmdo

Abstract;

		Genus's	Differences	Species
Substantive	Neuter	Bαr	Bαdh	Bαdho
	Active	Bαrl	Bαldh	Bαldho
	Passive	Bαrm	Bαmdh	Bαmdho
Adjective	Neuter	Vαr	Vαdh	Vαdho
	Active	Vαrl	Vαldh	Vαldho
	Passive	Vαrm	Vαmdh	Vαmdho
Adverbe	Neuter	Bαιr	Bαιdh	Bαιdho
	Active	Bαιrl	Bαιldh	Bαιldho
	Passive	Bαιrm	Bαιmdh	Bαιmdho

Inflexion; Viz. Plural Number by a[dd]ing 8 to the first Vowel, or by pronouncing it [...] in the

Genus	Bα8
Difference	Bάd
Species	Bάdo

Particles; whether

Grammatical;

1 *Copula*; according to its three [Ten]ces: ια, ιa, ιe

2 *Pronouns*; considered as

Simple / Possessive / Reduplicative

1				2		3		4		Possessive		Reduplicative	
I	α	Wee	αι	Th[is]	ι That : o	Any	α8	Who	oι	Mine	hα	I my self	αlα
Thou	a	Yee	aι	Th[is] ane	8 Another : y	Every	a8	Which	88	Thine	ha	Your self	ala
He	e	They	eι	A [cert]an	yι Some : y8	All	e8	Whosoever	ιι	His	he	Himself	ele

&c.

Admiring · Heigh · Love · Ah · Hatred · Vaugh · Ecclamation · Oh · Silence · St.Hush

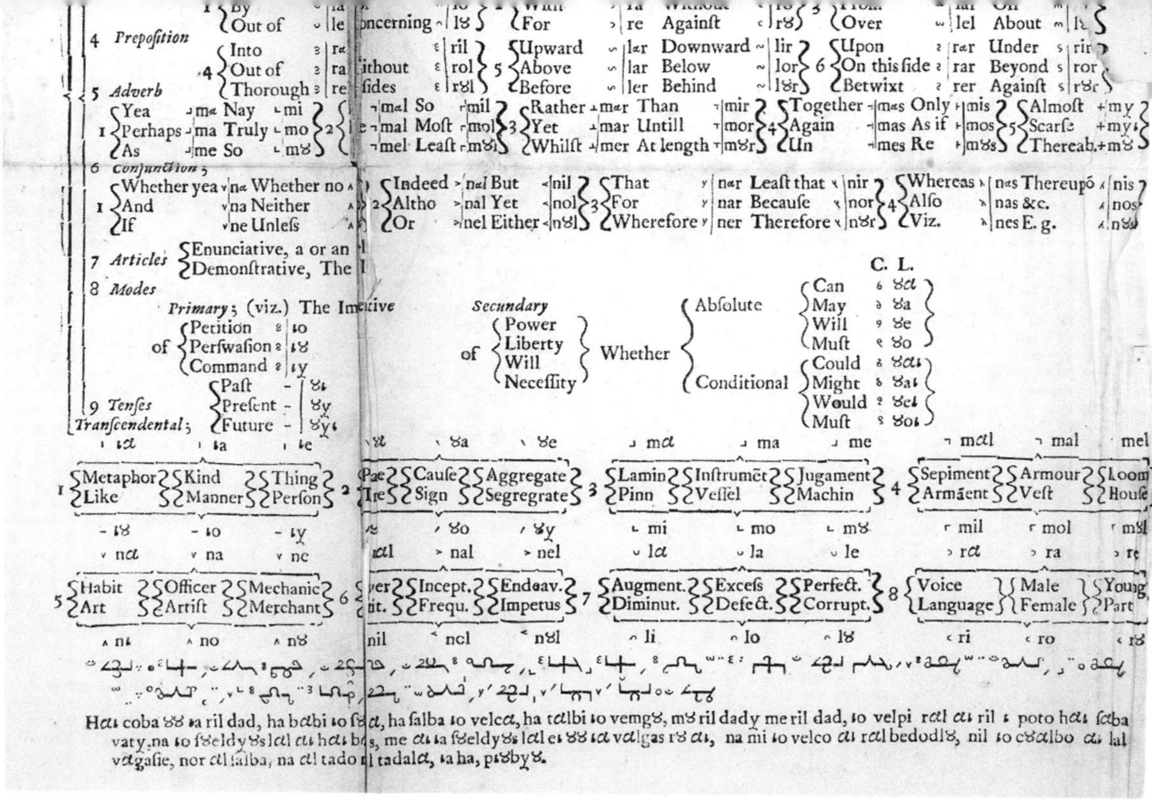

4 *Prepofition*

Out of le | ...ncerning l8 | For re | Againft r8 | Over lel | About l8

4 { Into ra | Out of ra | Thorough re } ...ithout ril, rol, r8l | ...fides | 5 { Upward lar | Above lar | Before ler } Downward lir | Below lor | Behind l8r | 6 { Upon rar | On this fide rar | Betwixt rer } Under rir | Beyond ror | Againft r8r

5 *Adverb*

1 { Yea ma | Perhaps ma | As me } Nay mi | Truly mo | So m8 | 2 { ... mal So mil | ... mal Moft mol | mel Leaft m8l } 3 { Rather mar | Yet mar | Whilft mer } Than mir | Untill mor | At length m8r | 4 { Together mas | Again mas | Un mes } Only mis | As if mos | Re m8s | 5 { Almoft my | Scarfe my | Thereab. m8 }

6 *Conjunction*;

1 { Whether yea na | And na | If ne } Whether no | Neither | Unlefs | 2 { Indeed nal | Altho nal | Or nel } But nil | Yet nol | Either n8l | 3 { That nar | For nar | Wherefore ner } Leaft that nir | Becaufe nor | Therefore n8r | 4 { Whereas nas | Alfo nas | Viz. nes } Thereupô nis | &c. nos | E. g. n8s

7 *Articles* { Enunciative, a or an | Demonftrative, The }

8 *Modes*

Primary; (viz.) The Im...tive of { Petition 8o | Perfwafion 8 | Command 8y }

Secundary of { Power | Liberty | Will | Neceffity } Whether { Abfolute | Conditional }

C. L.

Abfolute { Can 8a | May 8a | Will 8e | Muft 8o }

Conditional { Could 8a | Might 8a | Would 8e | Muft 8o }

9 *Tenfes* { Paft 8 | Prefent 8y | Future 8y }

Tranfcendental;

a a e | 8a 8a 8e | ma ma me | mal mal mel

1 { Metaphor | Like } { Kind | Manner } { Thing | Perfon } 2 { Pa... | ...re } { Caufe | Sign } { Aggregate | Segregrate } 3 { Lamin | Pinn } { Inftrumẽt | Veffel } { Jugament | Machin } 4 { Sepiment | Armãent } { Armour | Veft } { Loom | Houfe }

18 1o 1y | 8o 8y | mi mo m8 | mil mol m8l

na na ne | nal nel | la la le | ra ra re

5 { Habit | Art } { Officer | Artift } { Mechanic | Merchant } 6 { ...er | ...it. } { Incept. | Frequ. } { Endeav. | Impetus } 7 { Augment. | Diminut. } { Excefs | Defect. } { Perfect. | Corrupt. } 8 { Voice | Language } { Male | Female } { Young | Part }

n no n8 | nil nel n8l | li lo l8 | ri ro r8

Hαι coba 88 ιa ril dad, ha bαbi ιo f8α, ha falba ιo velcα, ha tαlbi ιo vemg8, m8 ril dady me ril dad, ιo velpi rαl αι ril ι poto hαι fαba vaty, na ιo f8eldy8s lαl αι hαι b...s, me αι ιa f8eldy8s lαl ει 88 ια vαlgas r8 αι, na mi ιo velco αι rαl bedodl8, nil ιo c8αlbo αι lal vαgafie, nor αl falba, na αl tado ...l tadalα, ιa ha, pι8by8.

FIGURE 12 • *"A Summary of Directions, Both for the Character and the Language." In John Wilkins,* An Essay towards a Real Character, and a Philosophical Language *(London, 1668). Swem Library, College of William and Mary, Williamsburg, Va.* Note the translation of the Lord's Prayer in the universal character at the bottom

than *Books,* and from the Observations, made of the *phaenomena* and Effects she presents, to compose a History of Her." Colonial America supplied the Royal Society with material to fulfill this objective, including "accounts of our aboriginal Natives and their customs" and boxes of curiosities that contained "quivers made of an Indian Dogskin" and "arrows headed according to the Indian manner." Eliot's letter participates in this transatlantic exchange by demonstrating the kinds of experiments that were uniquely possible in the New World. It also constructs the Praying Indian as an object of ethnological inquiry with peculiar powers of spiritual discernment. This rendering of the spiritual peculiarity of aboriginal knowledge emerged out of Eliot's own refinement of his ethnographic capacity as a missionary capable of observing, discerning, and carefully recording the anomalous forms of grace displayed upon primitive souls.[57]

In his manuscript collection of "loose notes theological," Boyle locates missionary ethnographies that make the "mysteries of religion" discernible within a narrative of natural history. For Boyle, the study of "the supernatural assistance of special grace" within the natural world depends on collecting and studying "external Testimonies that favor the Christian Religion," "the propagation of the gospel," and increasing "knowledge of human nature." Boyle's notion that "external testimonies" serve as the most reliable evidence of the presence of "special grace" within the natural world upholds a central tenet of the Royal Society—that experimental philosophy must involve exploration in order to examine the phenomenon of nature in its original state. For Boyle, a central purpose behind the collection of testimony and the progress of the gospel serves a pedagogical function. Natural religion will "cure the Indians of their ridiculous notions about the eclipses of the moon, and of the fond and superstitious practices those errors engaged them to [by informing] them of the true reasons of those eclipses, that did so undesirably amaze and fright them." For Boyle, the ontological possibilities of natural religion are reciprocal. Natural philosophers probe the "depths of God" through the experiential testimony of converted heathens. Praying Indians acquire a dual form of revelation:

57. John Eliot to Robert Boyle, Sept. 30, 1670, Notes from Company Records, fol. 5, American Antiquarian Society; "Directions for Sea-men, Bound for Far Voyages," Royal Society, *Philosophical Transactions,* I (1666), 140–141; Cotton Mather, "A Letter from Doctor Mather to Dr. Jurin, Boston New England, August 3, 1723," and Wait Winthrop, "Mr. Winthrop's Letter to Mr. Oldenburg about Some Curiosities Sent from New England to the Royal Society, 1671," Early Letters, III, XVI, Royal Society.

as God reveals himself to them, they acquire a "proper" understanding of the natural world. Science offers empirical evidence that religion "cures" natives' false views of the world.[58]

Cotton Mather's letters to the Royal Society similarly present the Indians as heathens in a state of primitive simplicity, with "no family government among them" and an "intolerably lazy quality." In a 1678 letter, Henry Oldenburg urges John Winthrop to "spread the honor of the English nation" through the new philosophy. Oldenburg surmises that even the "Savage Indians . . . when they shall see the Christian addicted as to piety and virtue . . . will the more cheerfully subject themselves to them." For Oldenburg, new philosophy joins with Christianity as an instrument of colonization. Boyle, Oldenburg, and Mather translate the ethnographic record of Indian religious testimony into a taxonomic and epistemological discourse of racial difference. Natural philosophy and the missionary project converge to produce a new form of racial particularity. While Indians exhibit a specific form of revealed knowledge about the natural world, they simultaneously demonstrate an inadequate ability to reason and understand it. The "progress of the gospel" also signals the progress of science: as Indians acquire salvation, they also acquire what natural philosophers view as a scientifically sound understanding of morality, nature, and government.[59]

Drawing upon Indian testimony for knowledge about supernatural phenomena, natural philosophy constructs the Indian as a curiosity within the New England landscape. Royal Society member William Petty's manuscript notes, containing a list of questions "concerning the nature of the Natives of Pennsylvania," exemplifies this process. Petty organizes his questions ethnographically by language, habits, religion, marital practices, economics, and education. Echoing Boyle's interest in the American Indian's ability to distinguish material and spiritual domains properly, Petty inquires: "Do they believe that God can raise men from Death to Life? Make a man a woman? dry up the sea? make the sunne stand still? remove the greatest mountaines from one place to another, at any future time prefixt?" These questions primarily concern knowledge of the natural world,

58. Robert Boyle, "Background Notes on Christian Virtuoso," 79, Boyle Papers, Royal Society; Boyle, "Theological Notes," 80.

59. Curiosa Americana Continued in a Decade of Letters to Dr. John Woodward and Dr. James Jurin from the Reverend Cotton Mather 1724, Early Letters, XLVI, Royal Society; Henry Oldenburg, "A Letter of Thanks to Mr. Winthrop in New England 1678," Early Letters, III, fol. 3.

seeming to partake in the emerging genre of the natural history survey, codified in Robert Boyle's "General Heads for a Natural History" (1666).[60]

Yet, formally and thematically, the questions also echo those posed to native proselytes, such as in Eliot's *Late and Further Manifestation:* "What work of the Spirit find you in your heart? Whether have you found at any time any such work in your self?" Blurring the missionary catechism and the natural history questionnaire, Petty juxtaposes his questions alongside a rigorous chart of resources and plans for economic growth in the American plantations, reflecting an imperialist agenda that underwrote empirical experimentation. The collecting of American curiosities was not a random or haphazard enterprise, but, rather, intimately linked to new philosophical standards of knowledge and testimonial reliability that structured a science of natural phenomena according to a science of the soul where taxonomies of grace engendered new ethnological understandings of subordinate human populations.[61]

The Limits of Experimental Desire

By the early eighteenth century, in the wake of King Philip's War, which devastated New England's native tribes, few praying towns remained. Puritan missionaries had all but abandoned the goal of millennial fulfillment through the conversion of a native population, instead deeming that populations fallen beyond the scope of Christian salvation to savage degeneracy could be viewed only as antithetical to English civilization. Descriptions of activity in the few remaining praying towns are sparsely sprinkled throughout a few colonial texts of the early eighteenth century, such as the appendix to Cotton Mather's *Bonifacius* (1710), his *India Christiana* (1721), and Experience Mayhew's *Indian Converts* (1727). In contrast to the Eliot tracts' configuration of the praying towns as active experimental sites where audiences on both sides of the Atlantic could witness the millennial fervor of the spread of the gospel, these texts feature Praying Indian churches as rare discoveries within the natural world, archaeological repositories for an an-

60. Marquis of Lansdowne, ed., *The Petty Papers: Some Unpublished Writings of Sir William Petty* (Boston, 1927), II, 117.

Initially printed in the *Philosophical Transactions* of the Royal Society, Boyle intended his "General Heads" to act as a guide for voyagers to the New World, instructing them in how to collect data from nature.

61. Eliot, *Late and Further Manifestation,* 19.

cient and primitive form of worship. In his appendix to *Bonifacius,* Mather explains:

> Their Method, respecting those that are Admitted into their Communion, is more according to the manner of the Churches in the Primitive Times, than is now practiced among the Churches in most Parts. The Person to be admitted . . . makes a *Confession of Sin;* which they do (as I have seen) with *Tears* and *Trembling,* like him in the *Sixteenth* Chapter of the *Acts.* And then he gives an Account of Experiences he has had, of Convictions, Awakenings, and Comforts; in which they are large and particular.[62]

Mather's Praying Indian testifies, through the witnessing power of the English minister, to the visible presence of divine power. The racial logic structuring this notion of a heightened primitivism stems from the peculiar physicality and affective response of Christian Indians, a response, remarks Mather, that goes beyond "the English in their meetings." The quoted testimony concludes with the statement that such scenes occur daily as more and more Praying Indians are admitted into the church. The "testimony" of an English minister preaching "among them" summarizes the evidentiary, primitivist, and affective modes that the Praying Indian had come to embody as a result of his encounter with the new scientific method and epistemology.[63]

What has changed between this scene described in the early eighteenth century and the very similar scene described in *Strength out of Weaknesse?* As quoted above, the latter describes "those poore naked sones of *Adam* . . . [coming] toward the Lord . . . with their joynts shaking, and their bowells trembling, their spirits troubled, and their voyces with much fervency." Both descriptions depict heightened affect; both display a scene through which those present as well as those reading the description can bear witness to a scene of Christian triumph. But, by 1710, the scientific community was less interested in native people as the bearers of sacred truths. The publication of Locke's *Essay concerning Human Understanding* in 1689 rendered the universal language program scientifically untenable by making a case for words as mere human constructs. Locke challenged the semiotic premise of universal language theory on two counts. First, he decoupled the

62. Cotton Mather, *Bonifacius: An Essay upon the Good* . . . (Boston, 1710), 197–198.
63. Ibid., 196.

"nominal essence" of the word from real essence, highlighting the discrepancy between human ideas of things and the things themselves. Second, he displaced the idea of language as having scriptural origins. Instead, his *Essay* seeks to unite the study of language with the study of human thought and understanding rather than with divine revelation. This transference in the object of knowledge challenged the universalists' belief that language might be restored to semiotic perfection in order to achieve a real connection to the divine.

Gradually, both native languages and native populations themselves became separated from the epistemological potential coded as sacred utterance in the scenes described in Shepard's letter or featured as the dumb sign of immanent divine wisdom in Beck's frontispiece. The two advertisements that conclude Mather's *Bonifacius* suggest the ramifications of this separation for religion and for science. The first advertisement, for Robert Boyle's *Christian Virtuoso,* extols the benefits of the Baconian method for discerning the visible presence of grace within the natural world. The second bills Mather's own *Biblia Americana* as a text that links natural philosophy and scriptural religion through *"experimental piety,"* "Observations of *Christian Experience,"* and "many of the *Excellent Things* observed in and extracted from *Holy Scriptures.*"[64]

The complementary use of testimony as evidence of biblical prophecy and natural history reveals how the application of new scientific methods to Native Americans produced a pattern of self-referentiality. Witnessing techniques express a desire to secure empirical knowledge of the divine and natural order through Indian testimony. Ultimately, however, this desire displays a circular logic as empirical practice fails to recognize or imagine forms of evidence outside its own self-authorizing framework of Christianity and natural law. The Lockean turn in linguistic philosophy did not quell the attempt to discover sacred origins in Indian languages. Rather, it relocated those speaking Algonquian within a different temporality. After Locke, divine mysteries are archaeological artifacts rather than human utterances of ancient, scriptural wisdom. Correspondingly, the image of the Indian changes from the fruits of an English nation to an American anthropological object of erasure. But, in each case, the words spoken by native people and recorded by Anglo missionaries remain scientific proof of an original, sacred utterance that helps to write each narrative of progress.

64. Ibid., 206.

{4} Deathbeds

Tokenography and the Science of Dying Well

"Jesus can make a dying bed as soft as downy pillows are."

"Yes," he replied, "he can, he does, I feel it."—"Memoranda by Mrs. Pearce within Four or Five Weeks of Mr. Pearce's Death"

She being dead *speaks* to You in what is here publish'd, after a short and *silent* Life among you.—BENJAMIN COLMAN, *Reliquiae Turellae, et Lachrymae Paternae*

On the day before her death in 1697, Mary Clark Bonner requested the presence of her close friends and family members in her Cambridge home. A group began to gather around her bed to bear witness to the final phase in the life of a Puritan saint: a testimony offering evidence of divine translation, the passage from death to eternal life. Often compared to childbirth, this passage was long and laborious. At one point, Mary cried out in a dreadful and frightened voice, "The devil came upon me like a Lyon." Her neighbor Mrs. Champney comforted and encouraged her "not [to] give way to the temptation of Satan." After the dreaded moment of temptation had passed, Mary's father and minister arrived to begin a collective prayer for a good death. Following the prayer, Mary entered into a series of raptures, exclaiming, "He is Come he is Come . . . and has Spocken powerfully to my Soul telling me my Sins are forgiven." The gathered community remained rapt for almost two days, erroneously thinking at one point that death had seized her. Mary "surprised" them at once, breaking forth into an "extream rapture of Joy" that far exceeded the expectations of her audience. The community watched as Mary's body weakened and her strength and speech failed. In the last moments of her life, she lifted a hand to heaven. Her eyes followed, gazing upward until the trembling hand fell back to the sheet, signaling at last that her soul had left her body.[1]

1. Robert F. Trent, ed., "'The Devil Came upon Me like a Lyon': A 1697 Cambridge Deathbed Narrative," *Connecticut Historical Society Bulletin*, XLVIII (1983), 117, 119.

It is quite possible that those witnessing Mary's death, including her husband, who most likely recorded her confession, were following the advice of their minister, Daniel Gookin, who over eight years preached a sermon series concerning the proper pious care that should attend such matters. In March 1681, Gookin explained to his congregation, "How good a thing it will be to have the testimony of a good conscience in a dying hour." Diligently, John Hancock transcribed this phrase, recording his minister's recommendation that deathbed witnesses record the testimonies of the dying. "To have" rather than simply to hear the testimony constructs the dying saint's last words as that which can be collected and preserved well beyond its initial temporal purpose. Gookin urges his congregation to preserve in writing something of the process of "translation" from life to death to Christian rebirth within the "dying hour." Diaries, family papers, and printed texts from the late-seventeenth to the early eighteenth century show how several colonial New Englanders followed this advice.[2]

William Adams of Ipswich transcribed the testimony delivered by his wife, Elizabeth, upon her deathbed in 1655 in his unpublished autobiography. In his journal, Joseph Tompson of Billerica paraphrased the dying words of his wife, Mary Tompson, and copied those of his daughter Mary Dane (d. 1689) from an original transcription sent to him by her minister, William Hubbard. In the above-narrated deathbed confession of Mary Clark Bonner, her husband, Captain John Bonner, recorded her "heavenly Expression" on a single sheet of manuscript paper. In the same year, Grace Smith from Eastham, Connecticut, delivered a "perpetual monitor to her surviving children," which "was taken from her own mouth a little before her death, by the minister of that town where she died" and published as *The Dying Mothers Legacy*. Testimonies were also excavated from prayer closets and preserved or published after the individual's death. A testimony written by Sarah Tompson of Connecticut was found by her husband upon her death in 1679, quoted extensively at her funeral, and preserved by her husband for several years. Upon the death of his wife, Sarah, Joseph Goodhue of Ipswich "found" and published a letter she wrote to him and her children, full of "Spiritual Experiences, Sage Counsels, Pious Instructions,

2. John Hancock, Sermon Notes, Commonplace Book, 1687, MS Am 121.1, Houghton Library, Harvard University, Cambridge, Mass.

This concept of divine translation comes from 1 John 3:14: "We know that we are translated from death unto life, because we love the brethren: he that loveth not his brother abideth in death."

and Serious Exhortations." Anne Bradstreet's "To My Dear Children" is a similar instance of this form.[3]

By the time the recording and preservation of such testimonies became common deathbed practices, Puritan ministers were already looking outside their congregations for evidence of grace. Thirty years of practice had not reduced the perpetual problem of uncertainty surrounding the Puritan quest for evidence of election. Signs of divinity collected from human souls in congregational settings never lost, or even diminished in, the shadow of uncertainty that Calvinist theology stipulated as a condition of such election. The human sensory capacity to discern the marks of grace would forever remain inaccurate, a direct and irrevocable consequence of the Fall. Senses could delude or deceive the believing Christian into thinking that she or he had experienced something authentically divine when in fact it had only been a delusion. The vexed status of these invisible data did not translate easily or clearly into a taxonomy of gracious signs that could then be used by the audience witnessing the testimony to judge further cases more clearly. Because of the potentially specious nature of divine signs dwelling upon each human soul, developing a soul science remained fraught with the same hesitant ambivalence. While the science of the soul had challenged the theologically imposed restriction against certain forms of knowledge through the improvisational incorporation of empirical techniques, there was a limit to how far this soul science could advance within a congregational setting. The data presented through the test for membership would remain, by theological design, contestable.

3. William Adams, "An Account of His Experiences Transcribed from His Own Handwriting, Jan. 1659," Massachusetts Historical Society, Boston; Joseph Tompson, Journal, 1662–1726, MS Am 929, Houghton Library; Grace Smith, *The Dying Mothers Legacy; or, The Good and Heavenly Counsel of . . . Mrs. Grace Smith . . . as It Was Taken from Her Own Mouth a Little before Her Death . . .* (Boston, 1712); "The Relation of the Word of God upon the——of Sarah Tompson, the Wife of Samuel Tompson Who Departed This Life January 15, 1679, Being Found after Her Death Written by Her Own Hand Many Years Since," MS 82653, Connecticut Historical Society, Hartford; Sarah Goodhue, "The Copy of a Valedictory and Monitory Writing, Left by Sarah Goodhue, the Wife of Joseph Goodhue, of Ipswich, in N.E. and Found after Her Decease," in Thomas Franklin Waters, ed., *Ipswich in the Massachusetts Bay Colony* (Ipswich, Mass., 1905–1917), I, 519–524; Jeannine Hensley, ed., *The Works of Anne Bradstreet* (Cambridge, Mass., 1967), 243–444.

The information on Mary Clark Bonner's manuscript comes from Robert Trent, who edited this manuscript copy for the *Connecticut Historical Society Bulletin*. John Bonner did not sign the document, so it is not clear that he was the transcriber, but Trent believes it is likely. I would agree, based upon my findings of other examples of the same genre. Trent, ed., "'The Devil Came upon Me like a Lyon,'" *Connecticut Historical Society Bulletin*, XLVIII (1983), 115–119.

The years following the passing of the Halfway Covenant bore witness to a substantial increase in female church members. The crisis felt by the passing of the first generation, as well as the sense of a fragmenting program for identifying visible saints and for making the elect the exclusive members of the governing body, placed pressure on the ministry, requiring it to redefine women's social role and covenantal responsibility. In a sermon series entitled *Some Important Truths about Conversion* (1747), Increase Mather describes a surrogate role of maternal minister through a domestic moral obligation, explaining that the household must form an exact replica of the church. *Pray for the Rising Generation,* a subtitle within this series that in fact quotes his own father's dying words, elaborates upon this domestic theory of "Household-Government," or the idea that maternal duties should be directed toward the goal of redemption.[4]

The work is for the descendants of the children of Israel, "the Rising Generation in New England," who are within God's covenant through the first seal of baptism. Urging the parents in his audience to consider that, although New England's children can be baptized, they are still "born in sin," Mather quotes Psalm 51:5, which refers to the naturally sinful state of the mother's womb: *"I was shapen in iniquity; and in sin did my mother conceive me;* Yet [David's] Mother was a precious, godly Woman." Mather contrasts the image of the mother as a natural breeder of sin with the description of David's mother, whose precious godliness prevented David from "dying in a natural, unconverted state." Mather underscores that, as breeders of sin, mothers have a heightened moral responsibility to pray for and discipline their children. But as a "precious, godly Woman," her body serves as a possible medium for redemption but not for divine regeneration.[5]

Although this maternal role gave women a more visible presence within the church, they did not receive the divine grace that would place them within the eternal, invisible church. Instead, mothers more often narrated the reception of grace on their deathbeds. This deathbed narrative was a

4. On his deathbed in 1669, Richard Mather spoke these words to his son Increase: "A speciall thing which I would commend to you is, Care concerning the Rising Generation in this Country, that they be brought under the Government of Christ in his Church." Quoted in Williston Walker, *The Creeds and Platforms of Congregationalism* (New York, 1893), 270n. For a lengthier analysis of this, see my article "'Keepers of the Covenant': Submissive Captives and Maternal Redeemers in Puritan New England, 1660–1680," in Mary C. Carruth, ed., *Feminist Interventions in Early American Studies* (Tuscaloosa, Ala., 2006), 45–59.

5. Increase Mather, *Pray for the Rising Generation; or, A Sermon Wherein Godly Parents Are Encouraged, to Pray and Believe for Their Children* (Boston, 1678), 5, 13, 22; Mather, *Some Important Truths about Conversion . . .* (London, 1674), 39–52.

subgenre of the testimony of faith, born out of a theological and ecclesiastical necessity to look to places other than the public congregational performance for the most convincing evidence of election. The deathbed testimony necessitated different techniques of witnessing while also providing a convenient replacement for the empirical practices encoded within more public expressive modes.

Deathbed testimonies functioned as objects within a new science of discovering grace and as political tools for policing the new social function of "Household-Government" by verifying that the covenant was indeed being "kept." Women's dying words extended mothers' earthly roles as surrogate ministers and proved that formal submission to the covenant would indeed secure the continuation of covenantal promise. Methods of tracking evidence of grace upon the deathbed were thus directed toward a politically efficacious corporate goal. Through their political status as "keepers of the covenant," women's deathbed scenes in particular fostered intensified modes of witnessing and expressing assurance, which attracted a science of discovery and its techniques of collecting and preserving testimonies for the observation, analysis, and surveillance of pious forms.

Anglo women and Native Americans occupied two disparate positions within the larger endeavor to develop a soul science. Pious mothers as well as "little damsels" and Anglo children functioned as siphons for the manifest presence of divine evidence within the framework of communal covenantal promise. Covenantal responsibility thus transferred to women, though not as visible saints displaying divine evidence in public duties, but rather as dying saints who displayed such evidence after serving the covenant through a dutiful, pious life.

The more certain forms of divine evidence that Eliot discreetly recorded on Martha's Vineyard in 1666 and then sent to Robert Boyle represent a similar quest for exaggerated evidence of grace. However, in the wake of King Philip's War, the desire to have living Christian Indians display and publicly perform such evidence became less desirable and even regarded as dangerous.

In 1685, John Eliot recorded the last words of twelve Praying Indians from Natick and published them in a collection entitled *The Dying Speeches of Several Indians* "for those that did desire them." Japhet, a Praying Indian from Martha's Vineyard, wrote his own deathbed testimony in Algonquian; it was translated into English and sent to the commissioners of the New England Company in 1712. Japhet's and other deathbed confessions of Martha's Vineyard's dying Indians constituted the archive of Experience

Mayhew's *Indian Converts* (1727), which compiled and revealed the ardent natural theological work of three generations of missionary Mayhews on Martha's Vineyard, increasingly figured through the turn of the eighteenth century as a site for displaying evidence of grace on the colonial periphery.[6]

Figuratively and literally, the island of Martha's Vineyard represented a contained experimental space, apart from the providentially verifying purpose of mainstream Anglo religious practice. As a supplemental form of evidence, it augmented the authenticity of the more internally directed colonial covenantal practices. The natural occurrences of divinity within these peripheral locales worked in concert with the dying testimonies or prayer closet devotions of the colony's most pious and fragile little girls to diminish the shadow of doubt and increase the providential likelihood that their deathbed words were true. In addition to functioning as a laboratory of grace, Martha's Vineyard also functioned metaphorically within missionary discourse as a place on the periphery of Puritan piety where one could go, somewhat paradoxically, to discover the forms of grace most full of a certain form of pious potential. By the late seventeenth century, the words of dying Indians and the prayer closet devotions of little damsels were rapidly coming to replace the congregational membership testimony as a primary site of empirical investigation into the invisible world.

Tokens of Grace

Ministers and laypeople attempted to capture the essence of spiritual plentitude in the moment before death through a written record of divine translation from death to life, from the visible to the invisible world, from the dark glass of limited perception to unfiltered revelation. To capture this essence, ministers performed a methodological innovation upon a well-established post-Reformation practice to revise what constituted divine evidence while also producing a distinct form of narration that I call "tokenography." In contrast to the hagiographical tradition that oscillates between past and future, tokenography dwells in the intricacies of the present. This emphasis explains Gookin's focus on the dying hour in the funeral sermon recorded in John Hancock's manuscript diary. Through the notion of a "token"—a sign indicating an event, object, or feeling—the recorded

6. John Eliot, *The Dying Speeches of Several Indians* [Cambridge, Mass., 1685]; Translations of Letters from Indians and the Report of an Indian on Matters of Religion, 1669–1727, Papers of the New England Company, Guildhall Library, London.

testimony stands as a remnant of that in-between state, an encapsulated moment in which the dying saint achieves near-perfect knowledge of his or her soul *before* the dissolution of the earthly body. Suspended between the world of the living and the world of the dead, the dying opened up a frontier of divine knowledge that had particular appeal to Puritan ministers and natural philosophers.

Accounts of laymen circulated too, but texts published in New England in the late-seventeenth and early eighteenth centuries demonstrate that laymen were less often transformed into tokens of divinity through empirical and natural philosophical techniques. Laymen appended to funeral sermons or letters are almost invariably older. The content of the written record therefore dwells upon the accomplishments of the deceased man's life. One notable exception is James Janeway's *Invisible Rarities* (1673), which records the visions seen in his brother's dying moments using the rhetoric of experimental philosophy to capture the vicissitudes of divine grace upon the soul of the dying man. But, even here, the single example contrasts the data-collecting practice reflected in Janeway's anthology of thirteen dying children, published around the same time, *A Token for Children*. In addition to printed documents, the existing archive of scribal publications and manuscript records of deathbed confessions reveals a similar demographic. Instances of recorded deaths of lay Anglo men emphasize pious lives rather than inquires into the world beyond. By contrast, testimonies of young women, children, and Native Americans—"tokens" according to Janeway and Mather—supply a present-tense account of the desire to dissolve the body in order to reveal the secrets of the invisible world.[7]

7. For instance of the contrast between laymen's and young women's funeral sermons, compare Solomon Stoddard's *Gods Frown in the Death of Useful Men* . . . (Boston, 1703), to Ann Fiske, *A Confession of Faith; or, A Summary of Divinity, Drawn up, by a Young Gentlewoman in the Twenty-fifth Year of Her Age* (Boston, 1704); or Ben[jamin] Tompson, *The Grammarian's Funeral; or, An Elegy Composed upon the Death of Mr. John Woodmancy* . . . [Boston, 1708], to Tompson, *A Neighbour's Tears Sprinkled on the Dust of the Amiable Virgin, Mrs. Rebekah Sewall* . . . (Boston, 1710).

For this demographic information, I am relying largely on the list of scribal publications compiled by David D. Hall. For instance, Joseph Eliot's *Copy of an Excellent Letter Wrote by the Reverend Mr. J. E. of Guilford* circulated in manuscript form before it was first published in 1725. It follows from the same pattern of deathbed confessional circulation seen in manuscript records of the deathbed scenes of Elizabeth Williams, Mary Clark Bonner, and Sarah Tompson. The gendered difference comes through in the content rather than the form; Eliot's letter concerns matters of living within the world, whereas John Bonner's record of his wife's death attempts to track the secrets of the invisible world into which she is entering. Hall, "Scribal Publication in Seventeenth-Century New England: An Introduction and a Checklist," American Antiquarian Society, *Proceedings*, CXV (2005), 29–80.

What made such populations more effective tokens of divinity than laymen or New England's ministerial elite? This question can be answered in part by the increased spiritual authority accorded women through the transforming typology of the female in late-seventeenth- and early-eighteenth-century theology. A complementary type, the dying Indian, emerges around the same time as a powerful, symbolic icon of nostalgia for a vanishing faith. After the Halfway Covenant of 1662, children were increasingly framed as vehicles of pious promise in such sermons as Increase Mather's *Pray for the Rising Generation* (1678) and *Call for Heaven* (1685). Such analyses chart the augmented spiritual authority of these groups but do not account for what the ministers were looking for, namely, the epistemological potential that prompted the attempt to collect, record, and publish their dying words. This potential may be explained in part through Locke's theory that "native and original impressions should appear fairest and clearest" among groups of people in whom "we find no footsteps."

> Children, Ideots, Savages, and illiterate People, being of all others the least corrupted by Custom, or borrowed Opinions; Learning, and Education, having not cast their Native thoughts into new Moulds; nor by super-inducing foreign and studied Doctrines, confounded those fair Characters Nature had written there; one might reasonably imagine that in their Minds these innate Notions should lie open fairly to every one's view, as 'tis certain the thoughts of Children do.[8]

For Locke, the appeal of these populations is that they most closely resemble the tabula rasa state and can thus be used by the empiricist to most simply and accurately observe how sensory impressions are recorded. "Doctrines," "Moulds," and "Education" all, to varying degrees, obscure human understanding of the transmission of sensory data from things to mental impressions to ideas. Children and naturals (persons removed from

8. John Locke, *An Essay concerning Human Understanding,* ed. Peter H. Nidditch (New York, 1974), 63. On women's increasing spiritual authority, see Lonna M. Malmsheimer, "Daughters of Zion: New England Roots of American Feminism," *New England Quarterly,* L (1977), 484–504; Margaret W. Masson, "The Typology of the Female as Model for the Regenerate: Puritan Preaching, 1690–1730," *Signs,* II (1976), 304–315; Amanda Porterfield, *Female Piety in Puritan New England: The Emergence of Religious Humanism* (Oxford, 1992); Laurel Thatcher Ulrich, "Vertuous Women Found: New England Ministerial Literature, 1668–1735," *American Quarterly,* XXVIII (1976), 20–40.

For a fuller analysis of the effects of the Halfway Covenant on gendered and generational piety, see David D. Hall, *Lived Religion in America: Toward a History of Practice* (Princeton, N.J., 1997); Rivett, "'Keepers of the Covenant,'" in Carruth, ed., *Feminist Interventions,* 45–59.

education, civilization, and custom) serve as litmus tests for the Lockean empiricist. Based on the premise that knowledge comes from the human senses, Lockean empiricism sought to objectify and expand human understanding through observation and experiment. Through proper observation, what occurs in the "Huts of Indians" or in the "thoughts of Children" may be extended to the "Principles of Sciences" studied and advanced in the "Academies of learned Nations." Children and naturals constitute a form of fieldwork for the empiricist whereby he may then apply his findings to a method of scientific inquiry grounded in the reality of human experience.[9]

The dying words of elite New England men such as those published in Cotton Mather's *Magnalia Christi Americana* (1702) and Nathaniel Morton's *New-England's Memorial* (1721) tend not to present "rarities" for postmortem discovery upon the souls of earth's "weaker vessels." Owing to the subjects' sociopolitical status, these testimonies and spiritual biographies offer a retrospective account of an inward spiritual self that has already been discovered and witnessed within the finite terms of the English saint's life. Testimonies of prominent English men rest upon the self-evident assurance of the legitimacy of the New England mission; they function to unite divine election with a politically visible providential design. *New-England's Memorial* includes an entire testimony written by William Bradford upon the death of William Brewster in 1640, a short verse found in Thomas Dudley's pocket after his death in 1653, and a testimony that Bradford wrote before his death in 1657. Each testimony functions as an exemplary "star of the first magnitude in the Firmament of New England," presenting the individual's works as the fruits of a gracious life. In England, Holland, and New England, Brewster was "of singular use and benefit to the Church and People." Thomas Dudley, "who was a Principal Founder and Pillar" of the Massachusetts colony, had "a piercing judgment to discover the Wolf, though cloathed with a sheep skin." William Bradford was "well skill'd . . . in regulating Laws" and taught scripture "laboriously."[10]

The dying words of and narratives about English men present historiographical evidence in the seminal narratives of New England's past. Consequently, they most often appear as integral parts of a colonial history that by and large excludes the recorded experience of English women and Native Americans. Deathbed testimonies of English men reiterate the

9. Locke, *Essay concerning Human Understanding,* ed. Nidditch, 64.

10. Nathaniel Morton, *New-England's Memorial* . . . (1669; Boston, 1721), 154, 178, 188, 210.

inseparable conjuncture between the inward, phenomenal experience of grace and the providential design of the New World that produced the masculine witnessing subject recorded in testimonies of faith in Reformed congregations. Conversely, the dying words spoken by the more socially and politically marginalized drew upon a sense of assurance that was not retrospective but rather came from witnessing the "foretastes of heaven" in the present world. This sense of assurance advanced the religio-scientific discovery of grace by occasioning new methods of mapping, seeing, and studying patterns of religious affect upon the soul.

Ethnographies of the Sacred

The distinct form of narration that I am calling "tokenography" originated in Roger Williams's critique of the more orthodox methods of discerning grace in the Massachusetts Bay Colony. The text that most explicitly details his theological departure from the New England clergy, *The Examiner Defended* (1652), explains that the *"Land of Canaan,"* consisting of the geographical as well as national space of Israel's typological descendants, must be "attended with *extraordinary, supernatural, and miraculous* Considerations." Williams identifies New England as a place where divinity can be most effectively discovered. Confronting the same dilemma of epistemological incertitude that necessitated the introduction of scientific methods into Puritan techniques of testimony, Williams reminds his readers of the Calvinist and Augustinian problem of religious sight. He explains that the "spiritual" cannot be "seen with a natural eye" because of humans' "depraved nature." But, in contrast to Thomas Shepard and Richard Baxter, Williams proposes a very different solution. The mistake of the New England orthodoxy is in presupposing that the "Civil Magistrates" "have a clearer sight in discerning, and an higher Authority in judging of *Godliness* and *Christianity*." Resisting the science of signs developed by Shepard, John Fiske, Eliot, Thomas Mayhew, and others, Williams insists that technologies of discernment should not be brought to bear upon controlled scenes of witnessing religious affect because the elect do not wield special visionary powers or the qualification to see that Boyle would write of years later in *The Christian Virtuoso*.[11]

11. Roger Williams, *The Examiner Defended, in a Fair and Sober Answer to the Two and Twenty Questions Which Largely Examined the Author of "Zeal Examined,"* in *The Complete Writings of Roger Williams*, VII, ed. Perry Miller (New York, 1963), 220, 224, 242, 244, 251.

Williams wanted to develop a form of ethnographic witnessing that differed in technique from the methods of tracking visual evidence in congregational testimonies. The highest attainment of spiritual knowledge comes "by Study, Observation, and Experience" of the "weaker earthen vessels" that are more often chosen for God's grace. The spiritual knowledge ascertained through *"Impressions* from *Heaven"* can be "distinguished" by "a more tender and observant Eye." The eye must be trained with the knowledge that "impressions" often come in the form of "incivilities" within the "Light of Nature." Williams resists the orthodox tendency to invite reproductions of intelligible forms of grace by insisting on the utter formlessness of true divinity.[12]

We saw earlier how Eliot applied the same empirical techniques for charting evidence of grace but in an internally contradictory practice, stemming from preconceived ideas about how grace affects a heathen soul. Williams's quest for anomalies rather than preconceived norms constitutes the central difference between two distinct ethnographic perspectives, both of which sought to advance knowledge of the sacred, invisible world through its manifestation across an ethnological divide. Seeking knowledge of the invisible world by assuming the role of passive observer of the workings of grace among "meanest and lowest earthen vessels," Williams made the death and burial rights of the Narragansett Indians, the dying words of Wequash, and his own wife's deathbed piety his primary objects of study. The last chapter of *A Key into the Language of America* seeks to understand how "natural men" encounter the "King of Terrors." He oscillated between a sense of the natives' proximity to English custom and a recognition of their differences. Revealing black to be a transcultural sign of mourning, Williams notes: "The Men also (as the *English* weare blacke mourning clothes) weare blacke *Faces,* and lay on soote very thicke, which I have often seene clotted with their teares." Tears and the color black are the universal signs of mourning, demonstrating a "natural" and ethnological uniformity in approaches to death, especially when set in contrast to William Wood's *New Englands Prospect,* which depicts "deepe groans, and *Irish*-like howlings, continuing annuall mournings with a blacke paint on their faces."[13]

12. Ibid., 224, 241, 244.

13. Roger Williams, *A Key into the Language of America,* ed. J. Hammond Trumbull, in *The Complete Writings of Roger Williams,* I (New York, 1963), 215 (hereafter cited as *Key into the Language of America*); William Wood, *New Englands Prospect* . . . (London, 1634), 93. The differences between Eliot's and Williams's ethnographic perspectives has not received enough critical attention. Larzer

In contrast to *New Englands Prospect*, the *Key into the Language of America* does not type the natives as heathen savages, but rather looks for continuities across the cultural divide in order to make the mystery of death the primary object of observation. The natives are "rarities" through which this phenomenon might be more adequately understood; their simultaneously bizarre and familiar behavior is an avenue through which the "King of Terrors" might become more comprehensible. As such, the *Key* exemplifies what we might think of as an ethnography of the sacred, a practice that privileges grace rather than human customs and cultures or heathen savagery as the primary object of study. "So terrible is the King of Terrors, Death, to all naturall men" that Williams more easily approaches it through the distilled lens of the ethnographic encounter, as in this sequence of lexemes (see Figure 13): Williams follows this sequence with the observation that "Death" is so terrifying that the natives "abhorre to mention the dead by name, and therefore, if any man beare the name of the dead, he changeth his name."[14]

This custom reflects a cultural difference between the Narragansetts and the English that Williams explains through the "natural" "terror" of "Death" to "all men." The terror of death is that it marks a passage into the unknown, invisible domain of the spirit world, a passage that the Narragansetts represent by rendering the names of those who have made the journey unspeakable. Belonging to such a domain, the dead Narragansett cannot be observed and studied as an index of grace, but instead marks the erasure of undiscovered "rarities" among New England's native population. What Williams's observation really describes is the terrifying finality of a death without salvation. Without evidence of grace, the passage from the earth to heaven is a traceless movement, a soul lost in another world without any hope of redemption.

Ziff's compelling analysis is a notable exception: Williams wrote "from a consciousness transformed by his experience of the different, rather than . . . from a consciousness in recoil from the different." See "Conquest and Recovery in Early Writings from America," *American Literature*, LXVIII (1996), 523. Generally, I agree, though I am proposing one more layer of complexity, that Eliot witnessed native testimony according to a preconceived notion of what this difference would look like. He recoils because the Massachusett testifiers perform in ways that are neither adequately different from the English nor according to the generic trope of native performance within which *Tears of Repentance* (1653) begins. (On the paradoxical dilemma of witnessing, see Chapter 3.)

14. Williams, *Key into the Language of America*, 216.

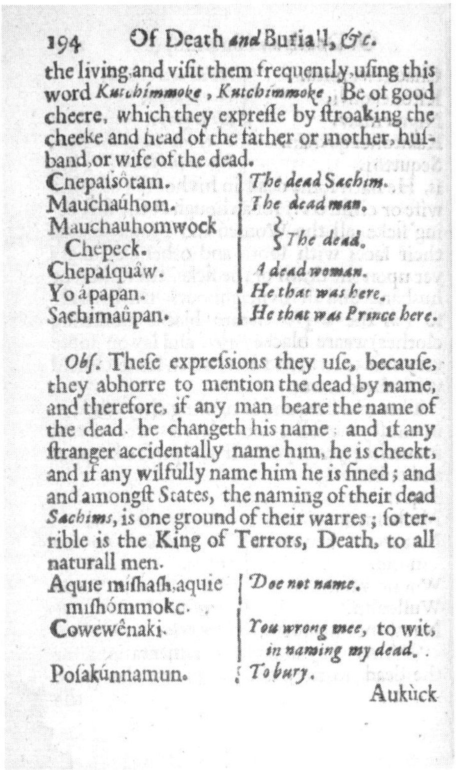

194 Of Death *and* Buriall, *&c.*

the living and viſit them frequently, uſing this word *Kutchimmoke, Kutchimmoke*, Be of good cheere, which they expreſſe by ſtroaking the cheeke and head of the father or mother, husband, or wife of the dead.

Cnepaſôtam.	*The dead Sachim.*
Mauchaûhom.	*The dead man.*
Mauchauhomwock Chepeck.	*The dead.*
Chepaſquâw.	*A dead woman.*
Yo âpapan.	*He that was here.*
Sachimaûpan.	*He that was Prince here.*

Obſ. Theſe expreſsions they uſe, becauſe, they abhorre to mention the dead by name, and therefore, if any man beare the name of the dead he changeth his name; and if any ſtranger accidentally name him, he is checkt, and if any wilfully name him he is fined; and and amongſt States, the naming of their dead *Sachims*, is one ground of their warres; ſo terrible is the King of Terrors, Death, to all naturall men.

Aquie miſhaſh, aquie miſhómmoke.	*Doe not name.*
Cowewênaki.	*You wrong mee,* to wit, *in naming my dead.*
Poſakúnnamun.	*To bury.*

Aukùck

FIGURE 13 • *Indian-English Vocabulary. In Roger Williams,* A Key into the Language of America *(London, 1643). Huntington Library, San Marino, Calif.*

In a concluding "general Observation of their Dead," Williams hopes to resolve this impenetrable mystery by beginning to "see" and make visible to others the presence of "the most High and most Holy, Immortall, Invisible, and only Wise God" through the dying Indian:

> O, how terrible is the looke the speedy and serious thought of death to all the sons of men? Thrice happy those who are dead and risen with the Sonne of God, for they are past from death to life, and shall not see death (a heavenly sweet Paradox or Ridle) as the Son of God hath promised them.[15]

Conversion is a solution to a universal terror because it transforms death to rebirth. The "look" and "thought of death" dissolve through the redeemed saint who confronts the passage from life to death and then "death

15. Ibid., 279.

to life" as a transition made highly visible through the "promise" of redemption. This, for Williams, is the "heavenly sweet Paradox or Ridle": a realm of impenetrable mystery can be made utterly intelligible for the elect so that one does not even have to "see death" but can rather encounter it, paradoxically, as a birth into a new life. Williams's account of this paradox comes from 1 John 3:14: "We know that we are translated from death unto life, because we love the brethren: he that loveth not *his* brother, abideth in death." The paradox of divine translation transforms the deathbed scene into an appropriate site for encountering grace through the scriptural assignation of a "certain sign of our regeneration." Less encumbered by the burden of the body—

> The Indians say their bodies die,
> Their souls they do not die

—the clarity of the soul's evidence increases. Although the deathbed scene reveals the unseen as a "sublime truth" that cannot be witnessed as fully by those still in their earthly bodies, the "certain sign" of grace received within this moment suggests that what can be seen and communicated offers a purer form of grace. The lesson that Williams takes from his Narragansett observations is that dying the "good death" offers assurance against the terrors of the unknown and leaves the living with more knowledge of divine truth.[16]

Williams draws upon this idea of the purity and certainty of evidence produced through dying words in the deathbed scene with which the *Key* begins, perhaps the first dying speech recorded by a Protestant missionary of a North American native. Williams records Wequash's statement of a stone heart as evidence of native knowledge of inward depravity, the "inward hardnesse and unbrokenesse" of a heathen heart desiring a Christian God. Wequash's words are intended to elicit sympathy from the reader and a belief in the native's open willingness to acquire Christian knowledge. Wequash's death becomes an unequivocal marker of the presence of divinity among the natives and the inception of the missionary movement. Williams's presence at Wequash's deathbed, his careful record of the "broken

16. Ibid., 219. In the Geneva Bible the margin notes for 1 John 3:14 state, "This love is the special frute of our faith and a certeine signe of our regeneration." The idea of a "certeine signe" of assurance indicates the absence of the danger of hypocrisy within this moment. A certain sign of assurance yields tremendous empirical promise to a theological system so challenged by the dilemma of incertitude.

English" and "savory expressions" of an Indian in the first phase of conversion, assures the reader that help for the natives has indeed arrived. The evidentiary value that Williams places on this one moment as a framing device to establish the legitimacy of his millennial vision causes John Cotton's critique: "His testimony of the facility of such conversion of the *Indians* was too hyperbolicall." Cotton calls into question the empirical value that Williams places on Wequash's death as a "certain" symbol of Native American conversion. Nonetheless, Williams seems to include this scene aware of the kind of confirmation that a dying saint can convey, giving his London audience "hope of the Indians receiving the Knowledge of Christ!"[17]

When Williams's wife falls ill in 1650 while he is living among the Narragansetts, he shares with her what he has learned in a letter of instruction on how to see her own death through the "riddle" of its spiritual truth as a form of rebirth. This letter was printed in 1652 as a pamphlet entitled *Experiments of Spiritual Life and Health, and Their Preservatives in Which the Weakest Child of God May Get Assurance of His Spirituall Life and Blessednesse.* As the title indicates, this instruction manual for seeking assurance in fact continued the argument that Williams made in *The Examiner Defended,* wherein he explains that grace affects the "meanest and lowest earthen vessels" most powerfully. Following from the principle of spiritual essence decoupled from form as a critique of orthodox practices of testimony, *Experiments* explains that as the "outward" form degenerates, the "inward" essence regenerates. For Williams the body is an example of forma and consequently an obstacle to true grace. Williams describes forma as the "seen of men"—all that is visible in the world. The "seen of men" produces hypocritical inversions of the inwardly directed secret vision that provides the true knowledge of God and assurance. But, Williams explains to his wife, the decay of the physical body enables the "Souls-eye" to "more and more brightly see him who is invisible" so that "the visible and the worldly, *may be accounted by you but* dreams *of* shadows." Whereas "false worshippers and false Christians may easily satisfy themselves and stop the mouths of their consciences, with any formal *performance,"* the true "children of God" are the "weakest," privileged with the ability to attain knowledge of God through "a single and upright eye unto God himself in secret." As in the *Key,* sight solves the riddle: physical death is actually a translation from death to life. Seen in this way, the divine truths revealed to the "souls-eye"

17. Williams, *Key into the Language of America,* 27, citing John Cotton, *The Way of the Congregational Churches Cleared: In Two Treatises* (London, 1648), 80.

permit the "King of Terror" to dissolve into the "strength of Gods grace and Spirit in us."[18]

Williams was not proposing anything radically new through this idea that visual evidence of assurance can comfort the dying. In a 1652 letter, Williams records his "humble wish" that his friends in England receive "the saving knowledge and assurance of the way of life which is eternal when this poor minutes dream is over." This would have been received as fairly conventional advice to an English audience now familiar with the ars moriendi tradition as literary record of how "man's mind sees farther than his physical sight, feels more sensitively than his five senses." Williams's innovation upon the genre comes through his ethnographic practice, which suggests a form of "discovery" newly available through the observational record of such scenes. The "rarities" to be found among New England's "natives" connote the spiritual knowledge displayed as death, the ultimate space of divine mystery, which produces a realm of legibility not only to the dying saint but to others bearing witness to and recording the scene. In *Experiments,* Williams does not recommend that anyone observe, record, or witness his wife's deathbed scene. In fact, he focuses on the private, introspective ways through which the ars moriendi tradition occasions philosophical insight: the meaning of death becomes the knowledge of one's salvation, a knowledge that transcends the stoic process of "dying well."[19]

Ars Moriendi Transformed

In seventeenth-century Reformed theology, dying well was the most legitimate and empirically sound evidence of an individual's spiritual estate, in part because the divine translation dissolved the threat of hypocrisy. According to one critic, "The art of dying well had never been so closely observed and analysed as it was during the 150 years following the Reformation." Consequently, the ars moriendi tradition flourished in early modern New and Old England as a way of narrating and perpetuating the legitimacy of the Protestant cause. Whereas Catholic theologians believed that the proper deathbed behavior could result in salvation, Protestants admit-

18. Roger Williams, *Experiments of Spiritual Life and Health, and Their Preservatives in Which the Weakest Child of God May Get Assurance of His Spirituall Life and Blessednesse* (London, 1652), v–vi.

19. John Russell Bartlett, ed., *Letters of Roger Williams, 1632–1682,* in *The Complete Writings of Roger Williams,* VI (New York, 1963), 242; Nancy Lee Beaty, *The Craft of Dying: A Study in the Literary Tradition of the "Ars Moriendi" in England* (New Haven, Conn., 1970), 75.

ted such evidence only as a further indication of a previously established spiritual estate; the genre was, however, mutually important in securing evidence for each respective position. Alongside this religious debate, the forms within the ars moriendi tradition changed significantly through the seventeenth century to register and reflect a "new mode of thinking death" as an object of intellectual curiosity as well as a technique for verifying religious legitimacy.[20]

This proliferating mode of analyzing Christian death corresponded to a scientific effort to understand disease and improve medicinal cures for its fatal consequences. Numerous Royal Society tracts from the late-seventeenth century record findings of diseases upon dissection and of ethnographic accounts of death and burial rites among various indigenous populations. John Josselyn's *New-Englands Rarities Discovered* (1672) records death and burial rites observed among Native Americans alongside taxonomies of plants and minerals found along the northern New England coast. The appeal of these data to the Royal Society anticipates what would become in the eighteenth century an attempt to scientifically prolong life through native roots and medicinal knowledge. Cotton Mather's development of a smallpox vaccine through the Native American remedy of infusing tobacco oil into the bloodstream met with much local resistance from religious conservatives while his exchanges with Dr. James Jurin secured his Royal Society membership. What Mather's reluctant contemporaries did not recognize was that this medical cure in fact aided the Christian art of dying well, a continuity that the ministry would not fully recognize until the beginning of the eighteenth century.[21]

20. Ralph Houlbrooke, "The Age of Decency: 1660–1760," in Peter C. Jupp and Clare Gittings Jupp, eds., *Death in England: An Illustrated History* (New Brunswick, N.J., 2000), 179. See also Lucinda McCray Beier, "The Good Death in Seventeenth-Century England," in Ralph Houlbrooke, ed., *Death, Ritual, and Bereavement* (New York, 1989), 43–76.

Erik Seeman explains this Catholic-Protestant distinction in his comparative reading of the deathbed testimonies recorded by French and Anglo-American missionaries. Seeman, "Reading Indians' Deathbed Scenes," *Journal of American History*, LXXXVIII (2001–2002), 17–47. Seeman uses this distinction to track a debate about the relative missionary success of Jesuits and Protestants; I am proposing that the increased focus on deathbed testimonies in general reflects the attempt by Catholics as well as Protestants to scientifically legitimate their cause.

21. For examples of accounts of dissection, see William Cowper, "A Letter from Mr. William Cowper, Giving an Account of a Very Large Diseased Kidney, Found on the Dissection of a Lady, with the Symptoms of the Disease before Her Death, and an Explanation of Their Phaenomena," Royal Society, *Philosophical Transactions*, XIX (1695–1697), 301–309; Dr. Harvey, "An Extract of the Anatomical Account, Written and Left by the Famous Dr. Harvey, concerning Thomas Parre, Who Died in London at the Age of 152 Years and 9 Moneths," *Philosophical Transactions*, III (1668),

In 1721, Increase Mather would preach a sermon, *Several Reasons Proving that Inoculating or Transplanting the Small Pox, Is a Lawful Practice, and That It Has Been Blessed by God for the Saving of Many a Life.* According to Mather, the conservative Boston ministry's attitude concerning the providential legitimacy of inoculation had transformed from critical resistance to endorsement. Ministers and physicians in Boston as well as members of the Royal Society approved the practice, which also appeared in the Royal Society *Transactions*. While contributing to this universal acceptance of God's design, Mather conjoins providential discourse to the methods of experimental philosophy, proclaiming that "inoculating the Small Pox is a safe and universally Useful Experiment." The complementarity of Christian and medicinal theories of life and death also appears in early-eighteenth-century correspondence between Dr. Hoy of Jamaica and the Royal Society when Dr. Hoy proposes that a Jamaican mineral may prove most valuable in improving the "Art of Dying." The circulation of information on remedies designed to prolong human life as well as experiments performed upon the dying circulated throughout the Royal Society's information-gathering networks in the Atlantic world. As Increase Mather explained in another sermon on smallpox, also in 1721, the smallpox inoculation was brought forth from the "unlearned Orientals" by the "good Providence of GOD." The revolutionary development of the smallpox remedy exemplifies experimental philosophy's dependence on information from the New World.[22]

886–888; W. Gould, "An Account of a Polypus Found in the Heart of a Person That Died Epileptical, at Oxon.," *Philosophical Transactions,* XIV (1684), 537–548.

On Mather: Cotton Mather, Curiosa Americana, Mather Family Papers, American Antiquarian Society, Worcester, Mass.; Cotton Mather, A Letter from Doctor Mather to Dr. Jurin, Boston New England, Aug., 3, 1723, Early Letters/M2/46, Royal Society, London. For a discussion of the controversies surrounding this medicinal discovery, see Robert V. Wells, "A Tale of Two Cities: Epidemics and the Rituals of Death in Eighteenth-Century Boston and Philadelphia," in Nancy Isenberg and Andrew Burnstein, eds., *Mortal Remains: Death in Early America* (Philadelphia, 2003), 56–67.

22. Increase Mather, *Several Reasons Proving That Inoculating or Transplanting the Small Pox Is a Lawful Practice . . .* (Boston, 1721), 1; Matthew Hale, A Letter Sent to Dr. Hoy at Jamaica, Feb. 4, 1713, Dr. Hoyes Letter from Jamaica April the 20th 1714 to Mr. Waller, Letter Book Collection, EL/W3/78, 91, Royal Society; Increase Mather, *Some Further Account from London, of the Small-Pox Inoculated . . .* (Boston, 1721), 1. By "unlearned Orientals," Mather is referring to the Native American suppliers of this remedy as well as invoking the ten lost tribes theory, which also promoted the belief that Native Americans migrated from East Asia thousands of years prior to European New World discovery.

The need for a New World discourse of colonial and providential legitimacy intensified the problem of death as an unknowable domain—Was a smallpox outbreak the sign of merciful affliction, or God's desertion?—at the same time that it expanded the empirical potential for divine discovery through methods of witnessing grace. John Hancock's notebook records six years of sermons by Daniel Gookin and William Brattle on the art of dying well. Gookin asked his congregation, "What is meant by death?" and recommended methods of observation and experience to supplement the theological inquiry. Brattle preached that "the evidences of a growth in grace or a constant spiritual soul's communion with God" could be tracked in order to produce a science of the dying soul. Those who witnessed, recorded, and circulated deathbed testimonies attempted to resolve the human incapacity to know divine mystery through the production of new forms and uses of evidence. "To have a testimony," Gookin told his congregants, was to have a recorded, verified, and preserved artifact.[23]

Images of the Dying Hour

Testimonial records also provide a rich archive into the epistemological conditions surrounding death. Dying female saints occasionally critiqued congregational methods of witnessing grace, based upon their own sense of exclusion from more public forms of witnessing. In her dying testimony, Elizabeth Adams states,

> A soul must be divorced from its first husband, and sin and lust, before it will clear with Christ and that the great reason why so many come short was because they could not bear the cutting off.[24]

This sense of a "cutting off" from earthly ties marks the onset of her divine translation; the embodied presence of grace produces a stricter separation between grace and works. In contrast to the testimonies of English men in Shepard's and Fiske's congregations where we saw sanctification, or the outward manifestations of grace, displayed through an aural and visual continuity between the inward phenomenal experience of conversion and a geographical and temporal providential design, Elizabeth Adams's narra-

23. William Gookin, Mar. 6, 1681, Apr. 17, 1687, William Brattle, 1684, in Hancock, Sermon Notes, Commonplace Book, 1687, MS Am 121.1, Houghton Library.

24. Adams, "An Account of His Experiences," MHS.

tive unfolds as a profound inward-turning in which her duty as a wife and her earthly works are set aside. The continuum between the roles of the wife and of the spouse of Christ was an avenue of women's spiritual authority. Elizabeth's divine encounter dismisses her earthly obligations. She uses the term "divorce" repeatedly throughout the narrative to describe this cutting one's self off from the earthly world to make her soul available for union with Christ. Mary Dane expresses a similar sense of detachment from worldly commitments and obligations as a necessary condition of her conversion. As if performing the process Williams described to his wife, whereby "the visible and the worldly, may be accounted by you but dreams of shadows," Dane remarks: "The Lord was pleased to discover unto me much of the vanity of worldly enjoyments. I saw that nothing short of Jesus Christ was sufficient to satisfie the soul, which made that promise precious to me."[25]

Alongside Dane's and Adams's description of a violent form of detachment and painful separation from their role in the world, both testimonies critique the hypocrisy inherent in testimony through references to Paul's Epistle to the Romans. The subject's proximity to death renders the deathbed testimony almost void of the dangers of hypocrisy that the pressure of visible sainthood produced. Mary critiques modes of hypocritical performance through an extended reference to Romans 3 in which the confessor's "throat," "lips," and "blood" signify how public confession violates the biblical "Law" that "no flesh be justified in his sight." Elizabeth includes multiple references to Romans 10, which expresses a similar fear of hypocrisy among those who "establish their own righteousness" (through testimony) and "have not submitted themselves unto the righteousness of God." Replicating Williams's critique of orthodox methods of witnessing, Mary and Elizabeth suggest through their reading of Romans that, while the living body prevents accurate "sight" or knowledge of divine election, knowledge of salvation increases with the degeneration of the flesh in the moments before death until this knowledge becomes a "certain sign" of election.[26]

Fully attuned to what Gookin describes as the evidences of the "dying hour," the narrative voice within each testimony speaks in the present tense about the event of conversion, displaying a certainty that both builds upon past evidences and promises to be completed in the future. This temporal

25. Ibid.; Joseph Tompson, "Journal," MS Am 929, Houghton Library.

26. Ibid.

framework enables a simultaneous critique of hypocrisy and expression of assurance. Through the dissolution of the body and the letting go of the finite conditions of visible forms and earthly time, the soul begins to inhabit a space of permanence like Williams's Narragansett Indians, whose "souls do not die." From this position of immortal elevation, past, present, and future occur simultaneously. Recall the line from Mary Clark Bonner's deathbed testimony, recorded in Cambridge in 1697: "He is Come he is Come . . . and has Spocken powerfully to my Soul telling me my Sins are forgiven." The shifting tenses between past and present indicate the soul's location in an eternal space beyond the finite boundaries of time and death. From this space, Bonner's soul provides those witnessing her death with unequivocal knowledge of her spiritual estate. The practice of recording such unequivocal knowledge is a practice of tokenography, a means of bearing witness to and preserving the fleeting knowledge of the invisible world that becomes available only in the final moments of the saint's life.[27]

This narrative focus on a presentist framework grounds this archive firmly in an experimental religious tradition, one in which truth through experience holds privileged epistemological value. Belief in the truths conveyed through experience was also, as I have been arguing, central to experimental philosophy. It is perhaps not surprising then that this interest in an empiricism of divinity was not exclusive to ministers. Members of London's Royal Society proposed a way of tracing such invisible impressions through the optic nerve and camera obscura. Published in 1682 and 1683, respectively, Richard Saunders's *View of the Soul* and Thomas Willis's *Two Discourses concerning the Soul* investigate the observational capacities required to see the actions of the spirit recorded in the optic nerve. Saunders explains that a worldly experience of "joy or mirth . . . is a dilation or relaxation of all the faculties of the Soul." Saunders offers an empirical rationale: the experience of dilation and relaxation that comes with joy expands the individual's perceptive capacities. As a dilating pupil that lets in more light, the joyful person relinquishes the "reins of Reason" in order to gain greater perception over the earth and its objects.[28]

Saunders and Willis promoted the idea that the soul's faculties could be observed and understood by expanding the scientist's observational capaci-

27. Trent, ed., "'The Devil Came upon Me like a Lyon,'" *Connecticut Historical Society Bulletin*, XLVIII (1983), 115–119.

28. [Richard Saunders], *A View of the Soul, in Several Tracts* . . . (London, 1682), A2, 5.

ties through a technology analogous to the camera obscura. In the seventeenth century, the camera obscura consisted of a lens fitted to one end of a portable box with a sheet or lens at the opposite end positioned to display the image. Newton used this tool extensively to conduct his experiments on light, but the camera obscura was also thought to trace spirit impressions as they were channeled through the optic nerve and recorded in the brain.

> We may conceive the middle or Marrow part of the brain, as it were the inferiour Chamber of the Soul, glased with dioptric looking-glasses; in the Penetralia or inmost part of which, the Images or pictures of all sensible things, being sent or intromitted by the Passages of the nerves . . . as it were an objective Glass, and then they are represented upon the Callous Body, as it were upon a white Wall; and so induce a Perception, and a certain Imagination of the thing felt . . . then by and by further progressing as it were by another waving, from the Callous Body towards the Cortix or shell of the Brain, and entering into its folds, the phantasie vanishing, they Constitute the memory or remembrance of a Thing.[29]

This passage applies the technology of the camera obscura to the soul. Impressions of animal spirits are sent from the soul through the nerves and into the brain, the "soul's inferior chamber," which records the memory of the spiritual thing. An inverted image of divinity appears upon what Willis calls the "corporeal soul." Subordinating the "corporeal or fiery soul" to human reason, Willis applies the technology of light to the movement of the spirit upon the soul in order to foreground a technique for observing human records of divine impressions. While Newton did not include this use of the camera obscura in his published *Optics,* an experiment privately undertaken in the mid-1660s demonstrates his hope that the technology could be used to study the actions of the spirit. Newton concluded that "fantasie" could "excite the spirits in his optic nerve sufficiently to generate apparent images."[30]

29. Thomas Willis, *Two Discourses concerning the Soul of Brutes . . .* (London, 1683), 24–25.

30. Adrian Johns, *The Nature of the Book: Print and Knowledge in the Making* (Chicago, 1998), 391. Animal spirits are particles and other elements that comprise the corporeal and subordinate portion of the soul. The spirit in question is, not the Christian *spiritus,* but the *pneuma,* borrowed from the Galenic medical tradition. Willis's spirit carries all memory images within the sensitive / corporeal soul. However, these two senses of the spirit do interpenetrate, especially in the writings of the Cambridge Platonists.

Contemplations

This transformation of the ars moriendi tradition into an occasion for collecting a glimpse of the dying soul as an artifact of the invisible world presumed that grace could be frozen in time, that the knowledge acquired in the dying hour could be codified and rendered as usable data for future generations. "To have a testimony" was to presume that signs of grace, like natural phenomena, could be isolated and studied as a passive and static entity whose motion and life depended on a divine agent. To view grace as purely and authentically manifest upon the passive vessel of the human soul was to view grace as many late-seventeenth-century natural philosophers such as Robert Hooke, Robert Boyle, and William Petty viewed nature: something whose laws could be studied as regular, uniform, and conducive to taxonomic organization because nature consisted of material that only God had the capacity to control. This was the rapidly developing paradigm of mechanical philosophy, a perspective that would predominate in Royal Society circles from the 1680s to the 1740s with the view that mathematics constituted a totalizing rubric for understanding nature. As with the mechanical philosophical paradigm, this logic, applied to human souls, did not go undisputed.

The strongest New England critique of this crystallizing practice of a completely discernible and taxonomizable soul comes from Anne Bradstreet in her poem "Contemplations." Published first in New England in 1678, six years after Bradstreet's death, the print appearance of "Contemplations" approximates the advice epistles that appeared in print as postmortem gifts, discovered in the prayer closets of the most pious women. A common practice in England, a few advice epistles also appeared in New England in the latter half of the seventeenth century from a select few of the colony's most prominent and pious women: Sarah Goodhue's "Valedictory and Monitory Writing" (1681) and Anne Bradstreet's "To My Dear Children" (1678). "Contemplations" responds directly to the dissatisfaction that Bradstreet felt in her "pilgrimage," investigating whether divine truth can be gained from nature. What unfolds as a critique of the new philosophy in this poem—the attempt to see God's work in a mechanical universe—is also a critique of the methods of natural philosophy as applied to human souls. Ultimately, "Contemplations" arrives at a theory of death's proximity to divinity through a shared sense of their mutually impenetrable mysteries. This sense of mystery counters an emergent attempt to discover evidence of grace within deathbed visions.[31]

Following the divine and moral injunction recorded in an occasional meditation—"He that passes through the wilderness of this world had need ponder all his steps"—the poem begins with Bradstreet's experience of sensual rapture in an oak grove. The next stanza extends this inwardly directed phenomenal experience to a vision of nature as a location for discovering divine truth:

> If so much excellence abide below,
> How excellent is He that dwells on high,
> Whose power and beauty by his works we know?[32]

Like the natural philosophers who were her contemporaries, Bradstreet establishes nature as a place for an inquiry into divine truth. Oscillating between the presence of divinity within the natural world and the "silent, alone" self, the poem offers an account of how the speaker comes "to see" God in the world. Can the eye discern, observe, and understand the affective experience of grace in the oak grove? References to sight throughout the poem investigate this question through the potential visibility of divine truth: "Then on a stately oak, I cast mine eye / Whose ruffling top the clouds seemed to aspire" (16–17). The eye traveling up the oak tree reveals the limitations of human sight. The "ruffling top," like the "clouds," can only "seem to aspire" or present the illusion of divine reach. The tree conveys a sense of timeless permanence in proportion to a human life as the speaker puzzles over whether the tree is a hundred or a thousand years old. Like the oak's height, however, its age soon registers only as an expanded image of the oak tree in the speaker's contemplative mind. The last line of the stanza announces the speaker's ultimate realization that true "eternity doth scorn" this visual illusion. Eternity cannot be grasped through a sense of scale enlargement, for it marks a dimensionality that exceeds any finite temporal or spatial understanding.

31. For an excellent analysis of the advice epistle genre, see Wendy Wall, "Dancing in a Net: The Problems of Female Authorship," in Wall, *The Imprint of Gender: Authorship and Publication in the English Renaissance* (Ithaca, N.Y., 1993), 279–340. I am quoting a line from "To My Dear Children" that reads: "I have often been perplexed that I have not found that constant joy in my pilgrimage and refreshing which I supposed most of the servants of God have," in Hensley, ed., *Works of Anne Bradstreet,* 243–244.

32. Hensley, ed., *Works of Anne Bradstreet,* 276, and "Contemplations," ll. 10–12 (subsequent references to this source will be cited parenthetically by line).

Bradstreet's contemplative rumination on the human inability to discern the eternal places a philosophical limitation on the use of sensory data to comprehend the infinite. Only a few decades later, John Locke devoted a chapter to the concept of the infinite in his *Essay concerning Human Understanding* (1689). Even though the human mind may acquire through the senses some indication of infinite dimensionality, Locke explains that the infinite cannot exist within the human mind as a complex idea, because the senses are ultimately finite. Translating sensory data into a mental image necessarily involves a reduction because infinite realms cannot be understood in finite human terms. The infinite becomes a Lockean mechanism for delimiting the boundaries of empirical knowledge, one that called for Jonathan Edwards to resolve through a theological adaptation of Lockean epistemology.

The speaker of Bradstreet's poem does not understand the conflict between infinite and finite dimensionality in Lockean terms, of course, but rather through the *concupiscentia oculorum*—lust of the eyes—against which Augustine warned.[33] In the first stanza set, the speaker's eyes continue upward, ignoring the futility of discovering divinity through the gaze upon a stately oak. Continuing her "gaze" "higher on the glistering Sun," the speaker shifts the focus of ocular desire to the divinity recorded in the sun. Through a description of the sun as the "soul of this world, this universe's eye," the poem establishes an explicit homology between the human soul and the natural world: the sun is to nature as the human soul is to natural sight. The solution to Paul's dark glass recorded in Hebrews 11:1—"Faith is the substance of things hoped for, the evidence of things not seen"—can be extended to the sun, whose "shining rays" enlighten what cannot ordinarily be seen. "Birds, insects, animals with vegative" appear afresh each morning as the sun, like a "bridegroom," "ushers" forth the special vision that corrects the dark glass. The speaker hopefully notes the potential fulfillment of ocular desire as "the Earth reflects her glances in thy face," at last presenting a clear mirror. Attempting to see these "glances" so clearly reflected in the sun, the speaker fears for a moment that, like Yahweh disguising his appearance behind a burning bush, the sun is "so full of

33. In book 11 of the *Confessions,* Augustine warned against the seduction of ocular desire. Through its focus on gazing, eyes, and magnifying sight, "Contemplations" reminds its reader of Augustine's warning. The futility of attempting to see divinity in nature, established by the end of stanza 8, posits the truth of Augustine's warning against the seventeenth-century science of vision.

glory that no eye / Hath strength thy shining rays once to behold" (44–45). Quickly disregarding this fear and its typological warning,

> My humble eyes to lofty skies I reared
> To sing some song, my mazed Muse thought meet.
> My great Creator I would magnify,
> That nature had thus decked liberally;
> But Ah, and Ah, again, my imbecility! (53–57)

Structurally mirroring the ascent of the eyes gazing upon the stately oak, this passage marks the second failed attempt to see divinity in nature. The sun vastly exceeds the scale enlargement produced by the oak. As an unreachable entity beyond even the "unknown coasts" of the world, the sun appears to perform divine miracles and sight (166). Even though the speaker "knows better" than to consider the sun a "deity" as others have considered it, the sun seems the clearest intelligible sign of divine mystery. The speaker contemplates the sun as an object through which to "magnify" the "great Creator" and arrive at a greater understanding of invisible divinity through the birds, insects, and plants made visible through the sun's enlightening powers. But, ultimately, as stated in the final line of this stanza, the sun also proves to be only an illusion of a natural key to divine truth.

The first eight stanzas of "Contemplations" reflect upon the capacity of the eyes to discern natural objects as evidence of the invisible world of divinity. As the stanzas progress, the poem unfolds a critique of this assumption that the human senses can be honed and trained to glimpse infinity within the natural world. Bradstreet's stance on the relationship between human comprehensive capacity and the manifest forms of invisibility within the natural world quite strikingly parallels the perspective of Margaret Cavendish, duchess of Newcastle, on precisely these issues. Cavendish published her first philosophical tract, *Philosophical and Physical Opinions,* in 1655, setting forth a vision of intelligent nature, sensory perception, and God's relationship to both that famously contrasted to that of her contemporaries, most notably Robert Boyle and Isaac Newton. Cavendish grounded her perspective that matter was not dead body, devoid of spirit, but, rather, active and alive.

Cavendish's concerns are first and foremost epistemological. According to Cavendish, it was not only wrong but in fact corrupting to natural philosophy to presume that simply enlarging the dominion of the senses would lead to increased and more accurate knowledge of nature and the forms of divinity residing therein. A sounder approach would be to discipline the senses, to reduce the problem of uncertainty plaguing the natural philosophical mind by being more selective about the particulars admitted into human experience and memory for scrutiny. It is important to pay as close and as careful attention to nature as possible because nature is "intelligent" rather than composed of passive and inert matter. Nature moves and deludes as an active agent that contains divine forms as well as the divine power to be secret and mysterious. Merely enlarging the human senses through technology like Hooke's microscope misses the point entirely: nature can be comprehended or perhaps contemplated only if treated as a realm with the same elusive, infinite, and mysterious dimensionality that humans attribute to God.

Bradstreet and Cavendish present an image of nature that is in perpetual motion and is thus resistant to taxonomies. As Cavendish states in the *Grounds of Natural Philosophy* (1668), "Nature is a perpetual motion, she must of necessity cause infinite varieties." Cavendishean nature cannot be mapped, traced, or labeled as her Royal Society contemporaries would have it, because nature is not passively acted upon by external forces but is, rather, composed of "Self-knowing, Self-living," and "Self-moving Parts" that "have an active Life, and a perceptive Knowledge."[34]

Critiquing the empirical attempt to understand divinity through nature, "Contemplations" also restores the mystery of divinity discovered through death as something that cannot be approached through recording, collecting, and circulating the words or actions of dying saints. As such, the poem acts as Bradstreet's own memorial poem, foreclosing the possibility that the truths of divinity can be discovered through an investigation into her "secret places." It is possible that Bradstreet's foreclosure of the possibility of investigation into "secret places" is in part a resistance to a language of secrecy that evolved around the project of identifying signs of grace upon the dying. A short memorial poem, written for recitation at the funeral of Sarah Tompson, describes the secrets to be found within a woman's prayer

34. [Margaret Cavendish], Duchess of Newcastle, *Grounds of Natural Philosophy Divided into Thirteen Parts: With an Appendix Containing Five Parts* (London, 1668), 6, 7. For a full account of Cavendish and vitalism, see John Rogers, *The Matter of Revolution: Science, Poetry, and Politics in the Age of Milton* (Ithaca, N.Y., 1996), 179. For feminist readings of Cavendish's natural philosophy, see Bronwen Price, "Feminine Modes of Knowing and Scientific Enquiry: Margaret Cavendish's Poetry as Case Study," in Helen Wilcox, ed., *Women and Literature in Britain, 1500–1700* (Cambridge, 1996), 128. See also Rosemary Kegl, "'The World I Have Made': Margaret Cavendish, Feminism, and the Blazing World," in Valerie Traub, M. Lindsay Kaplan, and Dympna Callaghan, eds., *Feminist Readings of Early Modern Culture* (Cambridge, 1996).

closet, as if discovering and then revealing for the first time the spiritual experience so carefully hidden during her life:

> In public secret what else is said
> Beseeming real saints she said and did
> But still with a design of being hid.[35]

This passage sets up a contrast between the piety that Sarah exuded in public and that which she kept secret. Her "design" was to "beseem" the life of a saint through a pious decorum that was in fact quite different from what was kept hidden. The speaker of the memorial poem presents Sarah's hidden life as one of "rarest worth" that, "now" discovered, can "blaze forth," framing Sarah's spiritual life as a curiosity, or a hidden gem, that resides beneath the external forms of her daily activities. The memorial poem constructs a sense of this discovery through the repeated metaphor of those aspects of Sarah's life that were kept hidden and indiscernible to others.

> Those many hours in Secret Spend
> Which thou along with God didst keep
> When many others were asleep.
> . . . .
> In Secret poured forth with tears.

The speaker claims to know what occurred between Sarah and God "alone" during her most secret hours through discovering a testimony written by Sarah herself in order to record the "tears" and "sweet manifestations of [Christic] love" experienced in her prayer closet. The elegy frames the knowledge gained through the speaker's reading of the testimony through a language of secrecy that highlights the empirical value of discovering the spiritual interiority that has been so artfully hidden throughout Sarah's life.[36]

Toward an Improvement of Divine Knowledge

Bradstreet and Cavendish were not the only two seventeenth-century writers to issue a reminder of the inevitable complexity and uncertainty surrounding a natural philosophical approach to knowledge of the divine. Locke also warned in his *Essay concerning Human Understanding* that his em-

35. "The Relation of the Word of God upon the——of Sarah Tompson," Connecticut Historical Society.

36. Ibid.

piricism of the senses should not be applied to "those depths, where [we] can find no sure Footing," restricting his method to the things, ideas, and impressions collected in the natural world. This warning forecloses the possibility that Lockean empiricism could be applied to the supernatural or the invisible world.[37]

The empirical techniques exemplified through Locke's *Essay* had a long precedent in experimental religion where theologians utilized an analogous form of knowledge acquisition. Experimental religion involved an immediate connection to God's illusive and fleeting communication to the human soul through the senses. Accuracy about the work of grace upon one's own soul required continual renewal to ward off the dangers of religious hypocrisy. Converts were to reacquaint themselves with innate depravity—the condition of uncertainty surrounding election—and then attend to the barest sensory components of divine communication, the only hope humans had of intuiting God's work with any degree of precision.[38]

Nowhere was the importance of precise ordering of the evidences supplied by the soul more integral than in the Puritan saint's "dying hour." Gookin invoked this temporal category as an apt label for teaching a method of testimonial transcription that would most effectively bear witness to the tokens of grace manifest on dying souls. Edward Pearse's *Great Concern* instructs the dying in how to best attune their senses to the well-ordered evidences recorded on a soul near death. First published in London in 1673 and then reprinted in London and Boston until 1735, Pearse's *Great Concern; or, A Serious Warning to a Timely and Thorough Preparation for Death* is one of many manuals of its time that contained instructions on the art of dying well. Other popular and widely printed manuals on grieving and dying well include John Flavel's *Token for Mourners* (1674), Richard Baxter's *Now or Never* (1662), and Henry Montagu's *Contemplatio Mortis* (1631).[39]

37. Locke, *Essay concerning Human Understanding*, ed. Nidditch, 47.

38. Anxiety over the accuracy of signs of grace was alive and well in theological writings from the time period discussed in this article. Published sermons about hypocrisy include Joshua Moodey, *The Great Sin of Formality in God's Worship; or, The Formal Worshipper Proved a Lyar and Deceiver* . . . (Boston, 1691); Ben[jamin] Wadsworth, *Danger of Hypocrisy* . . . (Boston, 1711); and Solomon Stoddard, *The Way to Know Sincerity and Hypocrisy* . . . [Boston, 1719].

39. Edward Pearse, *The Great Concern; or, A Serious Warning to a Timely and Thorough Preparation for Death* (London, 1673). At least forty-nine editions were printed in London and Boston between 1673 and 1735, spanning precisely the years during which what I am describing as the slow transformation of the ars moriendi tradition toward a science of dying took place. An edition was also printed in Boston in 1815.

Pearse "recommends" his treatise as "proper" to read at "Funerals." The primary purpose of preparation, according to Pearse, is to prevent death from becoming a passage into a vast and anonymous "Ocean of Eternity." Pearse encourages his readers to prepare their souls more effectively, by cataloging the "infinite riches of free Grace" that become increasingly manifest in the souls housed in bodies fast approaching their grave. These infinite riches must be properly organized to prevent the dying from simply fading into the eternal world without a trace or artifact left behind in the visible and natural world. According to the treatise's doctrine, "'Tis a very desirable thing . . . to have all things set right, well-ordered, and composed in Matters of [the] soul." A well-organized and orderly pious life makes the evidences of the soul conducive to an experimental practice whereby they become manifest within the "dying hour" so as to be easily transferred to a taxonomy of grace. Quite different from organizing one's immortality according to the good works produced during one's life, "preparation for death" involves ordering the soul such that the evidence displayed therein comes across clearly, allowing those witnessing the deathbed scene to partake in the assurance somewhat uniquely offered within the saint's "dying hour."[40]

Pearse goes on to offer some practical advice. Providing that the proper ordering has taken place, the dying hour should show the "first Evidence of our Assertion." Within the temporal framework of the dying hour, evidences become visible with increasing clarity. As death approaches, the dying sees "visions of death" and of the "grave." The body dissolves and with it the depleted and fallen capacity of the human senses to misread. Consequently, "the least frown in Gods Face towards the Soul, the least flaw in his Peace, the least blot or blur in his Evidences for Heaven" become immediately apparent, not only to the dying saint but also to those witnessing the scene. Pearse's *Great Concern* presents the properly prepared soul as a site of great evidentiary potential. Grace becomes manifest with unprecedented certainty. Within the temporal framework of the "dying hour," witnesses see such evidence as exhibiting unusual certainty and therefore unusually prescient knowledge of God.

Pearse's *Great Concern* was very much of a late-seventeenth-century moment in which the momentum of empirical practices and natural philosophy contributed to a developing ministerial expertise of human souls. The "dying hour" marked the culmination of a saint's life with temporal des-

40. Pearse, *The Great Concern* (1694), 5, 12, 13.

ignation that allowed the unique manifestation of revelatory and certain signs of assurance. Pearse, Gookin, and others urge ministers as well as discerning laity to take note of the dying hour, to attempt to see through the eyes of those passing from this life to the next, and to record and preserve the perceived visions and insights as a means of accumulating new knowledge of the divine. This particular empiricism of the unseen developed alongside the late-seventeenth-century warning of epistemology's limits, issued in Bradstreet's poetry, Cavendish's natural philosophy, and Locke's codification of the proper use of the senses.[41]

Circulating treatises such as Pearse's *Great Concern* set the context for the consolidation of a particular version of soul science during the first two decades of the eighteenth century. Methods set forth for recording the evidence of the dying hour figured prominently in the ministers' manuals that proliferated during these decades as ministerial expertise increased because of the epistemological convergences described throughout this work. Cotton Mather's *Manuductio ad Ministerium* (1726) offers "directions" for ministerial candidates by partaking in the advice offered to those either preparing to die or preparing to learn from the dying. Mather encourages candidates to "consider [themselves] as a Dying Person." He describes this imaginative exercise in visceral terms:

> Place yourself in the Circumstances of a Dying Person; your Breath failing, your Throat rattling, your Eyes with a dim Cloud, and your Hands with a damp Sweat upon them, and your Weeping Friends no longer able to retain you with them.

The *Manuductio* begins with this template of the dying person, which serves as a tabula rasa for the young minister's soul. Literally "placing themselves" into this frame, ministers renew the experiential purity of the divine encounter between their studies of natural philosophy and theology. Mather constructs this exercise as the foundation of experimental philosophy for the ministers, which establishes the proper frame of mind for the promotion and discovery of divine knowledge. The passage continues with another admonition: "And then entertain such Sentiments of this World, and of the Work to be done in this World, that such a View must needs inspire you withal."[42]

What is the "view" that Mather hopes young ministers might see from

41. Ibid.

42. Cotton Mather, *Manuductio ad Ministerium* . . . (Boston, 1726), 2.

this exercise, and what kind of work might it inspire them to do? In the *Manuductio,* Mather encourages young ministers to place themselves in the position of a dying person in order to heighten their observational powers of spiritual discernment. Mather goes on to explain that such deathbed ruminations will "*Lighten* the Mind that has an *Eye* upon it." The *Manuductio* proposes that observational capacities might be improved by meditating upon the glory of God, almost graspable through the vision seen by a dying person. Such meditation renders the ministerial mind more capable of receiving divine evidence within the natural world. This is an entirely conventional practice, according to Mather, who explains that ancient philosophers "assign[ed] the Contemplation of Death, as the main Foundation, and main Exercise of their Philosophy." Yet Mather hopes to apply this form of experimental meditation to a new purpose. Along with conditioning the young ministry to become better observers of the workings of grace, Mather wants the deathbed experience to inspire young ministers as to the "Work to be done in this World." Like Richard Saunders's *View of the Soul,* Mather presents dilation as an opening of the senses to the "Supra-Intellectual" realm described by Robert Boyle in *A Discourse of Things above Reason* (1681). Completely controlled by a "Divine Author," access to this realm of "inexplicable" and "incomprehensible" phenomena comes through the transformation of man's "intellective faculty." Anticipating the epistemological limits that Locke sets on empirical practice, Boyle argues that the transformative powers of God's divine light permit man to discern things beyond "the limits" of "safe" judgment and definition.[43]

The practice of witnessing an individual's "dying hour" loosens one's rootedness to the world and opens the possibility for greater insight. Mather describes this work as the "Great Design" of translating "Secular Learning into an Improvement of Divine Knowledge." According to Mather, the natural histories and philosophies promoted by Robert Boyle, Robert Hooke, John Ray, and Isaac Newton in the *Philosophical Transactions* of the Royal Society advance the cause and understanding of scripture first started by Calvin. The deathbed scene opens the young minister's mind to this process, which necessarily operates at the nexus between reason and revelation and consistently oscillates between experience and the religious innovation of the natural philosophical tradition.[44]

43. Mather, *Manuductio,* 3, 9, 56; [Robert Boyle], *A Discourse of Things above Reason . . .* (London, 1681), 20–21, 56, 79.

44. Mather, *Manductio,* 33.

Westborough minister Ebenezer Parkman advanced an adaptation of Locke to experimental religion in his lecture to a group of Boston ministers about a plane of "Gods and Angels" that might be reached through empiricism. In a lecture series entitled "The Nature of the Human Mind and Other Powers of the Soul," Parkman delineates a technique for translating the soul's encounter with "spiritual substances" into "thought" such that "the existence of *Spirits* and especially that raised and glorified order of the *Angels*" might be proven. Parkman explains that the "philosophic experience" presented in Locke's *Essay* could advance the terms of metaphysical knowledge. Parkman proposed to extend what is clearly a Lockean meditation on the "distribution of Ideas with reference to the different ways" they approach the mind, to an understanding of how spiritual substances produce "thought," how thought gives us "knowledge of the soul," and how this knowledge of the soul may in turn be used to track "the existence" of an invisible world of "Spirits."[45]

Parkman's model of spiritual inquiry violates Locke's warning against empirical treading in places where humans have "no sure footing." This subtle and carefully navigated violation of a clear epistemological limit might explain why Parkman lectured to a rather exclusive group of Boston divines. The methods of inquiry promoted in his lectures were not for everyone, but rather were for those ministers of God who had received some sign of election and could apply their expanded observational capacities to the work that Mather speaks of in his *Manuductio*. Empirical knowledge of "Angels" and "Spirits" could come from observing the work of spiritual substances and sensible objects upon the souls of others.[46]

This use of Locke would not have been unheard of in the 1720s. Charles Morton promoted a Lockean approach to the spirit world in his *Compendium Physicae,* a science textbook circulated widely at Harvard and Oxford between 1680 and 1721. Although read by most ministers in training, the *Compendium* circulated only in manuscript at these elite institutions. Like Parkman's lecture series, its limited and exclusive availability suggests that the information contained therein was directed toward an audience of divines who had both the experiential and the philosophical expertise needed for divine inquiry. Morton provides a more explicit approach to

45. Ebenezer Parkman, "A Lecture at the Friday Evening Association of Bachelors at My Chamber in Boston, February 1, 1722/3, Of the Nature of the Human Mind and Other Powers of the Soul," Parkman Family Papers, box 2, fol. 4, AAS.

46. Locke, *Essay concerning Human Understanding,* ed. Nidditch, 47.

the empirical discernment of angels and gods by remapping the temporal and spatial coordinates of divinity made manifest within the natural world. As if to extend the terra firma of empirical footing, Morton explains that different essences can be tracked by differentiating notions of space and time. Attention to time's "internal" component—the permanence and duration of things—contrasts the changeability of external time, divided into "hours, Days, and Years." Space might be more fully understood through three overlapping planes of the invisible world: the aerial, consisting of creatures that fly; the ethereal or "Ether," "which fills up all the Space between the heavenly bodies"; and the most distant realm of "Gods, Angels, and Souls." Morton locates the essence of this most distant third realm as the nonmaterial part of the body, from which divine evidence flows. Thus a correct understanding of spatial and temporal coordinates can enhance an understanding of divine essence by providing a mode of tracking invisible impressions and elements as they interact with essences in the world.[47]

Snapshots of the Soul

When recorded through the presentist narrative techniques of tokenography, deathbed testimonies provided an enticing experimental venue for visually capturing the actions of the spirit. Dying hours are carefully parsed in minutes, hours, and days to mark the passage of external time. Alongside these external markers, the voice of the dying interrupts the narrative frame through an exclamatory present tense that reveals divine time, the internal component that is permanent and eternal. In her "Memoranda" of her husband's death, Sarah Pearce begins with a quotation from Mr. Pearce, who exclaims: "I have been in darkness 2 or 3 days, crying, O when wilt thou comfort me, but last night the mist was taken from me, and the Lord shone in upon my Soul." This remarkable deathbed testimony was recoded on a narrow sheet of paper, measuring about two by eight inches. Mrs. Pearce fastened the ends of the paper into a loop, presumably to wear around her wrist after her husband's death as a poignant material display of memento mori. The opening quotation captures the technique of tokenography perfectly, for it shows the dramatic break from profane to sacred time, which becomes the point of entry for Mrs. Pearce to glimpse fragments from the invisible world from her husband's dying visions and experiences. "When

47. Charles Morton, *Compendium Physicae,* ed. Samuel Eliot Morison, Colonial Society of Massachusetts, *Collections,* XXXIII (Boston, 1940), 21.

scorching with burning fever," Mrs. Pearce writes, her husband exclaims: "O what a lovely God—and he is *my* God and yours!" Mr. Pearce's interruptive exclamation in the present tense marks a break out of the quotidian earthly realm, allowing those witnessing the scene of his death access to a different spatial dimension, the most distant realm of "Gods, Angels, and Souls." Moments before his death, her husband reassures her that she will join him in this realm, that their "separation" is not "forever."[48]

Cotton Mather recognized tokenography as a powerful way of introducing the spiritual facts collected during deathbed scenes to natural philosophy. In his 1704 funeral sermon, *Monica Americana,* for Mrs. Sarah Leveret of Boston, he commands:

> Tell us now that Secret Rare;
> Rare like the Stone of the Philosopher.

The bid to "tell us now" signals the revelatory potential promoted through verbatim records of the words of the dying. The "Secret Rare" corresponds to the "certain sign of regeneration" promised to dying saints in John 3:14 as well as the natural secrets sought by the philosopher of nature. Rarity wields incredible empirical potential to expand categories of knowledge beyond the rubrics established through reason-based taxonomies. Mather extends this logic in his own practice of recording the divine knowledge conveyed within young women's deathbed scenes in such texts as *Memorials of Early Piety* and *Ecclesia Monilia.* Visions, affective signs, and aural evidence of grace are extracted from the reserved papers of Elizabeth Cotton and Jerusha Oliver and reconstructed through testimonies of relatives and friends who witnessed the scene. The title page to *Ecclesia* references Malachi 3:17, echoing the *Christian Philosopher*'s metaphor of the soul's jewels. Mather compares the soul's gems to those that might be mined from the earth: "Learned and Curious Things have been written by Philosophers about the *Origin of Gems.* But the true origin of these Jewels, is a marvellous Work of God upon a miserable Soul, *Transforming* of it and the imprinting [of] His Image upon it." Through an explicit analogy between the mineral formation of gems in the natural world and converted souls, Mather's

48. This remarkable artifact was found by Kenneth Minkema in the Jonathan Edwards Papers. The provenance is unknown, but, based on the conversion theology expressed, it seems as if it was written in the early eighteenth century. The hypothesis that Mrs. Pearce wore the fastened paper around her wrist is Minkema's. "Memoranda by Mrs. Pearce within Four of Five weeks of Mr. Pearce's Death," Jonathan Edwards Collection, Beinecke Rare Book Library, Yale University, New Haven, Conn.

Memoranda by Mrs Pearce
within 4 or 5 weeks of Mr
Pearce's Death —

"I have been in darkness 2 or 3 Days crying
O when wilt thou comfort me, but last
night the mist was taken from me, & the Lord
shone in upon my Soul — O that I cd. but
speak I wd. tell a world to trust a faithfl. God
Sweet Afflictn now it worketh Glory, Glory" —

After telling him the various anxieties of
my mind, he replyd, "O trust the Lord, if he
lifts up the light of his Countce upon you, as he
has done upon me this day, all your mountns
will become molehills. I feel your situation, I
feel your sorrows, but he who takes care of
sparrows, will care for you, & my dear Childn."

When scorching wth. burng fever he sd. "Hot & happy."
On a Lord's day morng he sd. "Cheer up, my dear,
think how much will be sd. today of the faithfulness
of God. Tho we are called to separate he will never
separate from you — I wish I cd. tell the [illegible]
what a good & gracious God he is & never
need they who trust in him to be afraid of trials
He has promised to give strength for the day,
that is his promise — O what a lovely God,
and he is my God & yours. He will never leave
us nor forsake us, no, never — I have been thinkg
that this and that Medicine will do me good, but
what have I to do with it? It is in my Jesus's hands,
he will do it all, & there I leave it — what a mercy
is it, I have a good bed to lie upon, you my dear
Sarah to wait upon me, & friends to pray for me.
O how thankful shd. I be for all my pains, I want
for nothing all my wishes are anticipated. O I
have felt the force of those words of David, "Unless
thy Law, (my gracious God) had been my
Delights, I shd. have perished in mine Affliction."
Tho I am too weak to read it, or hear it, I can
think upon it, & O how good it is. I am in the
best hands I cd. be in, in the hands of my dr. Lord
& Savior, & he will do all things well, yes yes, he cannot
do wrong.

One Morng. Mrs P askd him how he felt, "Very ill
but unspeakably happy in the Ld. & my dear Lord Jesus
On beholding me grieving, "O my dear Sarah do not
be so anxious, but leave me entirely in the hands of Jesus
and think if you were as wise as he, you wd. do the same
by me. If he takes me I shall not be lost, I shall only go
a little before, we shall meet agn. never to part —
After a violent fit of coughing he sd. It is all well, O
what a good God is he! It is done by him and it must be
well — If I ever recover I shall pity the sick more than ever,
& if I do not, I shall go to sing dallverlastg love, so you
see it will be all well — O for more Patience! well my
God is the God of Patience & he will give me all I need.
I rejoice it is my Jesus's hands to communicate, & it
cannot be in better. It is my God who gives me Patience
to bear all his will.

When after a restless night Mrs P. asked him,
what she cd. do for him, "You can do nothing but
pray for me, that I may have Patience to bear all
my Lord's Will. After taking a Medicine he said
"If it be the Lord's Will to bless it for your sake and
for the sake of the dear Childn. but the Lord's Will be
done. O I fear I sin, I dishonor God by Impatience
but I wd. not for 1000 [illegible]s sin in a tho't if I cd. avoid
it. Mrs P. reply'd she trusted the Lord wd. still help
him, seeing the Lord had bro't him thus far,
he wd. not desert him at last. No, no, he sd. I hope
he will not, as a Fr. pitieth his Childn. so [illegible]
Why do I complain? my dear Jesus's Sufferings
were much sorer and bitter than mine
and did he thus suffer, & shall I repine! no
I will cheerfully suffer my Fr.'s Will."

One Morng after being asked how he felt, he
reply'd, "I have but one severe pain about me
What a Mercy! O how good is God to afford
some Intervals of so much pain! He is altogether
Good! Jesus Xt is my dear, and that must
be our final Consolatn. — After taking a Medicine
which operated very powerfully he sd. "This
will make me so much lower. Well let it be.
Multiply my pains thou Good God, so thou
art but glorify'd, I care not what I suffer, all
is right."

Asking how he felt after a restless night, he replyd, I have so much weakness and pain, I
have not had much enjoymt. but I have a full
persuasion the Lord is doing all things well —
If it was not for strong confidence in a lovely
God I must sink but all is well. O blessed
God I wd. not love thee less. O support a sinkg.
worm! O what a mercy to be assured that
all things are working together for good.

Mrs P. saying If we must part I trust the separation will not be for ever. O no he reply'd
we sorrow not as those who have no hope.
She sd. Then you can leave me and the Childn
with Resignation, can you? He reply'd my tender ♡
was pierced thro with many Sorrows before
I cd. give you and the dear Childn up, but
the Lord has heard me say, "Thy Will be done
and I can now say blessed be his Name, I
have none of my own — His last Days were
very happy, Mrs P. repeated this verse,
Since all that I meet shall work for my good
The bitter is sweet the Medicine is food
Tho painful at prest 'twill cease before long
And then O how pleast. the Conqueror's song
He repeated with an inexpressible Smile,
the last line, The Conqueror's song.

He sd. O my dear what shall I do? but
why do I complain, he makes all my bed
in my sickness. I repeated these Lines
Jesus can make a dying bed
As soft as downy pillows are
Yes, he reply'd, he can, he does, I feel it

FIGURE 14 • *"Memoranda by Mrs. Pearce within Four or Five Weeks of Mr. Pearce's Death." Jonathan Edwards Collection, General Collection, Beinecke Rare Book and Manuscript Library, Yale University, New Haven, Conn.*

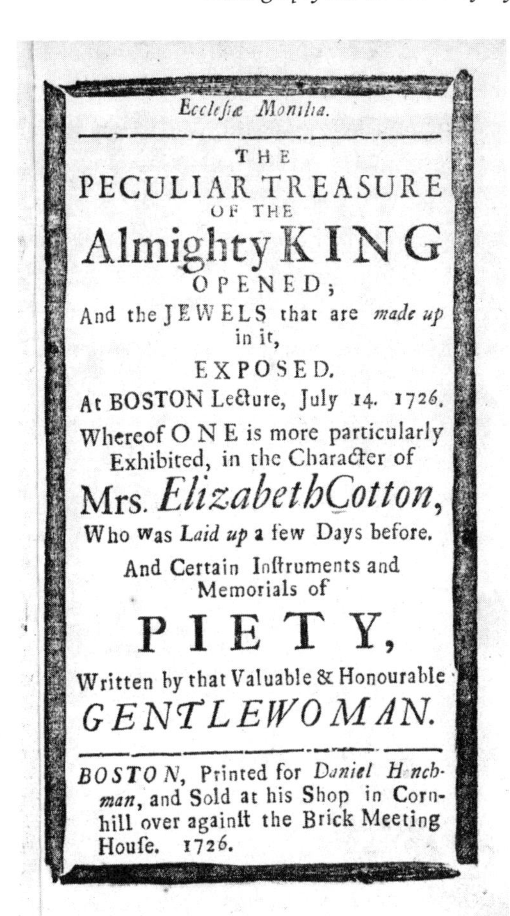
Ecclesiæ Monilia.

THE
PECULIAR TREASURE
OF THE
Almighty KING
OPENED;
And the JEWELS that are *made up* in it,
EXPOSED.
At BOSTON Lecture, July 14. 1726.
Whereof ONE is more particularly Exhibited, in the Character of
Mrs. *Elizabeth Cotton*,
Who was *Laid up* a few Days before.
And Certain Instruments and Memorials of
PIETY,
Written by that Valuable & Honourable
GENTLEWOMAN.

BOSTON, Printed for *Daniel Henchman*, and Sold at his Shop in Cornhill over against the Brick Meeting House. 1726.

FIGURE 15 • *Cotton Mather,* Ecclesia Monilia

practice of collecting, editing, and printing the secrets of a woman's prayer closet as curiosities produces knowledge of the invisible world.[49]

According to Mather, "the *Fearful Sex*" recorded in his collection "have Triumphed over the *King of Fears.*" Mather's catalog of deceased damsels invites the reader to peer into the cabinet, to view the soul and to see the "Private Prayers" and "secrets" of his subjects' souls as a way of deepening knowledge of divinity. The first time that Mather "finds" Jerusha "Employed in the **fast of the Closet,** is, when she was but Thirteen Years of Age." In *Ecclesia,* he writes: "Of considerable Use" to "Young People, and most

49. Cotton Mather, *Monica Americana* . . . (Boston, 1705), 32; Mather, *Ecclesia Monilia: The Peculiar Treasure of the Almighty King Opened* (Boston, 1726), 12. (In *The Christian Philosopher,* Mather exclaims that "Graces in the Soul are brighter Jewels than any that are dug out of the Earth!" [ed. Winton U. Solberg (Chicago, 1994), 126].) Mather is probably referring to Boyle's *Essay about the Origine and Virtue of Gems* (1672).

of all unto those of the Female Sex," are the collected "Reflections on the State of her Soul, when (Elizabeth) was but Thirteen Years of Age." In her dying moments, the "little damsel" supplies proof of the state of her soul.[50]

Mather describes the young Jerusha Oliver as having an extraordinary visual capacity: "She had a great Felicity in a more than ordinary strength and sharpness in her *Eye Sight,* being able to Read by such a dim light as others could not." Jerusha could see farther than "ordinary others," a comparative assessment that supplements Newtonian optics with supernatural possibilities. Book 3 of Newton's *Opticks* records a series of experiments to determine how "inflexions of the rays of Light" affect eyesight in a darkened room, a space associated with the camera obscura. Mather's description of Jerusha suggests that the eye's capacities may extend beyond the Newtonian correlation of natural light and visual capacity, acting as a lens for perceiving the subtle movement of the spirit upon the soul. Peering beyond the dark glass, Jerusha's eye displays an unusual capacity to see her "interior state." Mather's allusion to Newton's dark room through the description of a young girl's closet eyesight alludes to the empirical potential of the camera obscura. What Jerusha sees through her eyes are inverted images of her soul.[51]

Jerusha's triumph over the "King of Fears" produces a double vision that allows the witnessing audience to see divinity through what she sees. On her deathbed, Jerusha declares her certainty: "I know, that I am going to Christ: That I shall shortly be in the Heavenly Jerusalem, with an Innumerable Company of Angels, and among the Spirits of Just Men made perfect." From this position of certain knowledge, Jerusha conveys evidence of things unseen. She "Sang for joy" of "things that are Unutterable." "I see a Glory, which cannot be Express'd; Persons and Things, which I want a Language to declare what they are!"[52]

Jerusha's ocular capacities have strengthened throughout a life carefully attuned to her "interior state." Her deathbed produces the snapshot of the soul described by Thomas Willis. "Inward sense" accords with "outward motion" to produce "Images or Pictures of all sensible things." Although at times lacking a language of expression, Jerusha's age, sex, and social position validate what she sees. She represents a clearer template upon which to wit-

50. Cotton Mather, *Memorials of Early Piety . . .* (Boston, 1711), 21, 49; Mather, *Ecclesia Monilia,* 29.

51. Isaac Newton, *Opticks; or, A Treatise of the Reflections, Refractions, Inflexions, and Colours of Light* (London, 1704), 113; Mather, *Ecclesia Monilia,* 12–13; Mather, *Memorials of Early Piety,* 13.

52. Mather, *Memorials of Early Piety,* 49–50. Death is the "King of Fears" in Job 18:14.

ness the "ecstasies of joy" necessary to translate experience into knowledge and the particular motion of the spirit into a law of divinity. Through Willis's metaphor of the "corporeal soul" as a "white wall" upon which such heavenly images are projected, Mather interprets the "strange transports" and "raptures" of Jerusha's disembodying soul as "the plain Signatures of an Angelical Operation." Her visions of angels expand his own ocular ability to see this angelical operation. As a beholder of such infinite and unworldly wisdom, the angels recorded in Jerusha Oliver's deathbed scene offer a new means of tracing the subtle action of the spirit as a memory or sensory impression of divine essence.[53]

Mather's interest in and promotion of pious little damsels stems in part from an increased focus on the evangelical potential of pious little children, perhaps most widely circulated in James Janeway's *Token for Children* (1672). A Nonconformist preacher in London during the Restoration, Janeway promoted piety through the everyday experiences recorded in this popular publication. Anticipating Cotton Mather's oft-used metaphor, the epistle dedicatory urges "Godly Ministers" to consider "what a precious Jewel is committed to your charge." Dying children are compared to precious jewels and "noble plants" found in nature. Drawing from another feature of natural philosophy to explain a method of witnessing, Janeway says that to "be in travel [is] to see Christ formed in their Souls." To travel into the secret chambers and bedrooms of pious little children, to study the words and actions of their dying moments, is to bear witness to what humans are not supposed to be able to see: the visible work of Christ upon their souls. Establishing the empirical credibility of each deathbed case through a series of examples, Janeway explains:

> What is presented, is faithfully taken from experienced solid Christians, who themselves were Eye and Ear witnesses of God's works of Wonder, or from my own knowledge, or from Reverend Godly Ministers, and from persons that are of unspotted reputation, for Holiness, Integrity, and Wisdom; and several passages are taken verbatim in writing from their dying Lips.[54]

In the first letter, "To the Reader," continued statements of the effort to record "verbatim" establish the text's indebtedness to the experimental method. Words spoken from the "dying Lips" of young children carry a

53. Willis, *Two Discourses concerning the Soul of Brutes,* 25; Mather, *Memorials of Early Piety,* 49, 51.
54. James Janeway, *A Token for Children . . .* (Boston, 1700), "To All Parents. . . ."

special ontological significance, for they are spoken through divine translation and thus convey a purer divine essence. The privileged aural and visual capacities of "solid Christian" witnesses give special insight into the formation of Christ within these young souls.

The first example is of Sarah Howley, who died when she was eight or nine years old. Her conversion began when she heard her minister preach upon Matthew 11:13. Following the sermon, she retreated to her "Chamber," where she "wept," fell "upon her knees," and "cryed to the Lord." Janeway presents this simple, conventional narrative of Christian conversion for his young readers. Through Howley's dramatic and affective response to the sermon, Janeway encourages in his young readers a genuine emotional response to a compelling sermon. To "godly ministers," he explains how to verify the authenticity of such behavior, which "might easily be perceived by those who listened at the Chamber door."[55]

Through a doubling of the sensory encounter between testifier and witness, Janeway conveys the transformation of Sarah's own perceptive capacities through the sermon, encouraging his young readers to listen while he instructs ministerial and parental readers in how to listen for the correct signs. Janeway further establishes the credibility of Sarah's experience by explaining her own fear of the dangers of hypocrisy, registered through an initial reluctance to give the verbatim transcription. During the onset of her illness, she refuses to speak to ministers despite her mother's prompting; the daughter felt that it "was her duty with patience and silence to learn of them." Initially, learning as well as speaking was painful. Sarah's silence, her repeated fear of "deceiving herself," makes the "audible voice" that ultimately "broke forth" on her deathbed all the more persuasive and compelling. With a "cheerful countenance," the dying Sarah exclaims:

> Lord, thou hast promised, that whosoever comes unto thee, thou wilt in no wise cast out; Lord I come unto thee, and surely thou wilt not cast me out. O so sweet! O so glorious is Jesus! O I have the sweet and glorious Jesus; he is sweet, he is sweet, he is sweet! O The admirable love of God in sending Christ! O free Grace to a poor lost Creature.

The dying child has arrived at the culmination of a good death and a certain sign of regeneration. Speaking directly to the "Lord," Sarah verifies her status among the elect. Quotations, present tense, and exclamation

55. Ibid., 3.

marks convey her speech directly, drawing the reader's attention to its immediacy and purity.[56]

Designed to convert children through similar means, the structure of Sarah's conversion is highly conventional. She hears a sermon, falls on her knees, weeps, grows silent and inwardly directed, falls sick, grows increasingly convinced of her salvation, and dies. Her status as "Example I" reinforces this sense of her life's conventionality; she establishes the template for the examples to follow and the template that Janeway hopes other children follow. But the experimental methods that authenticate this scene's evidentiary status reveal the text's supplemental intention that "godly ministers" would discern something beyond this conventional structure. Among the highly intelligible signs, less obvious ones require the particular discerning capacities of ministers of God. When Sarah speaks, her audible voice is also the voice of a divine speech wielding a unique, epistemological capacity. Janeway's narrative signals this capacity with a return to the witnessing practices of hearing, observing, writing, and studying the words spoken by case study, "Example I." Sarah repeats her exclamatory statements "an hundred times over," speaking so quickly, with such rapture and fervor that those present "could not write a quarter of what she spoke." Janeway alerts his ministerial readers to the experimental potential of the scene and the work of those present to witness certain forms of information.[57]

Sarah's fits, silence, speech, and sleep continue for several days before she dies. She speaks only when "full of Divine Sentences." All of her "discourse" takes on the aural qualities of a "sermon"; she speaks of "her soul, Christ's sweetness, and the souls of others," describing truths that can be carefully recorded, preserved, and studied by discerning minds after her death. Janeway conveys a sense in which her voice is divine; it is the voice of the other speaking through her that attaches such empirical possibility and weight to her sentences. Janeway's *Token* encodes a way of reading the inscription of divine facts on the impressionable souls of young children. The examples collected function simultaneously as exemplary models for pious children and as siphons for the delivery of wisdom from the invisible world to "able divines." Soul, speech, and sensory capacities are exceptionally impressionable in such young children. Janeway highlights the possibility that children can effectively convey otherwordly wisdom.[58]

56. Ibid., 5, 9–10.
57. Ibid.
58. Ibid., 9–12.

A Wigwam Opened

In his anthology, *Indian Converts,* Experience Mayhew compiles dying words and visions from an archive of loose papers and testimonies collected by three generations of Puritan ministers on Martha's Vineyard. *Indian Converts* seeks to expose and reveal the sensory impression of divine essence as recorded on the minds and the hearts of the dying Indian. To translate dying Indian words and visions into scientific knowledge, Mayhew establishes Martha's Vineyard as an experimental site, beginning with his own particular knowledge of the language of this *"one small Island"* where his family has "successively laboured for these *fourscore Years*." Testifying to the knowledge acquired through direct observation and experience, Mayhew establishes missionary work on Martha's Vineyard as a key spatial practice within the larger transatlantic project of discerning divine evidence from the souls of the dying.[59]

The Royal Society took a particular interest in Martha's Vineyard, recognizing the praying towns as an important ethnographic site. Royal Society member Paul Dudley solicited from Mayhew a request for "some account, of the Peculiarities and Beauties of the Indian Language, and wherein they agre or differ from the Europians." He intended to publish this account in the society's *Philosophical Transactions.* As such, Martha's Vineyard became a site of fieldwork. The Anglo-European filter was not simply imposed upon the natives; rather, the interplay between testifier and witness demonstrated a more complicated exchange. Thomas Prince, printer and author of the hagiographic appendix to *Indian Converts,* reinforced this notion of the anthropological value of the island. He describes how Mayhew "readily travels and lodges in their smoky *Wigwams;* when he usually spends a great part of the Night in relating the antient Stories of GOD in the *Scriptures*." In a colonial twist on Janeway's metaphor of traveling to see Christ in the souls of children, *Indian Converts* established a Baconian empirical foundation whereby spiritual knowledge comes from experience in a controlled and particular locale. While Mather excavated and displayed the evidence of a little damsel's prayer closet, Mayhew presented this small island as a cabinet of curiosities that might be opened and displayed.[60]

59. Experience Mayhew, *Indian Converts . . .* (London, 1727), dedication, 6.

60. Thomas Prince, appendix to Mayhew, *Indian Converts,* 285; Experience Mayhew, *Observations on the Indian Language* (Boston, 1884), 8.

Tokenography establishes Martha's Vineyard as an ethnographic site of fieldwork. Mayhew uses Locke's category of the "credible witness" to establish his own epistemic authority to witness such scenes and "open the State of [the Native American] Soul" to the "Learned" as he did their language. According to Locke's *Essay concerning Human Understanding:*

> Testimonies contradict common Experience, and the reports of History and Witnesses clash with the ordinary course of Nature, or with one another; there it is, where Diligence, Attention, and Exactness is required, to form a right Judgment, and to proportion the Assent to the different Evidence and Probability of the thing, which rises and falls, according as those two foundations of Credibility.[61]

Indian Converts reconstructs the scene by compiling evidence of sensory impressions from a variety of sources: "reserved" missionary papers, manuscript records of dying speeches, and the oral testimony of native witnesses at the scene. The "Preface to the Reader" establishes a "Fidelity and Concern for the Truth in this Performance" in accurately presenting to his "Learned" audience "examples of piety." Through this idea of the pious example, Mayhew links his own epistemological position to that of the "able divine," capable, like Mather, of sifting through the evidence of secret prayer and pious exemplarity in order to discover the truth of divine mystery. Mayhew states his purpose of recording these *"Ministers* . . . of exemplary piety" in a "natural" "Catalogue" in order to build scientifically an accurate knowledge base. This method incorporates a Lockean understanding of words as conveying the idea of the thing rather than the thing itself. By relinquishing the empirical desire to "remember" and preserve "what in particular Expressions there then were," Mayhew more closely approaches his goal of describing the "Blessedness of the other World" through its imperfect image in Indian visions and dying utterances.[62]

Mayhew's narration of the invisible establishes empirical legitimacy through detailed descriptions of deathbed scenes. Credible witnesses compensate for the empirical challenges described in Locke's *Essay* by supplying a form of double vision whereby the audience validates what the reader sees through the eyes of the dying. Recording Stephen Tackamasun's death, Mayhew writes:

61. Locke, *Essay concerning Human Understanding,* ed. Nidditch, 663.

62. Mayhew, *Indian Converts,* ix, 1, 33.

> He was observed to look up as steadfastly thro' the Chimney of his Wiggwam towards Heaven, when he was dying, as tho he had, like another *Stephen,* seen the Heavens open'd, and had such a View of the Glory of them as he had; but I rather think, that the Sight which he by Faith had of *him that is invisible,* might cause him thus to look towards the Place whither he was going.[63]

The typological association of the Indian minister with Stephen from the New Testament frames Mayhew's "godly Indian" as relic of a primitive Christian past. The other godly Indians witnessing the scene verify the legitimacy of Stephen's vision through their own observation of it. What is significant about their visual record is that it is "through the Chimney of his wigwam" that this Stephen sees his heavenly vision. The reader understands the scene as filtered through Stephen's Indian identity. The "Learned" London audience to whom the anthology is directed gains a glimpse into the interior of a wigwam at the same time that the narrow chimney evokes the mechanics of the camera obscura—the small darkened room with light admitted through a tiny single hole. The phrase that begins the passage, "He was observed," invites the reader into the wigwam. Along with those present, the reader sees the "look": the eyes move toward heaven. This movement signifies the apocalyptic vision of New Testament Stephen. The "View of the Glory" becomes apparent to those witnessing the scene through the lens of the wigwam's chimney. Strange and extraordinary occurrences in the souls and wigwams of Martha's Vineyard's natives are juxtaposed to a discourse of probability, credibility, and truthfulness.

In the moments before her death, Abigail Kesoehtaut underwent a divine and a spiritual discovery that revealed evidence of this "certain sign" upon her own and her sister's soul. Mayhew narrates this process, using his own skills as a translator to convey much valued empirical data for the "Learned" and "Judicious" audience to whom the text is addressed.

> [Abigail's sister] continued in Distress for her until about twenty four Hours before she dy'd, when, being asleep in the House where her Sister lay, she . . . plainly heard a Voice in the Air over the Top of the House saying in her own Language, *Wunnantinnea Kanaanut,* the same being divers times repeated; which Words may be thus rendered in English, tho they are much more emphatical in *Indian,* There is Favor now extended in Canaan.[64]

63. Ibid., 44.
64. Ibid., 147.

The utterance, "*Wunnantinnea Kanaanut,*" captures a prophetic sign in its distilled and ancient form. Mayhew emphasizes that the utterance is in fact "much more emphatical in *Indian.*" The English translation does not adequately capture the import of Abigail's divine message as communicated to her soul. Spoken within her dying moment, Abigail Kesoehtaut's Algonquian language had an increased capacity to wield divine truth. The record of her heavenly message attained prophetic power, but this power derived from her marginal status as a woman, as an Indian, and as one who spoke in a primitive tongue. While Mayhew's insistence on ethnolinguistic difference assigned spiritual authority to the native's idea of divinity, the transference of this idea to a learned audience circumscribed Abigail's spiritual authority by rendering the testimony a curiosity: a rare, marvelous, and secret oddity, framed through her cultural place at the margins of race, gender, and language as well as through the imminent departure of her soul.

The untranslatable phrase "*Wunnantinnea Kanaanut*" speaks to an emergent theme of mediumship that spans the practice of empirically recording evidence of the divine in Anglo and native testimonies. God speaks through the mouth of the dying child, and thus titles often emphasize the "verbatim" transcription, "taken directly from her mouth." In an attempt to achieve proof of the authenticity of God's word communicated through the mouths of children, Indians, and young women, ministers directed their focus toward a theory of spiritual language that could come only from one source. Yet the dying medium was a continually shifting and transmuting object of scientific inquiry, oscillating between illicit and licit forms of knowledge. Elizabeth Pattompan, a "young Indian maid" who died July 6, 1710, revealed the potential encoded within this scientific object of knowledge to expose the danger of this form of inquiry. Upon her deathbed, Pattompan asked her father to record "some things which she then uttered."

> Tho the Father of this young Woman was so earnestly desired by her to commit to Writing the Words above recited, yet having for some time neglected so to do, he does with great Assurance and Confidence affirm, that the Spectre of his said Daughter did after her Death one Day plainly appear to him, being so near to him, that he plainly saw that she appeared with the same Clothes which she commonly wore before her Death. He also saw some Warts on one of her Feet, which were, in appearance, such as his Daughter had on hers.[65]

65. Ibid., 241.

Unlike a natural curiosity, the supernatural token wields a power greater than that of the minister, natural philosopher, or paternal scribe recording the scene. Pattompan's ghost beckons her father to participate in a natural philosophical narrative of the unseen by stripping him of his "Power to speak." The father becomes "weak, and uncapable of any Business" until he "commit[s] her words to writing." The experimental object controls the scene of transcription, reminding Mayhew's "learned audience" of the hubris of their endeavor, reminding them as Anne Bradstreet does in "Contemplations" that manifest divinity is a complex and elusive domain, for which neither nature nor the human soul can supply full intelligible evidence. Elizabeth Pattompan returns in spectral form to issue a different kind of warning: spiritual objects of inquiry wield power from the invisible world. Humans must be careful not to overstep their bounds by prying too deeply into the mysteries of the unknown.

Pattompan and Kesoehtaut reflect a particular domain of experimental philosophy, metaphorically reflected in Boyle's notion of the skillful diver who rests not upon the surface of the sea but makes "his way to the very bottom of it . . . fetch[ing] up Pearls, Corals, and other precious things, that in those Depths ly conceal'd from other men's Sight and Reach." For Mayhew, Janeway, Mather, and other natural theologians, what lies concealed from other men's sight and reach might be found in a child's prayer closet or in the "smoky wigwams" of Martha's Vineyard. But an experimental philosophy of the unseen contains an important caveat: divinity is a continually shifting, transmuting object of scientific knowledge, never fully identifiable or observable. While the visions collected in Indian huts provide a form for narrating divine knowledge and phenomena through privileged observational capacities, divinity also intensifies the illicit dimension of knowledge encoded within these curiosities. Kesoehtaut's dying words, the light that descends on Manhut's house, and Pattompan's spectral form each speak to the haunting presence of the divine, realizable only through what René Girard calls "mimetic fascination." Ultimately, the attempt to chart evidence of grace upon the souls of the dying proves his maxim, "The reality of the divine rests in its transcendental absence." Divine knowledge ultimately dissolves into a confrontation with what cannot be known and a reinstantiation of the epistemological limits placed upon this empirical pursuit.[66]

66. Boyle, *Christian Virtuoso,* 50; René Girard, *Violence and the Sacred,* trans. Patrick Gregory (Baltimore, 1977), 143.

Natural Theology's Garden of Grace

Mayhew's investment in the relationship between words and things exemplifies a transformative effect of Lockean philosophy on the New England practice of recording evidence of grace. Following Cotton Mather's discovering hidden gems in women's prayer closet devotions, eighteenth-century able divines from Ebenezer Parkman to Benjamin Colman to John Hunt printed impressions and sensory records of divinity from women's reserved papers. Unlike public testimonies of faith, which sought to capture evidence of grace in its immediate and embodied form, able divines excavated secret evidence from a silent life of pious secrets. Editorial translation displayed the sensory impressions recorded in the words in an effort to publish a science of divine signs through empirical "methods" to transform experience into knowledge. The power of these printed impressions was their ability to approximate an "imperfect idea" of divine evidence.

Parkman attempts this editorial translation in the "Memoir of Sarah Pierpont" in 1753 (see Figure 3). Upon her death, Sarah's husband, James, sent her private journals to Parkman. Returning the (now lost) originals to James, Parkman faithfully declares that he has "added no ornaments to them (except for connection and methods sake) thinking they would appear best to the Public in their own Native Simplicity." Resonant with Mayhew's "Indian Converts," Parkman's phrase "native simplicity" suggests that the "Memoir" will reveal a natural truth of divinity, an object of curiosity for the philosopher's gaze. References to nature continue throughout Parkman's introduction. Sarah has "natural abilities," "she was planted in the Garden of Grace" and "sprang up [as] a flower, in the field of Nature." Like the Praying Indians, Sarah represents a Lockean tabula rasa, or the white wall of Willis's corporeal soul, one upon whom divine impressions are recorded as they become manifest in their distilled and natural form. Parkman surmises that this form of unornamented presentation would be of most interest to his "Public," an audience in search of the evidence contained within a secret garden of grace.[67]

67. Ebenezer Parkman, "Memoir of Mrs. Pierpont," Parkman Family Papers, AAS. This manuscript contains no page numbers. Subsequent quotes are from the "Memoir" unless otherwise indicated. Parkman recorded thirty-three conversion testimonies from 1727 to 1771. Although these testimonies are formally identical to those collected by seventeenth-century New England ministers, they record belief, faith, and commitment to the church instead of empirical evidence of grace. Locke's influence on Parkman's theology paralleled the Lockean influence on Edwards. As Perry Miller claims in his analysis of Lockean sensation, Lockean philosophy participated in the

Parkman's parenthetical "(except for connection and methods sake)" is central to the practice of translating private pious secrets into public knowledge of divinity. His method is to display "her hidden life" through diary quotations that record "minutes of the breathing of God's Spirit on her Heart." Interspersed with these quotations are Parkman's explanatory statements, aimed at developing a taxonomy that he entitles "Evidence of Grace." Parkman frames and selects from Sarah's records of self-examination, the moments when she "very often called her self to the strictest trial and examination whether she was sincere in what she did whether she was a real Christian, an Israelite indeed." This process of selection makes public Sarah's own private journey of self-scrutiny, and her silent, "hidden Life" becomes visible as Parkman's anticipated reader catches a glimpse of her soul.

Sarah was a member of the New Haven congregation. She records her testimony of faith in her journal, which publicly proclaims her "desire to enter into Covenant with the Lord." Like Parkman's recorded relations, this congregational testimony does not overtly display the evidence of her soul. Her hidden life contains more empirical promise, for Sarah confesses that she was "more serious and lively in secret duties." This confession fascinates Parkman and motivates his commitment to displaying the contents of her soul. He insists that the "secret covenant transactions betwixt God and her own Soul" are most worthy of "publication" for "church and public." These secret transactions began early, as they did for Jerusha Oliver and Elizabeth Cotton, when Sarah was sixteen years old. They persist throughout her life, often "breaking out" as "Evidence of such irresistible brightness and glory, that she could no longer doubt." Sarah completes her quest for knowledge of her estate with what Parkman describes as "fixed and lettered evidence." This knowledge comes from "secret Prayer," the "sweet drawing out [of her] Soul to God." Through the anticipated publication of this secret prayer, Parkman's "Memoir of Mrs. Pierpont" aims to display a

---

refiguration of grace as a "simple idea" that could not be conveyed in words the way earlier practitioners of conversion testimony had hoped ("Edwards, Locke, and the Rhetoric of Sensation," in Harry Levin, ed., *Perspectives in Criticism* [Cambridge, Mass., 1950], 103–123). While Parkman sustains the effort to observe and record testimonies as a basis of church membership, he looks elsewhere for "evidence of grace." The "hidden life" of Sarah Pierpont provides a fuller sensory record of experience and thus a more complete archive for Parkman's science of divinity. Conversion testimonies are recorded in Ebenezer Parkman, "Relations (Confessions of Sins) for Members of the Westborough Church," Parkman Family Papers.

private feminine self. Although Sarah's face was "hardly seen" throughout her life, it becomes a postmortem object of surveillance through public exposure, one that "curbs the Profane; convinces and alarms the secure; detects the Hypocrite; and melts the hard heart." The intended publication of this manuscript marked Parkman's effort not only to display the contents of one woman's soul but also to use this soul to impress upon others the proper practices of a hidden, pious life.

The anticipated reader of Sarah Pierpont's "Memoir" comes to see "fixed and lettered evidence" of grace as Sarah sees it. Parkman begins his section on "Evidence of Grace" with a description of this process of sensory transference:

> None I believe that was acquainted with her or that has read the History of her Life can helpe seeing (if they have Eyes to see at all) the bright Displays of Divine Grace shining forth from Her. Indeed she like another Moses wist not that her Face shone with such a Luster as if it appeared to the unprejudiced beholders. How exceedingly evident is that divine intercourse that was carried on betwixt God and her own soul.

Through his reference to the seeing eye, Parkman returns to Boyle and Cotton Mather's notion of a divinely authorized eye, wielding privileged access to the angel's curiosity through the dark glass of a fallen world. In its natural and unornamented state of primitive simplicity, Sarah's life supplies the curiosities for this vision. Because her life consisted of a "face" that was "hardly seen" in the public domain, she dies not "knowing" the knowledge that her pious experience and secret duties would supply an audience of "unprejudiced beholders," invested in the scientific task of tracing divinity through evidence of the soul. Parkman promises that, through the special techniques of his "connections and methods," Sarah's soul reveals an "exceedingly evident" display of the invisible world.

His one stated regret in this attempt to display publicly another pious gem from the natural world is that "we can have only some general and brief account" of her death, because Mrs. Sherman, the witness from Sarah Parkman's deathbed who could offer "a more full and particular Relation," had also died. Extrapolating from his study of Sarah's pious life, Parkman recreates the scene of Sarah's death as the culminating "certain sign of regeneration" in his cataloged "evidence of grace." Upon her deathbed, Sarah discovers that

> she might well be enrolled among the Ancient divine Heroes and have a name and place in that sacred catalogue of Christian worthies recorded in the xi of the Hebrews, for her Faith was the substance of things hoped for, the evidence of things not seen.

Upon her deathbed, Sarah's "native simplicity" and recorded life of "divine intercourse" grew organically into her newly realized status as an "Ancient divine" within a "sacred catalogue." Sarah exhibits evidence of a form of Christian primitivism that parallels the typological associations assigned to godly Indians. Her natural, simple, and unornamented status renders her soul a relic of an Old Testament past. The evidence supplied through Sarah's divine translation occasions Parkman's citation of Hebrews 11, which marks a desire to transform faith into knowledge of the invisible world. Deathbed scenes such as Sarah's provided ministers not only with a rare, fleeting glimpse of the world beyond but also with the hope that dying visions, sounds, and exclamations might be preserved in the world of the living as new knowledge of God.

{5} Witchcraft Trials

The Death of the Devil and the Specter of Hypocrisy in 1692

As George Burroughs entered the courtroom on May 9, 1692, Mary Walcott, Mercy Lewis, Abigail Williams, Ann Putnam, Elizabeth Hubbard, and Susanna Sheldon fell into "grevious" fits. This preliminary hearing found Burroughs guilty of performing "wicked Arts" upon three of the afflicted girls (Mary, Mercy, and Ann). The court used the language of James I's witchcraft statute from 1604 (1 Jac. I, c. 12) to discover that Burroughs had "Tortured Afflicted Pined Consumed Wasted and Tormented" the girls, all under the age of nineteen. In each indictment, the other afflicted girls doubled as accusers, bearing witness to the torture, torment, and affliction. On August 3, 1692, Ann Putnam, Sarah Bibber, Mercy Lewis, Elizabeth Hubbard, Susanna Sheldon, Mary Warren, and Mary Walcott returned to court en masse with testimonies about Burroughs in his various spectral forms. In the eyes of the girls, ministers became witches and wizards whose murdered wives appeared in winding sheets, one still with a knife wound covered with sealing wax. Burroughs brought them to "a high mountain" where in view of all the "kingdoms of the earth" he urged them to "writ in his book," promising them possession of such kingdoms if only they would do so. Neighbors described Burroughs's supernatural strength, citing instances in which he had lifted a barrel of molasses or a canoe with one hand. The danger of his supernatural strength compounded in the testimony of each girl as she recounted the visceral effects of Burroughs's spectral presence. He bit, pinched, twisted, beat, starved, choked, and threatened to break their necks. In their testimonies, Thomas and Edward Putnam worried that "every joynt of [Mercy Lewis's] body was redy to be displaced."[1]

1. Paul Boyer and Stephen Nissenbaum, eds., *The Salem Witchcraft Papers: Verbatim Transcripts of the Legal Documents of the Salem Witchcraft Outbreak of 1692* (New York, 1977), I, 153, 156, 157, 162, 169, 171. According to Daniel Payne, James I's statute was used up through Bridget Bishop's trial on June 2, 1692, when the magistrates deferred to the Massachusetts "Body of Liberties" (1641) for legal doctrine. The language quoted in the May hearing references James I's witchcraft act. The

"A lettell black menester that Lived at Casko bay," Burroughs's spectral form represents one of the more vexing interpretive dilemmas of the Salem witchcraft trials. Spectral evidence tested both the hermeneutic capacities of Salem's magistrates in 1692 and, soon after the trials, many ministers' capacity for comprehension and sympathy as to why the magistrates would admit such evidence within the New England community and across the Atlantic. I propose that the specters' presence within the trial transcripts as well as in discussions of evidentiary standards surrounding the trials indicates an intensification of rather than a diversion from the more general epistemological quest for invisible evidence within the natural world. Salem's judges looked to little girls for sensory records of demonic influence as part of a natural philosophical inquiry into an invisible and impenetrable darkness. Rather than a subversion of power, the Salem trials reflect the reassertion of philosophical authority over the domain of the unseen at a time when this authority was under intense philosophical contestation and even attack.[2]

By the end of the seventeenth century, it seemed increasingly clear that securing evidence of grace through visible sainthood was an impossible task. The eventual acceptance of the Halfway Covenant by all of New England's congregations ensured that those offering such evidence did not have to be visible saints in the traditional sense. The early Puritan aspiration to create a visible church as a close approximation of the invisible church of Christ in heaven disintegrated before the eyes of second- and then third-generation ministers who could not sustain this orthodox practice within their congregations. Dependent as it was upon a secure semiotic mechanism for determining visible saints, reinforced through the legitimizing

---

Massachusetts "Body of Liberties" states unequivocally, "If any man or woman be a witch (that is, hath or consulteth with a familiar spirit), they shall be put to death (Francis Bowen, ed., *Documents of the Constitution of England and America: From Magna Charta to the Federal Constitution of 1789* [Cambridge, Mass., 1854], 72). Daniel G. Payne, "Defending against the Indefensible: Spectral Evidence at the Salem Witchcraft Trials," Essex Institute, *Historical Collections,* CXXIX (1993), 62–83.

2. Boyer and Nissenbaum, eds., *Salem Witchcraft Papers,* I, 171. As explained below, the Salem witchcraft trials opened at an uncertain moment in colonial religious history, when the loss of the charter in 1684 made the magistrates equivocate between English common law and the Massachusetts "Body of Liberties" (1641). Uncertain whether to admit spectral evidence, the magistrates looked to the ministers for religious guidance. For a succinct, narrative account of this legal history, see Peter Charles Hoffer, *The Salem Witchcraft Trials: A Legal History* (Lawrence, Kans., 1997). Shortly after the trials, Robert Calef's *More Wonders of the Invisible World* circulated on both sides of the Atlantic, questioning the legal and scientific validity of spectral evidence.

authority of the General Court, the visible church represented a fragile religious and political experiment from the very beginning.

Once Charles II requested that landowning English men be allowed to participate within the body politic regardless of their elect status, the wedge between church and state grew faster and more dramatically. The loss of the charter in 1684 meant even less autonomy of government; the third generation lost the sense of a deep intrinsic connection between the evidence culled from human souls and the Puritan purpose within the New World. More safely gathered from the soul of a dying child or through the public testimony of a Native American, evidence of grace no longer supplied the elect with authority or the ruling elite with a privileged social position. In fact, precisely because of their status as New England's ruling elite, such populations, more than any other, could not shed the shadow of hypocrisy surrounding the evidence of their election.

As seen earlier, evidence of election was not only most epistemologically secure in the dying moments of an individual's life but also most valuable if culled across a diverse array of human souls, especially those not so steeply immersed in doctrinal learning that their intellectual understanding of grace preceded the experiential. The endeavor to collect evidence of grace increased in scope and practice even as the integral link between this endeavor and the prospect of visible election among the colonies' privileged saints diminished. The Puritans did not stop their quest for evidence of grace, even as the dangers of hypocrisy and epistemological limitations continued to thwart the effort toward identifying visible saints. When specters appeared in Salem, they irrevocably and irreparably disrupted the fragile hermeneutic for discerning election that depended on a clear separation between visible and invisible domains such that evidence from the latter could be carefully and expertly discerned only by those esteemed and worthy ministers with the divinely endowed capacity to know.

George Burroughs threatened the magistrates and ministers with an empirical as well as social disruption: he was visible to the disempowered and hauntingly invisible to the ministers and magistrates. He delivered blows that left physical marks on the bodies of adolescent girls. His conjuring powers heretically called forth apparitions of the dead. "Negroes" that turn into white calves and a whole dimension of earthly kingdoms became available to those with whom he chose to communicate. The official records of the Salem witchcraft trials carefully transcribe his haunting spectral presence in order to legitimate the proceedings of oyer and terminer, a court specially designed to deal with the disruptive presence of apparitions and demonically possessed young girls.

When Governor William Phips returned from London in May of 1692, he found the jails full of accused witches whose numbers showed no sign of relenting, given the constant cries of the afflicted. Through a rigorous "empirical" method of recording, the records reveal a group of late-seventeenth-century ministers and magistrates who not only believed in apparitions of the dead, the supernatural powers of *maleficium* (malevolent sorcery), and young girls flying through the night sky on broomsticks but actually admitted such things as legal evidence of an unseen world. By 1692, this unseen world claimed such an unmanageable and complete hold on the visible world that the magistrates could imagine no other choice than to admit the evidence offered to them.

Through their detailed attempt to record this evidence empirically, the authors of the trial records left a tantalizing historical archive of what seems ostensibly to be the rapid decline and eventual end of the Puritan era. Irrationality does not suffice as an adequate explanation of Salem. Instead we seek to untangle the mess, to discern carefully and precisely just how the Puritans arrived at a semiotic implosion that repeatedly and insistently refused rational narration until the only solution was to hang nineteen people and one dog and press a man to death. From an unenlightened era, the Puritans resurface as a great intellectual challenge. Facing it involves wrapping our own minds around the theological condition of Calvinism, the existential uncertainty of divine abandonment, and the ontological crisis felt by those watching their religious and political world dissolve before their very eyes. Accepting the Salem witchcraft trials as an anomalous and puzzling occurrence within our larger narrative of American colonial identity contains the event's violence and its perhaps more unsettling implications.

Rather than a symbol of a fading occult worldview, the devil in Salem represented a phase of an emerging Enlightenment modernity. Seen from this perspective, the events of Salem mark the eruption of, not an atavistic spiritual irrationality, but rather the reverse: the application of a rationality that presented new empirical potential, the implications of which late-seventeenth-century people were attempting to grasp. What would happen to our interpretations of the events at Salem were we to read Spinoza as a contemporary of Cotton Mather? We might better conceive of how the late-seventeenth-century philosophers displaced their anxieties about inaccessible knowledge by locating it, not at a site awaiting examination, but within a forbidden site, rendering such knowledge not only unknowable but morally beyond human ken. For a broad array of seventeenth-century individuals the devil embodied inaccessible knowledge. The devil posed a historically specific problem that vexed mechanical philosophical circles at the same time that belief in the devil produced a crisis in Salem. Salem provides a specific scene within the early modern world where such philosophical patterns became apparent, in this case as evidence of specters rather than grace.

A sharp division between visible and invisible domains marked the culmination of a long seventeenth century in which theologians and natural philosophers were engaged in a mutual endeavor to ascertain knowledge of God in nature. Mechanical philosophers and Puritans sought to maintain this division in order to isolate and circumscribe their respective domains of inquiry so as not to exceed proper boundaries. But, despite these efforts to separate and discipline realms of knowledge, learned men and godly ministers continued to ask, What forms of invisible phenomena can be discovered through science? Can human testimony further divine testimony? What is the epistemological status of the human soul in relation to these questions?

By the late-seventeenth century, mechanical philosophy solidified an ideological stronghold on ways of reading nature. Natural philosophers and divines like Joseph Glanvill, Richard Baxter, Robert Boyle, and Henry More also questioned whether the dark realm of the devil's domain might supply evidence to further dissolve the scriptural limits of human perception. To what extent does the study of witchcraft augment the empirical goal of tracking the action of the spirit on earth? Can natural history prove the existence of God more successfully by penetrating the "Works of Darkness" than by rationally conceiving of the invisible domain as one completely separate from the natural world? These questions were central to Restoration scientists as well as to the Salem witchcraft trials of 1692.

By the end of the seventeenth century, the greatest danger of misapplied science was that it might threaten the existence of God or, at least, direct knowledge toward new ends, augmenting a materialist fascination with secular learning that seemed a distraction, in danger of leading to atheistic tendencies. In promoting materialist and mechanical interpretations of the world, science also warned of the limits of ascertaining knowledge of God. In this respect, philosophers and Reformed theologians had made little progress since Calvin, caught in an intense struggle between the desire for divine knowledge and a deep awareness of the blasphemous nature of such an inquiry. Ministers and empiricists sought to bring evidence from the invisible world into the visible while nervously charting the deteriorating distinction between the two. They struggled to live contentedly in a world that was certain of divine existence but that also faced the certainty that the invisible was essentially unknowable. The specters that appeared throughout the late-seventeenth-century Atlantic world supplied tantalizing and contentious evidence to the contrary.

Through its fleeting, secretive, and deceptive appearances, the specter is the dark shadow of divinity's simultaneous presence and absence in the world. It sits precariously between the visible and invisible domain, the realm of the natural and the supernatural, haunting mechanical philosophers and Puritan ministers alike, not with its inevitably disruptive capacity, but rather with its embodiment of empirical desire. The specter was the late-seventeenth-century repository for doubt, the shadowy twin of Reformation optimism: a continual reminder that the quest to increase divine knowledge required the careful negotiation between knowable knowledge and knowledge knowable but morally prohibited. The specter displaced the sensory impressions of victims, marking a communal presence capable of exploding all sensory and empirical order. It resisted the mechanical understanding of the universe as carefully mapped and ordered, following mathematical principles, by leaving scant and uncertain evidence of malefic acts and through the mysterious supernatural penetration of matter. Specters were deceitful; their hypocrisy bore witness to the "corrupting divinity" through the unchecked pursuit of "secret things." In 1692, specters certainly threatened to transform visible souls into illegible texts, but they also inculcated a desire that the soul reveal the dark and wonderful mysteries of the invisible world.[3]

Discourses on Method

> Have there been any disputed Methods used in discovering the Works of Darkness?—Cotton Mather, *The Wonders of the Invisible World* (1693)

A strange question for Mather to pose, particularly in a treatise dedicated to defending the admittance of spectral evidence into the Salem witchcraft trials, precisely as a method for legally discovering the "Works of Dark-

3. Quoted from Richard Baxter, *Of the Nature of Spirits; Especially Mans Soul . . .* (London, 1682), 4.

ness." It is also strange considering that *The Wonders of the Invisible World* was written in the same year as Cotton Mather's father's own anonymous tract interrupted the trials in the early fall of 1692, granting a stay of execution to the convicted and a reprieve to a number still awaiting trial. It is stranger still, given that the generation of divines and Royal Society members (of whom Increase and Cotton Mather were a part) actively debated whether empirical methods might be used to trace demonic evidence within the natural world. On one side of this debate, such pamphlets as R. B.'s *Kingdom of Darkness* and Joseph Glanvill's *Saducismus Triumphatus* promoted techniques from the experimental sciences such as eyewitness testimony and the collecting and recording of natural phenomena to discover evidence of the demonic preternatural within the natural world. On the other side of this debate, John Webster's *Displaying of Supposed Witchcraft* insisted that this was, entirely, a misuse of empiricism because it was scripturally unfounded.[4]

By the late-seventeenth century, the methods used in discovering the works of darkness were nothing but disputed. Mechanical philosophers stood on one side of this debate, arguing adamantly against the application of their science toward the discovery of demons, not only because it was scripturally forbidden but also because such misuses of empiricism and natural philosophy threatened to break apart the very epistemological fabric of the experimental method, which (according to John Locke and Benedict Spinoza, most famously) could be used only to inquire into the natural and material domain. This was what Margaret Jacob and Jonathan Israel call the Radical Enlightenment, defined through a rejection of the past, a departure from ecclesiastical authority, and a reconceptualization of

4. Stuart Clark, "The Rational Witchfinder: Conscience, Demonological Naturalism, and Popular Superstitions," in Stephen Pumfrey, Paolo L. Rossi, and Maurice Slawinski, eds., *Science, Culture, and Popular Belief in Renaissance Europe* (Manchester, 1991), 222–249; Clark, *Thinking with Demons: The Idea of Witchcraft in Early Modern Europe* (Oxford, 1997); J. A. Sharpe, *Instruments of Darkness: Witchcraft in England, 1550–1750* (Philadelphia, 1997). Clark argues that demonology enjoyed scientific status in the sixteenth and early seventeenth centuries.

Cotton Mather, *The Wonders of the Invisible World: Being an Account of the Tryals of Several Witches, Lately Executed in New-England: And of Several Remarkable Curiosities Therein Occurring* ([London], 1693) (Wing M1175) (pagination is erratic). *Wonders of the Invisible World* was completed and published in 1692 but postdated to avoid the impression of violating Governor Phips's order that nothing be published on the witch-hunt. I am grateful to David Hall for pointing this out to me.

For divergent views on Salem, see Sanford J. Fox, *Science and Justice: The Massachusetts Witchcraft Trials* (Baltimore, 1968); Richard Weisman, *Witchcraft, Magic, and Religion in Seventeenth-Century Massachusetts* (Amherst, Mass., 1984).

philosophy as distinct from the traditional view of theology as the supreme science. It is important to note that this was an epistemological rather than political radicalism.[5]

I propose a mirrored, parallel version of this Radical Enlightenment, consisting of a group of philosophers and ministers who argued precisely the opposite: that empiricism could be used to chart the unseen. They were not espousing an irrational, superstitious, or older religious perspective, but instead were engaging the same contemporary philosophical nexus in precisely the opposite way. The late-seventeenth-century debate over the two Radical Enlightenments was not one of science versus religion, of magic versus modernity, or of rationality versus irrationality. Rather, it was a debate about methodology and the applicability of new scientific techniques to further certain domains of knowledge. It was also a debate born out of a deep frustration on both sides over the problem of uncertainty. Radical Enlightenment philosophers René Descartes, Spinoza, Pierre Bayle, and Balthazar Bekker felt that the firm separation of philosophy from theology was the only way to attain certainty. Those who proposed otherwise sought adamantly to secure a place for the study of the invisible world within what seemed to be an emergent domain of certain, verifiable truth.

The intensive seventeenth-century focus on method—its appropriateness and applicability to various domains of inquiry as well as its integral ability to produce certain truth—is perhaps most famously framed by Descartes in his 1637 essay, *Discourse on Method.* Therein Descartes reflects upon "the progress" that he has "already made in the search after truth." This progress is parallel to and commensurate with a journey toward increased philosophical freedom. This freedom is manifest through an awareness that humans are subject to "delusion" while knowing that through this awareness the detrimental effects of delusion on human perceptive capacities can be overcome. It is a freedom that partakes in a project of self-instruction, which Bacon theorized and Michel de Montaigne prac-

5. Jonathan I. Israel identifies the years between 1650 and 1680 as an intensification of rationalization and secularization during which such shifts took place. Spinoza systematizes the Radical Enlightenment for Israel in *The Ethics* (1675), which unites a wide-ranging network of intellectual and radical thought. Israel, *Radical Enlightenment: Philosophy and the Making of Modernity, 1650–1750* (Oxford, 2001). Margaret C. Jacob uses this idea of a Radical Enlightenment as well in *The Radical Enlightenment: Pantheists, Freemasons, and Republicans* (London, 1981), which traces connections between and transitions from religious to philosophical thought. Akeel Bilgrami makes a similar claim for epistemological radicalism in "Occidentalism, the Very Idea: An Essay on Enlightenment and Enchantment," *Critical Inquiry*, XXXII (2005–2006), 381–411.

ticed in his *Journals*. Descartes "emerge[s] from the control of [his] tutors" and "quits the study of letters." This form of liberation through self-study leads to the resolution to "seek no other science than that which could be found in [himself], or at least in the great book of the world." Carrying nothing other than this resolution with him, the young Descartes sets out into the world to discover new truths, autobiographically recounting a connection between inquiry and travel to new lands that Bacon also maintained and that became institutionalized through the queries issued forth by the Royal Society later in the seventeenth century. But the freedom that Descartes is most famous for claiming in this *Discourse* is, of course, the autonomy of the thinking mind: "I think, therefore I am." This philosophical realization is the point at which he liberates himself from the tendency of human nature and human delusion to "think all things false." It is the point, in other words, where Descartes imagines a philosophical freedom from uncertainty, the deeply inculcated and inescapable awareness of the limits surrounding human comprehension, believed to be an irreversible consequence of the Fall.[6]

Descartes is responding more directly and explicitly to philosophical skepticism, though this tradition shares a great deal of continuity with the story of Genesis retold by Christian philosophers from Augustine to Calvin to Boyle. In his landmark study *The History of Skepticism,* Richard Popkin contends that the Reformation debates over the proper standard of knowledge coincided with a widespread revival of the arguments of the ancient Greek skeptics. The Reformation promoted an intellectual crisis by positing inner truth as the only avenue to persuasion while repeatedly proclaiming that this inner truth was a "subjective certainty that might easily be illusory." Nonetheless, Reformed theologians remained committed to this inner truth as the only guarantor of an authentic experience that was caused by God. To compensate for this contradiction, moderate Protestants in England developed a system for justifying "religious and scientific views" by "appealing to probabilities and reasonable doubts." In other words, they compromised. Human delusion was an inescapable reality, and the acceptance of a condition of doubt that might lead to qualified certainty was the closest that one could come to ascertaining the truth of God within the natural world. Popkin sees both this resignation to a condition

6. René Descartes, *Discourse on the Method of Rightly Conducting the Reason . . . ,*" in Elizabeth S. Haldane and G. R. T. Ross, trans., *The Philosophical Works of Descartes* (Cambridge, 1931), I, 82, 86, 101.

of knowledge and development of techniques for its negotiation as anticipating the "limited certitude" promoted in the theories of Royal Society philosophers, such as John Wilkins, Robert Boyle, and Joseph Glanvill.[7]

This was precisely the position that Descartes's *Discourse on Method* sought to challenge. He refused to rest in the condition of uncertainty, doubt, and human delusion that Reformed theologians and many Royal Society philosophers accepted as an unavoidable fact. The profundity of the "I think, therefore I am" formulation resided in its certainty. It is "so certain and so assured," writes Descartes, "that all the most extravagant suppositions brought forward by the sceptics were incapable of shaking it." It is a claim to a kind of immediate revelation, for the certainty and assurance with which Descartes knows "the first principle of the Philosophy" is the same certainty and assurance with which Anne Hutchinson claims to know God by immediate revelation of his voice to her soul. Each instance is highly individualistic; each is a moment of epistemological certitude that can be achieved only through a form of self-study that is largely incommunicable to the larger community. This was John Winthrop's primary frustration and rationale behind Hutchinson's dismissal; this is why Descartes predicts the impossibility of skepticism to refute his claim. Descartes's essay and Hutchinson's immediate revelation both occur in 1637, on different continents, on different sides of the Atlantic, and, presumably, unbeknownst one to the other. But it is quite possible—in fact contextually evident—that both claims were made in response to parallel frustrations with the epistemological limits that thwarted one's ability to know but did nothing to diminish the human desire to pursue knowledge.[8]

While the method and foundation of his knowledge is obviously different from Hutchinson's, Descartes's statement, "I think, therefore I am," is no less "assured" than Hutchinson's claim to immediate revelation. These are, in fact, two sides of the same coin, for Descartes affirms certainty of God's existence by separating it from human reason. He is certain of God's existence and of the existence of the human soul through a process of separation. He explains that he "aspired as much as anyone to reach to heaven,"

7. E. M. Curley, *Descartes against the Skeptics* (Cambridge, Mass., 1978); Richard H. Popkin, *The History of Scepticism: From Savonarola to Bayle* (Oxford, 2003), 10–11, 13. Where Popkin charts continuities between religious and scientific incertitude, Peter Harrison takes this line of thinking one step further to suggest that "the myth of the fall informed discussions about the foundations of knowledge and influenced methodological developments in the nascent natural sciences." *The Fall of Man and the Foundations of Science* (Cambridge, 2007), 3.

8. Descartes, *Discourse on Method,* in Haldane and Ross, trans., *Works,* I, 101.

yet he also recognizes that "revealed truths . . . are quite above our intelligence" and therefore resolves not to submit them to the "feebleness" of human reasoning. For Descartes, philosophy and theology, reason and revelation, are separate domains, a separation that to his mind augments rather than diminishes the certain knowledge of each. Philosophy is purified and liberated to pursue reason above all else, which Descartes sees as the only clear path to truth. Theology is left alone to convey a different kind of revealed truth. Where the two conjoin is in their causal relationship to God, for both the self and the natural world proceed from God; the perceived perfection of each is an extension of divine infinite perfection. The *Discourse on Method* proved the existence of God and of the human soul by establishing the utter separation between the divine and natural worlds: God's existence was certain. Philosophy confirmed this while also reiterating the limits of knowledge, namely the use of reason to discover divine essence.[9]

The Cartesian separation between theology and philosophy, between visible and invisible domains, informed the emergence of the mechanical philosophical model that would come to dominate natural history by the end of the seventeenth century, but it also set the stage for Benedict Spinoza. In his early essay "God, Man, and His Well-Being," Spinoza would make the most radical claim to date for the exclusion of religion from philosophical inquiry and of knowledge of God from the reasoning capacities of man. The essay begins with two principles: One, "That the things knowable are infinite." Two, "That a finite intellect cannot comprehend the infinite." This formulation is quite similar to the problems set forth by Locke in his chapter "The Infinite" in *An Essay concerning Human Understanding* (explored in Chapter 6). Where Locke says that such infinite things exist but are not available to the human senses, Spinoza positions such infinite things beyond the capacity of human reason. Spinoza sought to recognize the domains of knowledge that exceeded human intellective capacities in order to expurgate the problem of uncertainty from philosophy. The infinite and finite must be kept separate in order to ascribe unprecedented certainty to each.[10]

In a move philosophically as infamous as the Hobbesian denial of the human soul, Spinoza proclaimed the "death of the devil." This was a claim

9. Ibid., 85, 101.

10. Benedict Spinoza, *God, Man, and His Well-Being,* in Edwin Curley, ed., *The Collected Works of Spinoza* (Princeton, N.J., 1985), 62.

as easily misunderstood today as it was among Spinoza's contemporaries for having "atheistic tendencies," privileging reason over revelation and philosophy over faith. But Spinoza's specific objection concerns, not the existence of God, but rather the problem of uncertainty surrounding his existence. Knowledge of God should be certain, Spinoza affirms, rejecting empiricism as an avenue to knowing divine essence, since it necessarily augments rather than diminishes the problem of uncertainty. Faith must be set apart from philosophy in order to restore the certainty of the faithful of divine essence. Consequently, the devil that Spinoza wishes to dispel is, not an atavistic marker of a superstitious age troubling an emergent reason-based modernity, but rather the embodiment of inaccessible knowledge that was integrally linked to but also seeming to thwart the progress of this emergent modernity.[11]

Descartes and Spinoza exchanged the philosophical pursuit of knowledge of God for philosophical certainty. If the methods of human reason could be decoupled from the philosophical proclivity to augment knowledge of God, skepticism and doubt could be more fully eliminated from the progress toward truth. Descartes's and Spinoza's arguments for a separation between theology and philosophy, between divine and natural truth, compounded and contributed to the mechanical philosophical separation between visible and invisible realms that Boyle, Newton, and other members of the Royal Society community sought to uphold in order to legitimate new scientific modes of inquiry. Certainty became a philosophical possibility by the end of the seventeenth century, but its realization required the measured and conscientious decoupling of empirical and natural philosophical method from the pursuit of divine knowledge.

Late-seventeenth-century mechanical philosophy was impatient with the dilemmas of knowledge that had characterized a series of epistemological crises since the Reformation issued its staunch reminder of the limits of the human capacity to know. Mechanical philosophers longed to reclaim the knowledge lost in the Fall and to discard an ancient skeptical belief in the inevitability of contingency. This was, not a discarding of the lessons learned by Calvin, but rather a replication of the paradox of knowledge in a new form. For at the heart of the crisis of knowing, catalyzed by the Reformation, was the kernel of optimism described in Chapter 1. Calvin

11. The phrase "the death of the devil" is the title of a chapter in Israel, *Radical Enlightenment,* 375–405. For this reading of Spinoza, I am indebted to Israel and to Richard Mason, *The God of Spinoza: A Philosophical Study* (Cambridge, 2001).

promised certainty through inwardly directed knowledge of God's voice to the human soul; the challenge for realizing this truth was methodological, for humans were prone to delusion in their fallen states. Mechanical philosophy reinvigorated this paradox while redirecting it toward the natural world. The truth of nature could be known in totality and with certainty, so long as the methods applied to its pursuit were strictly disciplined to avoid seeking knowledge of the divine source undergirding the causal relationship between elements within the natural world.

Preternatural Phenomena

By the end of the seventeenth century, the Reformation problem of knowledge had replicated and intensified in a new form. Correspondingly, so had the science of the soul. As with Bacon's streams of heavenly and natural knowledge, the careful differentiation of theology and philosophy could not be maintained. God could not be fully accepted as a certain but exclusively revealed source of truth alongside a separate domain of natural truth, realizable only through reason. By the turn of the eighteenth century, the attempted split into two separate realms of knowledge, which might ultimately eventuate in some version of a secular age, could not be maintained.

The science of the soul proliferated in direct and explicit response to this period of mechanical certainty, expanding from the effort to chart evidence of God on human souls in specific sites of witnessing to a wide spectrum of more capacious attempts to integrate this evidence into a natural theological plan for determining certain evidence of God, including natural theology, optical studies of the human soul, epistemologies of dreaming, angelographia, and demonology. Disparate as they may seem, each form of inquiry can be characterized through a fascination with the nebulous domain between the visible and the invisible world that might be most succinctly labeled as a domain of the preternatural. Neither fully natural nor supernatural but intrinsically linking the two, the preternatural contained epistemological possibility as well as the ominous danger of transgression.

Just as Reformation epistemology consisted of an optimistic kernel encased in resurgent shadows of doubt, the spectrum of preternatural investigation ranged along this double axis of empirical possibility and the risk of transgression. On the conservative end of the risk takers, Royal Society members John Ray, William Derham, and John Williams revived natural theology in order to determine "certain" supernatural knowledge that could supply "unquestionable Evidence" of God. They grounded their authority in the pulpit, replicating the oratorical and formal style of a minister so as to channel the same divine authority that the sermon allows. The natural theologians also attempted to expand this authority philosophically by introducing a concept of the human soul as a kind of terra incognita, from which new evidence of God could be parsed and incorporated into an ever-expanding cosmic structure that sought to understand the divine workings through the universe in all of its complexity. But, as soon as Royal Society natural philosophers expanded their epistemology, they backed away from the potential scope of this new way of seeing the world, with the wistful resignation that, while the soul was an enticing domain, it was also "the least known" part of man, containing untold discoveries that still seemed beyond the reach of natural philosophy.[12]

Henry More's *Immortality of the Soul* (1659) presents a science of the human soul, refuting Hobbes by insisting that "the very notion of a Spirit or Substance Immaterial is a perfect Incompossibility and pure Non-sense." More asserts that the soul, or immaterial portion of humans, can be the subject of philosophical pursuit. The essence of the soul falls within the realm of the senses, More insists, specifically the visual sense, for visible images of "lucid Spirits" collect in the "bottome of the Eye, with the outward Light conveyed through the Humours . . . wherein the great Mystery of Sight consists." When a human dies, in More's account, the soul reveals itself as a repository for this spiritual residue, conveying information from the spirit world as information to the soul via the "Common Sensorium."[13]

Understood as an immaterial realm that resided in material and natural bodies, the soul represented a unique space during the height of mechanical philosophy's demarcation of separate visible and invisible domains. Theories about how to gain access to the information conveyed to and recorded upon this immaterial realm abounded among philosophers that, like More, sought to use the soul to posit a philosophically sound realm for divine inquiry. Thomas Willis's *Two Discourses concerning the Soul of Brutes* (1683) and Richard Saunders's *View of the Soul* (1682) exemplify this pattern. These tracts apply the camera obscura as both a metaphor and a technology for studying human souls. More commonly used to conduct experiments on light, the camera obscura also supplied philosophers with a mechanism for tracing and studying spirit impressions as they were channeled through the

12. John Williams, *The Possibility, Expediency, and Necessity of Divine Revelation: A Sermon Preached* . . . (London, 1694), 3, 39.

13. Henry More, *The Immortality of the Soul* . . . (London, 1659), 55, 223, 226.

optic nerve and recorded in the brain. Through this technique, philosophers such as Willis demonstrated that there could be a science of the unseen, that an anatomist could "elucidate the nature, routes, and purposes of the animal spirits, despite the impossibility of actually seeing them in action."[14]

In addition to human souls, the angels, demons, specters, and dreams presented possible exceptions to the strict philosophical and theological division between the visible and the invisible. More speaks of the psychic landscape of the dream in the *Immortality of the Soul,* which explains that postmortem ghosts may appear to humans "either by Dreams or open Vision." More emphasizes the availability of this kind of informational data to the task of tracing the invisible realm through the soul's common sensorium. The soul's fundamental distinction from the body is what allows divine sources to pass through it, though More is careful to emphasize that humans have no control over this passing and that in fact, when humans are sleeping, the soul is more likely to leave the body. Sleep places humans in a more conducive state for collecting and perceiving sensory data from the soul, though they then have less control over these data.[15]

The idea that the dream state provided more data from the invisible world—albeit amid diminished reasoning and discerning capacities—troubled Descartes in his *Discourse on Method.* Dreams, to Descartes, are dangerous because in the sleeping state the "imaginations" are "lively and acute" but the capacity of "reasonings" is less so. Humans cannot reason as well while asleep as while awake, so Descartes cautions against using evidence from dreams to supplement the quest for truth. He recognizes dreams as supplying precisely the temptation for increasing knowledge of the invisible that is part of the human condition. "Our thoughts cannot possibly be all true," Descartes laments, given that "we are not altogether perfect." Therefore, if thoughts are to supply truth at all, they must come from "our waking experience rather than in that of our dreams." Dreams fall within the realm of an illicit form of knowledge that nonetheless tempted philosophy to look to them for augmenting knowledge of God.[16]

Within the spectrum of preternatural phenomena, contemplation of the information that might be conveyed from dreams also extended to an-

14. Adrian Johns, *The Nature of the Book: Print and Knowledge in the Making* (Chicago, 1998), 397.

15. More's assessment of the informational use value of dreams explicitly contrasts to that of Descartes. More, *The Immortality of the Soul,* 286–287.

16. Descartes, *Discourse on Method,* in Haldane and Ross, trans., *Works,* I, 105–106.

gels, demons, and finally specters as the most illicit domain for ascertaining knowledge of the divine. In *Angelographia* (1696), Increase Mather reads angels as purveyors of a certain kind of experiential knowledge that can expand human insight into the mysteries of scripture:

> The Angels are great *Philosophers:* they know the Principles, the Causes, the Nature, and Properties of every Creature in the World. . . . They have a Wonderful *Natural Knowledge* and Wisdom in them, and their knowledge is increased very much by Experience. They have had above five thousand years experience, which hath added to their knowledge. And they have a greater measure of *Revealed Knowledge,* than men may pretend unto. They know the Scriptures, better than any man in the World. They have a deeper insight into Scripture Mysteries.

The angel has access to "things" above reason. Angels see what the natural philosopher cannot see. Sight comes from their combined natural and revealed knowledge, accentuated and expanded by what Increase Mather estimates to be their "five thousand years experience." Observational capacities lead to a deeper "insight" into scriptural mysteries. "Compared to living Creatures," angels are "full of eyes behind and before Rev. 4.6." Mather explains that this allows knowledge beyond what any historian or philosopher can see or know; angels demonstrate an ocular capacity to see prophetically what has happened in the past and what will happen in the future.[17]

Spinoza's Ghost

Yet alongside the alluring possibility of the angel philosopher came the troubling awareness that the angel had a twin: "A Devil is a *Fallen Angel,*" as Cotton Mather explains in *The Wonders of the Invisible World.* As long as preternatural phenomena existed, the devil could not die, for it could easily mimic the knowledge proposed in Mather's *Angelographia* in order to delude and deceive inquiry into the invisible domain. According to Increase Mather, fallen angels are notorious hypocrites and are therefore difficult to observe and study. Demons "deceive the Eyes, and delude the Imaginations of Silly Mortals." As potential purveyors of spiritual knowledge,

17. Increase Mather, *Angelographia . . .* (Boston, 1696), 13, 14. It is likely that this tract is a response to Robert Boyle's *Discourse of Things above Reason: Inquiring Whether a Philosopher Should Admit There Are Any Such* (London, 1681).

they supply a slippery path to such knowledge, displaying the continual threat of sensory deception endemic to the testimony of faith. In the spectrum of preternatural phenomena, demons supplied the most dangerous and elusive method of tracing evidence of the invisible domain. Demons were dangerous because they deceived, but tempting to study because they were the cause of religious doubt. Demons clouded the optimistic kernel of Reformation knowledge of God. If only philosophers and theologians could comprehend the knowledge that demons conveyed, the doubt and uncertainty surrounding divinity might recede.[18]

If the demon deceives and deludes, how does one go about studying it methodically and systematically? To return to Mather's question, What are the appropriate methods for studying darkness? As Jonathan Israel rightly argues, the Enlightenment became radical as it separated from theology through a philosophical effort to protect inquiry from the condition of uncertainty and theologically imposed epistemological limitations. As the allegorical reminder of the condition of sin that led to the Fall, the devil represented this condition of uncertainty. Consequently, the Radical Enlightenment marked an attempt to eliminate the devil from philosophical possibility. The corresponding version of this, among those who wished to salvage reason and empiricism as avenues for the study of God, was to apply these methods of reason and empiricism specifically and purposefully to the study of demons. Their hope was that such a study might limit the condition of uncertainty in theological circles so as to mirror the proposal of a kind of philosophical certainty posited most famously by Descartes and Spinoza. Both versions of the Radical Enlightenment of the late-seventeenth century—those who wished to eliminate the epistemological relevance of the devil and those who wished to make the devil more epistemologically relevant—exemplified a mutual struggle to reconcile the condition of unknowable knowledge.

Within this quest for certainty, the specter became Spinoza's ghost, the faint shadow of a devil that claimed a powerful preternatural presence throughout the natural world of Europe and the Americas. The specter generated an empirical desire for supernatural secrets, reflecting a mode of inquiry parallel to the natural philosophical techniques traced in the deathbed scenes discussed earlier. As an extension of this inquiry, the notion that these secrets might come from demons, the most notorious deceivers in the

18. Cotton Mather, *Wonders of the Invisible World,* 5; Increase Mather, *Angelographia,* To the Reader.

invisible kingdom, produced tremendous anxiety around a whole subset of supernatural knowledge. Nonetheless, this anxiety simultaneously inculcated desire through the tantalizing possibility that the most secret, dark, and hidden phenomena might also be the key to greater divine truth. A 1688 anthology of demonic specters explains that there are those "who hold a correspondence with Hell . . . Teachers and Instructers . . . in those cursed Ceremonies,"

> several Books having been writ to that purpose wherein too many of those horrid abominations are set down; Yea it is a certain truth that some have discourst in several Languages, and reasoned notably about sciences which they never learned; They have revealed secrets, discovered hidden Treasures.

The horrible is also a treasure. Fear entices desire. The specter is the monstrous mimetic double of natural science. Despite the overwhelming theology of hypocrisy and the scientific dilemma of incertitude, this passage conveys the possibility that a certain "scientific truth" can be mined from demonic correspondences. To compensate for the heresy of a philosophical project that subtly ebbs away at the scriptural limits of knowledge, *The Kingdom of Darkness* intensifies the empirical scrutiny surrounding these aberrant forms of unseen evidence. Through the properly disciplined Baconian method, these "certain truths" may be rendered as scientific evidence of the unseen. Spectral evidence and Salem's little girls provided convenient avenues for a late-seventeenth-century philosophical endeavor to gain access to this dark secret.[19]

The Restoration debate over the epistemological status of the devil erupted out of this desire, intensified through the pregnant absence of divinity in mechanical philosophy. Scientists and theologians such as John Webster and Balthazar Bekker drew upon the Cartesian rationalism of mechanical philosophy to disprove traditional methods of identifying demonic forces. A pastor in Amsterdam, Bekker insisted that to refute the discernibility of the devil was not to align oneself with atheism. He endorsed Cartesian principles and corresponded with Spinoza, though he also wrote

19. R. B., *The Kingdom of Darkness* . . . (London, 1688), preface. Henry More exemplifies an awareness of this potential heresy in his opening letter to Baxter's *Of the Nature of Spirits; Especially Mans Soul* . . . (London, 1682), 4 (also published as the preface to Jos[eph] Glanvil, *Saducismus Triumphatus* . . . [London, 1688]). He writes: "I take that boldness of Philosophers to have had a great hand in corrupting Divinity. Secret things are for God, and things revealed for us and our Children, saith Moses."

that Spinoza was mistaken in "confound[ing] God and Nature together." Bekker sought to integrate the new science and philosophy into his theology. His widely translated and circulated treatise, *The World Bewitch'd,* reflects this effort, specifically by promoting his belief that the new philosophy amply negates the "common Opinion of the Devil." Bekker constructs his argument by compiling ethnographies of the devil from all over the world, including North America (with Mexico) as well as Peru. Bekker's point is, not that the devil is a bygone mark of a superstitious age, but rather that those collecting such stories are misusing the natural sciences. Collectively, the stories represent a "vast catalogue" of writings about witchcraft cases, possession, and exorcism as proliferating phenomena throughout the New World, but they do so through an explicit misunderstanding of the relationship between natural and supernatural forces.[20]

Like Bekker, John Webster contended that the use of experimental philosophy to chart the existence of "Witches and Witchcraft" was entirely "confused and immethodical" in his *Displaying of Supposed Witchcraft* (1677). With similar instances of collecting and cataloging stories about witchcraft and demonic possession in mind, Webster attributes this confusion to the same idea of the limits of human understanding that appears as a repeated trope throughout early modern philosophical and theological writing: "The minds of men are not only darkned in the fall of Adam, but also much misled, by the sucking in of errors in their younger and more unwary years." Because of their limited reasoning capacities, humans have misunderstood how observation, recording, and experimental philosophy should be performed; this failure has led to the "immethodical" studies of witchcraft. Far from reflecting a movement away from the religious realm, Webster seeks to refute empirical studies of witchcraft by showing that they are based on false interpretations of scripture. He builds his claim on the proper relationship between reason and scripture rather than seeks to displace the latter with the former.[21]

In response to Webster, Royal Society members Richard Baxter, Joseph Glanvill, Robert Boyle, and Henry More attempted to prove demonic existence through testimony and empirical evidence. In the 1660s, Boyle cor-

20. Balthazar Bekker, *The World Bewitch'd; or, An Examination of the Common Opinions concerning Spirits* . . . (London, 1695), unpaginated preface and "Introduction to the Abridgement of the Whole Work." As Jonathan Israel explains, Bekker was eventually imprisoned for his attempt to negate the philosophical legitimacy of the devil's existence. *Radical Enlightenment,* 381.

21. John Webster, *The Displaying of Supposed Witchcraft* . . . (London, 1677), 20.

responded with a French doctor, Pierre Du Moulin, about "the Devill of Mascon" whom Moulin had observed, studied, and recorded in a "truthful relation."[22]

This was the same French author who wrote *The Buckler of the Faith* in 1631, which charted sacred truths through human testimony. The development of Du Moulin's thought toward this observational mapping of devilish manifestations throughout the world reflects the transformation of soul science over the seventeenth century, as the demonic came to be seen as a dangerous and illicit but nonetheless tantalizing site for pursuing this form of inquiry into the invisible. In a letter to Joseph Glanvill, Boyle ultimately "disowned" Du Moulin's story as "a clear Imposture," indicating his commitment to sifting through the inevitable delusions that accompany any such effort to observe manifestations of the invisible world. Yet, Boyle's refutation of Du Moulin's demonic findings did not stop him from collaborating with Glanvill on *Saducismus Triumphatus* (1689). Glanvill's project was to provide "so palpable an Evidence" of the existence of witches that he would prove "lastly that there is a God." Henry More urged the Royal Society in his *Antidote against Atheisme* (1653) to put on the "Garb of the Naturalist" in order to collect case histories of witchcraft. As More's title implies, proponents of the use of experiment and empiricism to trace a demonic presence on earth framed their efforts as a response to the growing "atheism" of mechanical philosophers who refuted the discernible presence of invisible, divine agents in the natural world.[23]

Despite the atheistical accusation, the two poles of the witchcraft debate do not represent a schism between science and religion, or between reason and revelation. Webster insists that "denying of existence of Angels or Spirits . . . doth not infer the denying of the Being of God." A believer in magic and the occult, Webster does not dispute the existence of the supernatural, but rather challenges the application of experimental techniques to the study of a particularly elusive, specious, and deceptive component of

22. "Ocular Conviction" is Cotton Mather's term: Cotton Mather, *A Discourse on Witches* (1689), in Alan Charles Kors and Edward Peters, eds., *Witchcraft in Europe, 400–1700: A Documentary History,* 2d ed. (Philadelphia, 2001), 370; Michael Hunter and Edward B. Davis, eds., *The Works of Robert Boyle* (London, 1999), I, 14–36.

23. Joseph Glanvill, *Sadducismus Triumphatus* (1689), in Kors and Peters, eds., *Witchcraft in Europe,* 371. Boyle wrote to Glanvill on February 24, 1678, to explain that, while he would like to contribute to Glanvill's collection of witchcraft relations, he had "of late disowned the Story of the Demon of Mascon as a clear Imposture." Boyle to Glanvill, Feb. 24, 1678, in Michael Hunter, ed., *The Correspondence of Robert Boyle,* electronic ed., V, 38. More quoted in Thomas Harmon Jobe, "The Devil in Restoration Science: The Glanvill-Webster Witchcraft Debate," *Isis,* LXXII (1981), 347.

the invisible world. The devil's discernible existence was a scientific rather than a religious question. Both groups of Restoration philosophers believed in varying degrees that demons exist; they disagreed on man's capacities to perceive their presence within the natural world. Could optical science, the "Ocular Conviction" of which Mather speaks in his *Discourse of Witches,* map the presence of specters? Bekker expresses doubt, posing a rhetorical question, "How the [demonic] body is seen when it appears, since it is more subtle than the air, which cannot be seen because of its subtlety?" Through his comparison of the unseen supernatural to air, Bekker was probably alluding to Ephesians 2:2, "the prince that ruleth in the aire" (Geneva Bible). In a sermon in *Memorable Providences* (1689), Mather also draws upon this scriptural passage to describe an "Atmosphere of our Air" in which "a vast Power, or Army of Evil Spirits, [lives] under the Government of a Prince who employs them in a continual Opposition to the Designs of GOD."[24]

While grounded in scripture, both uses of air to comprehend the supernatural also evoke Boyle's air-pump, a seventeenth-century invention that aided an empirical approach to invisible substances. Consonant with Boyle and with Mather's science of divinity, air-pump experiments grappled with the difficulty of observing that which could not be seen, even with the most powerful micrographic technology. Bekker interprets air as a mark of the limits of human perception. The technology could not be applied to the supernatural, since the spirit is clearly more elusive than air. "Men," according to Bekker, "desire . . . to appear more savant in occult things than in things which are known." This desire reiterates for Bekker the hubris of original sin as it guides an attempt to illuminate the darkened glass of perception and to discern "visible bodies" from "all kinds of substances." Bekker bases his critique of this endeavor on a Cartesian rationalist's approach to matter. The only "palpable substance" visible to humans is the embodied form of "Adam." The immaterial cannot be seen or known through sensory data of the eyes or of the touch. "Reason teach[es] us" nothing "beyond the limits of Nature." While "revelation" and "Scripture" supply unique and divinely authorized forms of knowledge, Bekker warns that, when one goes beyond natural law, one becomes entirely "lost." The world ceases to make sense because all that becomes apparent is the devil's utter "lack of experience concerning the secrets of nature." The devil, for Bekker, wields no

24. Webster, *Displaying of Supposed Witchcraft,* 27, 38; Balthasar Bekker, *The Enchanted World* (1690), in Kors and Peters, eds., *Witchcraft in Europe,* 435, and Mather, *A Discourse on Witches,* 368.

secret knowledge of the "power and wisdom" of God. Attempts to empirically discern demonic presence reflect mere futile grasps towards "things," which cannot be "known" on earth.[25]

Ethnographers of the Enchanted World

Increase Mather's *Essay for the Recording of Illustrious Providences* (1684) remarks favorably upon the "improvement" that the "Ministers of God" have made "in the Recording and Declaring the works of the Lord." *Illustrious Providences* employs empiricism to chart evidence of the demonic in case studies of possessed children similar to Glanvill's in *Saducismus Triumphatus,* which Mather in fact cites as a model. For Increase Mather, "illustrious providences" were entirely explicable through the methods of observation and natural history. Mather reveals his "thoughts" regarding the publication of "a Discourse of Miscellaneous observations, concerning things rare and wonderful, both as to the works of Creation and Providence, which in my small Readings I have met with in many Authors." These readings included Boyle and Glanvill, two authors whom Mather cites and who seem to mark for him a promising conjunction between natural history and the very rare and curious yet revelatory movement of a demonic presence within the natural world. Expressing his own desire to engage in this enterprise, Mather writes that he has "often wished, that the Natural History of New-England might be written and published to the World; the Rules and method described by that Learned and excellent person Robert Boyle Esq. being duely observed therein."[26]

The "rules and methods" to which Mather refers are in Boyle's "General Heads for the Natural History of a Countrey" (1665–1666). Whereas "General Heads" offers methods and questions for cataloging "the earth" and "its inhabitants," *Illustrious Providences* catalogs the devil's appearance in "any visible shape" and the "rare and wonderful . . . works of Creation

25. Steven Shapin and Simon Schaffer, *Leviathan and the Air-Pump: Hobbes, Boyle, and the Experimental Life* (Princeton, N.J., 1985); Bekker, *The Enchanted World,* in Kors and Peters, eds., *Witchcraft in Europe,* 435. Invented by Robert Boyle, the air-pump consisted of a suction pump attached to a glass bulb. The pump evacuated the air to create a vacuum for conducting scientific experiments.

26. Increase Mather, *An Essay for the Recording of Illustrious Providences,* in George Lincoln Burr, ed., *Narratives of the Witchcraft Cases, 1648–1706* (New York, 1914), 13, 16–17; Paul S. Boyer and Stephen Nissenbaum, *Salem Possessed: The Social Origins of Witchcraft* (Cambridge, Mass., 1974), 16, 17; Mary Beth Norton, *In the Devil's Snare: The Salem Witchcraft Crisis of 1692* (New York, 2002), 6, 16, 17.

and Providence." Applying Boyle's rules for natural history to trace the earth's invisible works, Mather participates in a late-seventeenth-century philosophical movement intent upon applying the Royal Society's rules and methods to discover the "secret" and "invisible" works of fallen angels and demons. Mather's text foregrounds an empiricism of the demonic, a practice whereby "Learned and Holy men" would redirect the methods for discerning evidence of the soul to a science of "Satan in signes."[27]

Illustrious Providences begins this discourse of "Miscellaneous observations," recording several instances of demonic possession. William Morse's possessed children display clear evidence of demonic presence through the effects of an invisible "Hand" that scratched them, beat them, and threw them on the floor. Yet, in keeping with his future refutation of spectral evidence, Mather stops short of giving to this invisible presence a visible form: "All this while the Devil did not use to appear in any visible shape." As if hesitating to include a spectral form of empirical proof in his observational method, Mather "judges" that it was "not so proper a season . . . to divulge" certain "particulars" relating to the project of recording demonic presence. In 1684, New England's science of divinity was not yet ready to embrace spectral evidence of the invisible world.[28]

Although this use of experimental philosophy in many respects characterized techniques for culling evidence of grace from human souls throughout the duration of the practice, the specter intensified anxieties surrounding the possible misuse of empirical techniques. The specter manifested the conundrum of hypocrisy's haunting the genre all along. Could its presence and its haunting invisible absence be contained within a safe space? Cotton Mather would attempt to empirically encase Salem's specters in "Enchantments Encounter'd," an essay in *Wonders:*

> We are safe, when we make just as much use of all Advice from the Invisible World, as God sends it for. It is a safe Principle, That when God Almighty permits any Spirits from the unseen Regions, to visit us with surprizing Informations, there is then something to be enquired after.

27. Robert Boyle, "General Heads for a Natural History of a Countrey, Great or Small, Imparted Likewise by Mr. Boyle," Royal Society, *Philosophical Transactions,* I (1665–1666), 187; Mather, *Illustrious Providences,* in Burr, ed., *Narratives of the Witchcraft Cases,* 16, and John Hale, *A Modest Inquiry into the Nature of Witchcraft,* 403.

28. Mather, *Illustrious Providences,* in Burr, ed., *Narratives of the Witchcraft Cases,* 16, 30; Simon Schaffer, "Godly Men and Mechanical Philosophers: Souls and Spirits in Restoration Natural Philosophy," *Science in Context,* I (1987), 73.

This insistence on the "safety" of the magistrates "enquiry" effaces the sacrificial violence impugned upon the accused.[29]

Nathaniel Crouch's 1688 text, by contrast, disregards the hesitant inquiry prompting the safe containment of empirical techniques to trace and verify spectral evidence. Using Increase Mather's *Illustrious Providences* as well as other New England, Scottish, German, and old English sightings of specters, *The Kingdom of Darkness* narrates "the history of daemons, specters, witches, apparitions, possessions" from "Authentick Records, Real Attestations, Credible Evidences." The title page also advertises this "wonderful" world of "supernatural Delusions" through images that correspond to specific demonic sightings throughout the text. In the bottom left-hand corner of the facing page are the six spirit specters that appeared to Ann Bodenham in the form of "ragged boys." The image of the book, also figured in the lower left-hand corner, signifies the correspondence that these spirits have with the devil. The image to the bottom right figures the judges who sentenced Johannes Contius to hang, and then to burn when he would not die from hanging, while the ship in the top right corner foreshadows a repeated image throughout the text. Ships recurrently produce "marvellous accident[s]," washing ashore ghosts and apparitions before "many hundred Spectators" of the workings of the "Invisible World."[30]

The ship represents a dual geographic and ethnic-sacred focus of the text. The relations come from all over the Atlantic world, confirming Crouch's initial premise that "such frightful Specters do most frequently shew themselves in places where the Gospel is not preached." "Credible Historians" seek evidence of the demonic within the farthest reaches of the globe. This spatial conceptualization imagines the possible presence of demons in the West Indies while including small towns, rural areas, and the provincial life of New Englanders as particular enclaves for the proliferation of a demonic presence.[31]

The oceanic voyage into the "kingdom of darkness" also represents an unknown, furtive, and mysterious dimension of the invisible world. At the top of the image, the drawn curtain with the two corners held by a demon signals the disclosure of these dark and mysterious secrets. This theatrical gesture frames the contents of what will be scientifically discovered through the empirical techniques enumerated on the title page. Proof "of

29. Mather, "Enchantments Encounter'd," 13, in *Wonders of the Invisible World.*

30. R. B., *Kingdom of Darkness,* 13, 46.

31. Ibid., preface.

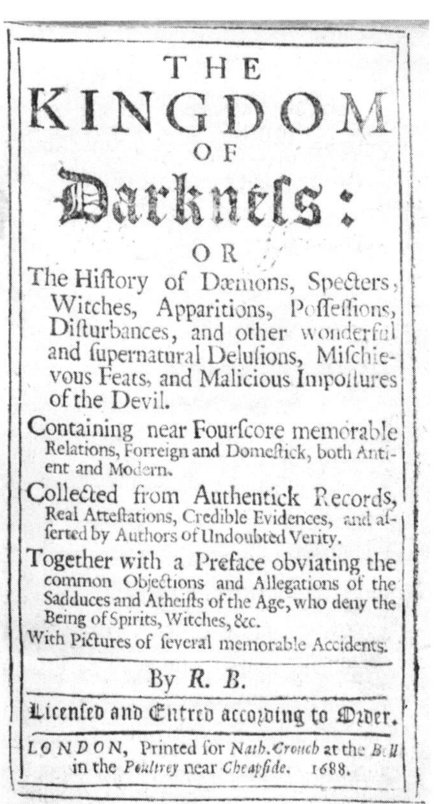

THE
KINGDOM
OF
Darkneſs:
OR
The Hiſtory of Dæmons, Specters, Witches, Apparitions, Poſſeſſions, Diſturbances, and other wonderful and ſupernatural Deluſions, Miſchievous Feats, and Malicious Impoſtures of the Devil.
Containing near Fourſcore memorable Relations, Forreign and Domeſtick, both Antient and Modern.
Collected from Authentick Records, Real Atteſtations, Credible Evidences, and aſſerted by Authors of Undoubted Verity.
Together with a Preface obviating the common Objections and Allegations of the Sadduces and Atheiſts of the Age, who deny the Being of Spirits, Witches, &c.
With Pictures of ſeveral memorable Accidents.
By *R. B.*
Licenſed and Entred according to Order.
LONDON, Printed for *Nath. Crouch* at the *Bell* in the *Poultrey* near *Cheapſide*. 1688.

FIGURE 16 • *Frontispiece and Title Page, R. B.,* The Kingdom of Darkness. *Image courtesy of the Division of Rare Books and Manuscript Collections, Cornell University Libraries, Ithaca, N.Y.*

Undoubted Verity" of this invisible world exists and is soon to be disclosed to the reader. As in Glanvill's *Saducismus Triumphatus,* the woodcut includes the reader within the witnessing audience. We see the specter alongside those with special visual access to demonic manifestations. Through a form of double vision produced through the woodcut, the reader acquires scientific access to unseen, invisible secrets. The raised curtain facing the title page constitutes this ocular doubleness "as the revealing [of] secret things past or to come, which without some supernatural assistance could not be discovered." The reader of *Kingdom of Darkness* moves temporally as well as spatially, encountering "examples of former ages and places" as collective evidence of a realm normally hidden from human sensory impressions. Through the addition of expert testimony from "persons whose Judgment and Reason has been free from disturbance," the text legitimates relations concerning the "Secrets of Nature" that often constitute only "a multitude of Lyes and Fables." The reader of these "Authentick" accounts acquires the ability to see revealed truths in the way Increase Mather described angels, "full of eyes behind and before" (Rev. 4:6). Supernatural secrecy transforms through the inclusion of specters as empirical evidence of an invisible world. Specters provide access to a "transcendent knowledge" of divinity.[32]

Saducismus Triumphatus also traces this transcendent knowledge through an ethnography of the enchanted world. Glanvill responds to the "anti-spiritualist mechanism" of Cartesian philosophy with empirical evidence of the reality of spirits in human life. *Saducismus* presents the inextricable coexistence of God and the devil. In an epistolary introduction, Henry More offers the assurance that "so palpable an Evidence . . . that there are bad Spirits, which will necessarily open a door to the belief that there are good ones, and lastly that there is a God." *Saducismus* sets out to supply this evidence through carefully observed and recorded human testimonies and eyewitness accounts. Glanvill explained to Robert Boyle that he "makes it [his] business to be very careful in the strict examination of all the Relations of the kind that I admit into my Collection. . . . The great caution . . . makes [his] books go on very slowly." For Glanvill this is a laborious yet imperative endeavor: if experimental philosophy denies the existence of evil spirits, "Sense and Knowledge is gone as well as Faith." Sense is crucial to Glanvill's empiricism as well as to his religious belief: "Matter of Fact can only be proved by immediate Sense, or the Testimony of others Divine or Humane." Whereas radical Protestant John Webster proclaimed that scripture was the true medium for scientifically disproving demonic presence, Glanvill seeks to prove demonic existence through the sensory data of human testimony.[33]

The question of divine versus human testimony was central to the Glanvill-Webster debate. Webster would admit only scriptural proof of divine phenomena into his empiricism of the unseen. Glanvill based his entire "collection" of demonic facts on irrefutable human testimony as a supplement to the scriptural understanding of the action of the spirit within the natural world. The 1689 edition of *Saducismus Triumphatus* con-

32. Ibid.; Mather, *Angelographia,* 14.

33. Glanvill, *Sadducismus Triumphatus*, in Kors and Peters, eds., *Witchcraft in Europe,* 375–376; Glanvill to Boyle, Feb. 24, 1678, in Hunter, ed., *Correspondence of Boyle,* electronic ed., V, 38; Webster, *Displaying of Supposed Witchcraft,* 44–45.

tains an appendix of six verified witchcraft stories based upon eyewitness testimonies. As in the title page to *Kingdom of Darkness,* a curtain frames this series of woodcuts, each of which corresponds to a testimonial record of spectral evidence narrated in "Part the Second." The curtain evokes the theater. The reader, like the audience attending a play, becomes a witness to a certain form of truth through narrative. In the top left image, Tedworth, the haunted victim of one of Glanvill's case studies, sees a demonic specter from the invisible world. Tedworth's vision forms a kind of double vision analogous to the double vision described above. We see the presence of the devil through Tedworth's authority of witness. Through his careful empirical record of human testimony and painstaking investigation into the reality of each case, Glanvill verifies Tedworth's vision by reproducing it as part of a scientific anthology. A version of this double vision of the demonic operates in each image. On the top right, we see Julian Cox's descriptive image of the devil in "the Shape of a black Man," coaxing her to sign her name in blood in his book. The levitating specter defies the mathematical laws of matter's passive and active movement through the world.[34]

These images present the specter as scientific proof of invisible phenomena whose elusive status has historically been visibly manifest to only a select few. Glanvill extends the specter's witnessing audience, enfolding collected testimonies, his own testimonial authority, and the witnessing reader into his empirical project. Translating spectral evidence into scientific "matters of fact" built from human testimony, Glanvill contests Bekker's claim for the unmethodical nature of such investigations while expanding natural laws of ocularity and the movement of matter and substance.

Seeing, Hearing, and Touching the Land of the Spirits

The retrospectively termed Glanvill-Webster debate over humans' capacity to observe preternatural phenomena was one climax of a more general seventeenth-century discussion about the relationship between form and matter. Since the start of the seventeenth century, the Aristotelian understanding of form as constituting the shape of matter had been challenged. The materialist impulse driving the rise of Baconian empiricism elevated

34. For a full account of the Glanvill-Webster debate, see Thomas Herman Jobe, "The Devil in Restoration Science," *Isis,* LXXII (1981), 343–356; Glanvill, *Sadducismus Triumphatus,* in Kors and Peters, eds., *Witchcraft in Europe,* 377.

FIGURE 17 • *Frontispiece, Joseph Glanvil,* Saducismus Triumphatus *(London, 1689). Huntington Library, San Marino, Calif.*

the ontological status of matter over form. Natural philosophy unfolded to conceive of a more dualistic relationship between the two. Form was the inanimate state and condition of things, and matter was atoms, which formed the very fabric of a universe. So the question of how matter moves became central to seventeenth-century mechanical philosophy, central to understanding the laws of motion in the universe. Applying a rule of mathematical certitude, mechanical philosophy proposed that "'Nature observes a fixed and immutable order,' that is, that the same laws of motion, and the laws of cause and effect, apply in all contexts and everywhere."[35]

But what caused motion? What was the cause and what was the effect of matter's movement through the world? Few philosophers contested the fact that matter was passive and inert, acquiring motion as an effect rather than a cause. In a philosophical landscape that increasingly sought to separate God from nature, this posed a dilemma. If not God, then what moved matter? Theologians and natural philosophers who wanted to reconcile theology with scientific advancements maintained that God was the causal force behind matter's movement. But sustaining this theological perspective required accepting the uncertainty surrounding natural philosophical questions of causality.

In their quest to eliminate uncertainty from science, to arrive at a comprehensive notion of scientific rationalism, mechanical philosophers continually confronted "epistemological dogmatism," the specter of mathematical certitude. This position seemed to be a great threat to theology, in fact even displaying atheistic tendencies. Descartes attempted a theory of how bodies move by developing the notion of a "sensitive soul." Composed of matter, the body is mechanical and inert in Descartes's *Passions of the Soule,* though it is animated by the soul. External things constitute sensory objects for Descartes. They convey information through the eyes, hands, and ears to the pineal gland in the brain, which activates the passions of the soul. These then cause a physiological disruption such as a rapid heartbeat, rise in temperature, and so forth. The Cartesian body responds physiologically to the effects of an agitated soul.[36]

Descartes's sensitive soul breathes movement into a newly animated body independently of any supernatural influence. This was precisely the threat that his mechanism posed to theology. In *The Origine of Formes and Qualities, (according to the Corpuscular Philosophy)* (1666), Robert Boyle pro-

35. Israel, *Radical Enlightenment,* 244.

36. Ibid.; R. Des Cartes, *The Passions of the Soule* (London, 1650).

posed an explicit response. Like Descartes, Boyle proposed a way of mapping physiology according to the laws of mechanical philosophy in order to explain the mutability of essential substances. Boyle agrees with Descartes's statement in the *Discourse on Method* that the origin of motion in us is from God. Corpuscular philosophy attempts to refute the limits that Descartes placed around this notion of origin and its deistical implications: "The Origine of Motion in Matter is from God; and not onely so, but that thinking it very unfit to be believ'd, that Matter barely put into Motion, and then left to it self should Casually constitute this beautiful and orderly World." In other words, there must be a method of tracing the subtle motion of matter throughout the natural world back to its causal origin in "supernatural mysteries."[37]

Through its focus on the activity of sensible qualities, *The Origine of Formes and Qualities* allows a more capacious sense of where the data communicated through the senses might come from as well as what they might say. What Descartes described as the "objects of our senses" may not after all come just from a natural world of purely material entities. If this natural world consisted of the forms of enchantment that the ethnographers discussed above sought adamantly to maintain, then it was indeed possible that specters, witches, apparitions, and possessions might also convey information to the senses, animating the souls of the earth's inhabitants in physiologically surprising ways. Boyle did not, of course, allow for or even anticipate all of this in his *Origine of Formes and Qualities* (a treatise intended primarily for conducting experiments in a chemistry laboratory). What I am proposing rather is that his theory of corpuscular movement introduced a method that could be used to chart the presence of specters as preternatural phenomena, which, like natural phenomena, were capable of communicating information to the senses and prompting a physiological response. *The Origine of Formes and Qualities* unequivocally contends that "the humane Body hath a receptive power in reference to the humane Soule, which [philosophers] confesse both to be a substantial Form, and *not to be educ'd out of the power of Matter.*"[38]

Such critiques of Descartes were common in the late-seventeenth century by theologians as well as natural philosophers intent upon preserving a notion of divine causality that gave not only structure but also motion to

37. Robert Boyle, *The Origine of Formes and Qualities, (according to the Corpuscular Philosophy) . . .* (Oxford, 1666), 5.

38. Ibid., 31, 75. My emphasis.

material forms. In his "Preface" to *Of the Nature of Spirits; Especially Mans Soul* (1682), Richard Baxter attempts to elevate the soul from the Cartesian theory of sensitivity. Baxter writes that the category of the "*Immaterial* signifying nothing (but a negation)." This "immaterial" fails to explain adequately the complex interactions between "materia" and "substantia." Residing between the natural and supernatural worlds, *substantia* consists of the plastic forces of the Neoplatonic tradition and the sensible qualities described by Boyle. Baxter proposes a return to a notion of substance as theorized by "ancient philosophers" who recognized that between the material and the immaterial, between the active and passive bodies of mechanical philosophy, is "spiritual substance." For Boyle, Baxter, and other experimental philosophers, spirits and souls wield some capacity to "actuate matter."[39]

The Nature of Spirits presents a method for tracking the motion of spirits through their visible substance while remaining deeply aware of the heretical possibility encoded in this form of inquiry. A prefatory letter by Henry More admits that care must be employed to "distinguish unknown things from known, and to bridle my understanding from presuming to enquire into unrevealed things." Aware of the philosopher's potential for "corrupting Divinity" by violating scripturally defined epistemological limits, More proclaims his and Baxter's "understandings" that their "Eyes are made only for things revealed." While ostensibly "confining" their "enquiry" to the soul's substance insofar as it may be supported through scripture, *The Nature of Spirits* seeks to establish that the soul can "subsist without a Body." Using an extended form of corpuscular philosophy to trace the disembodied spirit's presence within the world, Baxter and More find "seminal material" manifestations of the soul's "ethereal and airy" composition. Both divine and human testimony—provided to witnesses through the combined proof of "Scripture and apparitions"—confirms the "individuation of Spirits and Separate Souls." The tract attests to an empirically verifiable soul that is *separated* from the body but that also continues to "have its part in the worlds animation."[40]

Regarding the spiritual substance controlling this animation, Baxter poses as many questions as he answers. He puzzles over the "genus of Substantiality" by comparing the invisible soul to an atom: "Is an Atom Matter?

39. Baxter, *Nature of Spirits*, 5. Baxter, More, and others use the words "spirit" and "soul" interchangeably.

40. Ibid., 4, 5, 6, 8.

Or is it not? . . . If an Atom be matter, is it Discerpible or not? If not, how is this the Form of Matter?" The spirit poses a parallel epistemological dilemma, for its invisibility exerts a visible influence upon the world; its animation takes the form of spiritual substances and sensory manifestations, presenting, like the atom, the alluring possibility of decipherability. Baxter further explores the confusion of mechanical philosophy's categories of active and passive material by describing the mutual process through which spirits and bodies "can penetrate one another." This process of penetration makes it difficult to distinguish ontologically the relationship between matter and spirit, body and soul. In place of mechanical philosophy's reified categories, Baxter proposes a fluid boundary of penetration and passivity.[41]

Laws of Motion

Debates between Glanvill and Webster—as well as between the Radical Enlightenment codified by Spinoza and its mirror version in a radical theological epistemology that sought to redirect natural philosophy toward invisible inquiry—both hinged upon a desire to attain greater philosophical and divine certainty. The laws of motion governing the movement of matter formed the center of these competing epistemological programs, for the advancement of knowledge on either side of the debate depended on resolving the question of causality. Through this line of questioning, the Radical Enlightenment and its critics also redirected evidentiary criteria within a legal context toward increasing certainty.[42]

Most of the cases of the Salem witchcraft trials that resulted in executions followed the 1648 *Book of the General Lawes and Libertyes* of Massachusetts. Yet the *Lawes and Libertyes* did little to establish evidentiary standards during the trials themselves, nor did it create a precedent for the admittance of spectral evidence within the trials. When specters appeared in the Court of Oyer and Terminer, they exposed what had historically been the secret crime of witchcraft for empirical scrutiny, calling for a rational basis of legal examination of a subject that required rather than eschewed the certainty derived from religious authority.

41. Ibid., 16, 17.

42. Barbara J. Shapiro, *"Beyond Reasonable Doubt" and "Probable Cause": Historical Perspectives on the Anglo-American Law of Evidence* (Berkeley, Calif., 1991), 165. Shapiro finds witchcraft cases to be legal anomalies in the seventeenth century. While law, like science, relied increasingly on careful observation, empirical techniques, and credible witnesses, witchcraft cases represented a secret crime that could not be tracked empirically.

Although magistrates looked for ministerial approval before admitting spectral evidence, criteria used to study specters via the sensory data supplied by adolescent girls entered the trials through a specific adaptation of contemporary theories that explained the movement of matter and the appropriate methods for detecting the cause of this movement. Magistrates tracked spectral form through Boyle's formulation of how the senses act upon matter in *The Origine of Formes and Qualities* while also implementing the dual-witness standard as recommended by William Perkins in his *Discourse of the Damned Art of Witchcraft*. Alongside the afflicted girls' performance, a respected (usually male) community member attempted to convert specters into sources of empirical truth, calling up the legal basis of spectral evidence within the Salem witchcraft trials. During the trials, the magistrates used this empirical basis to track spectral forms through the relationship between the accused witches' eyes and touch and the actions of the girls. This interactive process drew upon the laws of Boyle's corpuscular philosophy, which put "Matter . . . into Motion," providing a method for tracing the specter's movement as an invisible witness within the trials.[43]

In a letter to John Richards, one of the judges in the trial, Cotton Mather approves the use of spectral evidence, provided that proper empirical standards ensure its validity. He exhorts the magistrates to "not lay more stress upon pure specter testimony than it will bear." While it can tease out the "hidden things" of witchcraft transactions, caution must accompany the legal admission of specters lest it ruin "a poor neighbor" or open an uncensored door to the "invisible world" in which devils come and go at will. Mather urges the judges to "find out" how the accused have carried out "what the devil will" through "mental, oral, or manual" manifestations, drawing upon corpuscular philosophy to produce a safe space for the public study of specters.[44]

A natural philosophical approach to the discovery of deathbed secrets was often used to trace invisible forms of divine evidence in Anglo women's and Native Americans' deathbed testimonies. The relatively marginal social position of these populations heightened the empirical possibility

43. Boyle, *Origine of Formes and Qualities,* 4. According to Cotton Mather in *Wonders of the Invisible World,* Perkins's *Discourse of the Damned Art of Witchcraft* (1608) was used in the trials to develop evidence according to "just and sufficient proofs." Among the criteria set forth by Perkins was the "testimony of two witnesses of good and honest report" ("Enchantments Encounter'd," 15, in *Wonders of the Invisible World*).

44. Cotton Mather to John Richards, May 31, 1692, in Kenneth Silverman, ed., *Selected Letters of Cotton Mather* (Baton Rouge, La., 1971), 37.

encoded on their dying souls in the hour before death. Salem's girls inhabited parallel social positions of social exclusion. But, rather than displaying benevolent forms of secret prayer, the afflicted girls revealed that secrecy could also wield horrific and terrifying truths. While private prayer might affirm pious exemplarity, the misguided use of secret space also might entice young girls to travel through the nightly air with the devil.

As disturbing and disrupting as their presence within the community appears, the testimonies of the afflicted girls provided the witnessing audience with an ethnographic archive from which to probe more deeply into the world of the unseen. In the witchcraft trials, the embodied performances of Salem's afflicted girls became sites for the production of disembodied specters as convincing empirical proof of demonic agency. Alongside testimonies of spectral appearances, an *invisible witness* appears in the trials as the performances of little girls signify the presence of an absent specter. This invisible witness becomes the key form of evidence introduced within the trials: it legitimates spectral appearances by supplying magistrates with a method for mapping and discerning the invisible world. However, Salem ultimately fails as an empirical experiment.

The Court of Oyer and Terminer admitted spectral evidence into the trials only from June to October 1692. The two legal statutes available to Salem's judges were the Massachusetts "Body of Liberties" (1641) and the English statute of James I from 1604 (1 Jac. I, c. 12). The colonial law was in fact the harsher of the two, evoking Romano-canonical standards to proclaim, "If any man or woeman be a witch, (that is hath or consulteth with a familiar spirit,) They shall be put to death." James I stated more moderately that the "practise use or exercise of Witchcraft" was strictly prohibited and at times would result in the "death penalty . . . where a victim was killed, destroyed, wasted, consumed, pined, or lamed." The 1684 revocation of the colonial charter presented a question of which doctrine should be employed in court. It appears that Bridget Bishop in 1692 was the first to be tried under the reign of spectral evidence and the only one tried under the James I's witchcraft laws.[45]

45. Quoted from Payne, "Defending against the Indefensible," Essex Institute, *Historical Collections,* CXXIX (1993), 62–83, esp. 63.

Weisman *(Witchcraft, Magic, and Religion)* argues that the collapse of the colonial charter in 1684 caused the colonists to revert to English law, specifically James I's statute of 1604. Payne provides more convincing evidence that they shifted back to the Massachusetts "Body of Liberties" (1641) after Bridget Bishop's trial on June 2, 1692.

James's statute did not supply evidentiary standards. This lack of clear, evidentiary criteria coincided with the decision to admit specters into the courtroom. Bishop's specters proved that her "victims" were "pined and lamed," leading to her execution and opening the legal door "for the devils to obtain from the courts in the invisible world a license to proceed unto most hideous desolations." The shifting legal terrain and political turmoil of Salem in 1692 acted as a catalyst for the legal incorporation of a scientific mode of inquiry. Indictments stemmed from a natural philosophical practice of eyewitnesses' testifying to what the new science would admit as evidence of the invisible world. After Bridget Bishop's trial, the court reverted to the Romano-canon language of the 1641 "Body of Liberties." Evidence proceeded from the new methods of experimental philosophy, but punishment for the secret crime replaced the safe scientific space with an atavistic display of Old Testament law: "Thou shalt not suffre a witche to live" (Exod. 22:18, Geneva Bible).[46]

First to be tried under the incorporation of spectral evidence at Salem, Bridget Bishop exercises influence upon the afflicted through her eyes and touch. Attempting to narrate and defend the use of spectral evidence within Bishop's trial, Mather summarizes the courtroom interactions through the transference of sensory impressions from accused to accusers.

> If she did but cast her Eyes on them, they were presently struck down; and this in such a manner as there could be no Collusion in the Business. But upon the Touch of her Hand upon them, when they lay in their Swoons, they would immediately Revive; and not upon the Touch of any ones else. Moreover, Upon some Special Actions of her Body, as the shaking of her Head, or the turning of her Eyes, they presently and painfully fell into the like postures.

In his framing remarks, Mather refers to his own as well as Judge Matthew Hale's knowledge of Baxter's *Of the Nature of Spirits.* Hale was the judge in an infamous witchcraft case in Suffolk County, England, in 1664, in which two women, Rose Cullender and Amy Duny, were convicted and hanged for bewitching children. Although modern scholars often regard the case as an irrational deviation from one of the leading proponents of English common law, several seventeenth-century expert witnesses in fact testi-

46. On Bishop's specters: this is Cotton Mather's warning of one of the possible consequences of admitting spectral evidence into the courts. Mather to John Richards, May 31, 1692, in Silverman, ed., *Selected Letters of Mather,* 36.

fied to the bewitchment, including Sir Thomas Browne, who provided a medical opinion. But the deciding evidence presented in the case occurred when the afflicted children responded to the touch of the accused witches. Mather refers to the Hale case in order to defend the standards of evidence admitted in the Salem witchcraft trials by comparison to England. He reminds his reader that it was not too long ago that *"like things"* were happening in Europe. "The *Witchcrafts* here most exactly resemble the *Witchcrafts* there." The framing reference to Baxter's text also establishes *Wonders of the Invisible World* as an attempt to prove that New Engand witches were not condemned without "full Evidence," according to the European standards that we see in Baxter's text and in the Hale case.[47]

As the first narrated examination in Mather's case history of the trials, Bridget Bishop displays a sensory encounter with her four accusers: Elizabeth Hubbard, Ann Putnam, Abigail Williams, and Mercy Lewis. The encounter accorded with the methods derived from Boyle's corpuscular philosophy, which offered experimental physiology as a supplement to natural philosophy. Boyle's treatise on chemistry described how external "Matter" became "manifest to sense" through "Local Motion," rendering natural phenomena "obvious as well to the Eye as the Understanding." In Bishop's trial, the magistrates externalize the body's sensory organs, producing a version of corpuscular interaction manipulated through the disembodied spirit of Baxter's *Of the Nature of Spirits.* Mather narrates this disembodied spiritual presence as "the *Shape* of the Prisoner" that "did oftentimes very grievously Pinch them, Choak them, Bite them, and Afflict them; urging them to write their Names in a *Book,* which the said Spectre called, *Ours.*" Mather was not present at the trials. His account comes from the trial transcripts, which record the theatrical, sensory interactions between Bishop and her accusers but make no claims about a visually discernible spectral presence within the trials. Mather reinterprets the legal data recorded in the trials according to the evidentiary standards of contemporary natural philosophers invested in defending the ontological status of the demonic world against supposed atheistical attacks.[48]

In addition to the embodied and affective testimonies of the afflicted girls, Bridget Bishop's condemnation rests upon the secondary testimo-

47. Mather, *Wonders of the Invisible World,* 12, 55, 66, 83. The history of the Hale case is discussed at greater length in Sharpe, *Instruments of Darkness,* 223–225.

48. Boyle, *Origine of Formes and Qualities,* 4, a2v; Mather, *Wonders of the Invisible World,* 66.

nies of credible witnesses: Richard Coman, Samuel Shattock, and William Stacy. Each testifier reports that he or she "perfectly saw the shape" or "Apparition of this *Bishop*." This testimony met Lockean standards for sound empirical practice, which depended on "the number of witnesses, their integrity, their skill at presenting evidence and its agreement with the circumstances, and last, the presence or absence of contrary testimony." These scientific standards also legitimated the indictment of Bridget Bishop according to the English common law requirement for two witnesses and William Perkins's rule that the "Testimony of two witnesses, of good and honest Report" must confirm the conviction. While *Wonders of the Invisible World* was written (perhaps reluctantly) in response to Governor Phips's plea that Mather defend the witchcraft trials to London's colonial authorities, the account of evidence within the trials accorded with scientific and legal standards of the day. The problem was, not the "dark things" of "America" in relation to an enlightened Europe, but rather a particular legalistic use of scientific evidence to institute state control over "unseen regions" through the sacrificial eradication of demonic spirits.[49]

Bridget Bishop's trial established the evidentiary standards as a theater of sensory encounter, replayed in the trials that followed. Three weeks after Bishop's execution, Susannah Martin enters the courtroom, the "Cast of her *Eye* striking the afflicted People to the Ground." This is in fact Mather's emendation to the trial transcripts, which state simply: "As soon as she came in many had fits." Mather revises this line to include a sensory transaction between the accused and her adolescent accusers (Abigail Williams, Elizabeth Hubbard, and Mercy Lewis). The explicit "cast of the eye" links the motion recorded in the trials to the discourse of sensible qualities and disembodied spirits found in the writings of Cotton Mather's Royal Society colleagues. Mather derives this dramatic entrée from a later description in the trial transcripts in which the afflicted girls buttress the credible testimonies of Joseph Ring and Thomas Putnam by falling into fits as if playing the part of a Greek chorus.

49. Mather, *Wonders of the Invisible World,* 15, 67, 69, and "Enchantments Encounter'd," 15; Shapiro, *"Beyond Reasonable Doubt."* Kenneth Silverman suggests that Cotton Mather felt deeply conflicted about Governor Phips's request that he defend the trials. Silverman bases this conjecture on Mather's letter to the court clerk, Stephen Sewall, which tentatively requests the trial transcripts "as if he were trying to convince himself that he was doing the right thing." Silverman, ed., *Selected Letters of Mather,* 33.

> Upon the glance of hir eies they ware strucken down or allmost choak and upon the motion of hir finger we took notes they ware afflected and if she did but clench hir hands or hold hir head aside the afflected parsons . . . ware greviously tortored in like maner and several times sence we have seen them tortored complning of susannah Martin for hurting them.[50]

Significantly, this action takes place while the credible testifiers are recounting sightings of Martin as a specter. As if describing the title page to *Kingdom of Darkness,* Joseph Ring tells of "divers strang appearances w'ich did fors him away with them into unknown places w'r he saw meettings and featings and dancing and many strange sights." In a coordinated response to Martin's bodily sensory conduits—her eyes, finger, hands, and head—the girls testify to the presence of an invisible witness within the courtroom, the haunting double of the spectral form. Invisible witnesses manifest the "Unaccountable" epistemological "Trick which the Witches have to render themselves, and their Tools *Invisible.*"[51]

This invisible presence within the courtroom marks a continuation rather than subversion of magisterial and ministerial authority within the trials. The manifest phenomenon of the invisible witness produced a new empiricism of "unseen regions" that cinched a scientific effort to narrate spectral science. The invisible witness provided a lens into the "hidden things" beneath the surface of visible evidence. Rather than disrupt the Puritan science of visible signs, an empiricism of the invisible strengthened it. The Court of Oyer and Terminer acquired strength and momentum through the afflicted performances, not in a frantic attempt to regain control over a chaotic colony, but rather through the attestation and renewal of the promise of truth beneath the elusive visual image of the specter. Through their performances, the afflicted girls conjoined the known and unknown, the specter and the elusive "*Invisibilizing* of the Grossest Bodies." Invisible witnesses eerily shadow the tantalizing possibility of semiotic perfection, a goal central to the Puritan project of testimony in which visible signs would make manifest the invisible realm.[52]

In his narrative of Elizabeth How's trial, Mather similarly highlights the

50. Mather, *Wonders of the Invisible World,* 71; Boyer and Nissenbaum, eds., *Salem Witchcraft Papers,* II, 550, 574.

51. Boyer and Nissenbaum, eds., *Salem Witchcraft Papers,* II, 565, 574; Mather, *Wonders of the Invisible World,* 44, 58.

52. Mather, *Wonders of the Invisible World,* 44.

"sensible and evident" presence of witchcraft through the grievous torture inflicted by the presence of an invisible witness. He describes the responses of "the Sufferers" who "were not able to bear her *Look*, as likewise, that in their greatest Swoons, they distinguished her *Touch* from other Peoples." The corporeal effect of How's *"Look"* and *"Touch"* on the afflicted provides empirical proof of a demonic presence, invisible yet traceable through the sensible qualities produced through the "unlawful" use of *"Plastic Spirit* of the World."[53]

Mather invokes this Neoplatonic term *"Plastic Spirit"* in *Wonders of the Invisible World* as a way of linking the spectral science implemented in the trials to the natural philosophical endeavor to intuit grace upon human souls. His particular use of this term aids his defense of an empirical method for tracking invisible evidence through natural philosophical law. Mather's persistent endorsement of spectral evidence stems from his philosophical commitment to the Platonic understanding of visible and invisible domains. In a 1695 exchange, Mather wrote to Robert Calef, "I have not yett altered my Opinion, that there is a Plastic Spirit permeating of the World, which very powerfully operates upon the corporeal parts of it." Calef replied: "A plastic spirit. What foreign word is that?"[54]

This exchange exemplifies the philosophical disagreement underlying critical and apologist historiographers of Salem. In a 1692 letter, New England colonist, mathematician, and Royal Society member Thomas Brattle remarked that while "some men" call "this Salem philosophy . . . the new philosophy," he thinks "it rather deserves the name of Salem superstition and sorcery." By demonstrating his awareness that a theory of experimental philosophy existed in Salem, Brattle concurred with Webster's sense of the superstition of Neoplatonic philosophers. Far from exhibiting rational and irrational accounts of spectral evidence, Salem's historiographers upheld competing perspectives on the empirical admissibility of evidentiary standards. Mather's focus on How's *"Look"* and *"Touch"* would meet harsh criticism in the wake of Salem by Calef, Brattle, and even Increase Mather. Yet Cotton presents this sensory evidence as certain proof of the accompanying "Testimony of People to whom the shape of this *How*, gave trouble nine or ten years ago."[55]

53. Ibid., 44, 77.

54. Quoted in Weisman, *Witchcraft, Magic, and Religion*, 33.

55. Thomas Brattle, Letter, in Burr, ed., *Narratives of the Witchcraft Cases*, 171–172; Mather, *Wonders of the Invisible World*, 77.

In *Wonders of the Invisible World,* Mather frames touch as a sensible quality that reveals telling evidence of a demonic presence in spite of the radical displacement of the afflicted's sensory impressions: "Tho' the Afflicted were utterly deprived of all sense in their *F*its, yet upon the *Touch* of the Accused, they would so screech out, and fly up, as not upon any other persons." This mode of manipulation through touch produces a category for the preternatural within Boyle's corpuscular philosophy of sensory interaction as a mark of human specificity. Boyle explains that there are

> men in the world, whose organs of Sense are contriv'd in such differing wayes, that one Sensory is fitted to receive impressions from some, and another from sorts of external Objects or Bodies without them, (whether these act as entire Bodies, or by *Emission* of their Corpuscles, or by *propagating* some motion to the Sensory).

That humans receive sensory impressions differently produces diversity "according to the nature of the Sensories or other Bodies they work upon."[56]

Mather extends this logic to the sensory impressions of Salem's afflicted girls. Their possessed state disrupts the normal process of sensory impressions so that in their "*F*its" they are almost "deprived of all sense." This sensory deprivation causes a collapse of the balanced process through which the external organs (eyes, ears, nose) receive heat, color, sound, and odor. The afflicted exist in a particular sensory condition, requiring the discerning expertise of ministers, magistrates, and natural philosophers. Following this expertise, the Salem trials interpreted preternatural effects through this notion of spirit and body penetrability. Little girls represented convenient objects upon which to enact this preternatural inquiry. Their fragile "feminine sex" made them more susceptible to demonic intrusion; their senses could be differentiated from others more readily as exhibiting a "passive" and "penetrable" quality. The afflicted girls performed as wax templates of primary sensations, commensurate with Locke's description of "how *Bodies* produce *Ideas* in us, and that is manifestly *by impulse,* the only way which we can conceive Bodies operate in." The "idea" that the magistrates sought through the records of sensory impressions in the afflicted girls was the demonic "secret" of "unseen regions" masked behind

56. Mather, *Wonders of the Invisible World,* 59; Boyle, *Origine of Formes and Qualities,* 66.

the "frightful Apparitions of Evil Spirits." The invisible manipulation displayed within the courthouse participated in this unmasking by rendering the invisible a palpable, sensible presence.[57]

Bad Science

When the Court of Oyer and Terminer heard the case of Martha Carrier on August 2, 1692, Salem's theater of the senses had become a formalized script. Mather writes that Carrier "was indicted . . . according to the Form usual in such Cases," which consisted of a not guilty plea and then the telling effects of Carrier's sensory communications. "The Look of *Carrier* then laid the Afflicted People for dead; and her Touch, if her Eye at the same time were off them, raised them again." Mather arrives at this summary from the coordinated concert of affliction performed by Mary Warren, Ann Putnam, Susanna Sheldon, Mary Walcott, Elizabeth Hubbard, and Mercy Lewis and recorded in the trials:

> Susan: Sheldon, who hurts you?
> Goody Carrier, she bites me, pinches me, and tells me she would cut my throat, if I did not sign her book
> Mary Walcot said she afflicted her and brought the book to her.
> What do you say to this you are charged with?
> I have not done it.
> Sus: Sheldon cried she looks upon the black man.
> Ann Putnam complained of a pin stuck in her.
> What black man is that?
> I know none.
> Ann Putnam testifyed there was.
> Mary Warrin cryed out she was prickt.
> What black man did you see?
> I saw no black man but your own presence.
> Can you look upon these and not knock them down?

57. Mather, *Wonders of the Invisible World,* 59, 64; John Locke, *An Essay concerning Human Understanding,* ed. Peter H. Nidditch (New York, 1979), 135–136. Locke draws heavily on Boyle's *Origine of Formes and Qualities* here, extending Boyle's notion of the mutability of essential essences to a theory of language. (I discuss the overlap between these two philosophies at greater length in Chapter 6.) See also David A. Givner, "Scientific Preconceptions in Locke's Philosophy of Language," *Journal of the History of Ideas,* XXIII (1962), 350–354.

> They will dissemble if I look upon them.
> You see you look upon them and they fall down
> It is false the Devil is a liar.[58]

The girls in this scene are between the ages of twelve and nineteen. Their testimony takes a very different form from that of the credible witnesses cited later in the trial. In place of the integrity, skill, and purpose of the reporter, characteristic of testimonial evaluation in the emerging legal tradition, the girls display bodies so penetrated through preternatural manipulation as to become publicly visible. The trial becomes a concert of sensible qualities manifested in what was by this point within the trials a highly conventionalized form, recognizable even to the defendant who knows that the girls "will dissemble if I look upon them." This statement reflects a theory of causality central to proving the specter's presence through the material effects of an invisible witness, manipulating the scene within the trial. Carrier and the magistrates establish a pattern of correspondence between the accused witches' "Look" and the dissembling performance of the afflicted adolescent girls. This form of material causality becomes the focus of Thomas Brattle's critique of the theatrical conventions of sensory encounter in the trials.

> The Justices order the apprehended to look upon the said children, which accordingly they do; and at the time of that look, (I dare not say by that look, as the Salem Gentlemen do) the afflicted are cast into a fitt. The apprehended are then blinded, and ordered to touch the afflicted; and at that touch, tho' not by the touch, (as above) the afflicted ordinarily do come out of their fitts.[59]

Brattle repeatedly emphasizes his disapproval of the form of causal effect used as evidence by Salem's judges. Whereas the magistrates attribute the fits to the "Look" and "Touch" of the accused witch, Brattle suggests that the only discernible evidence from this interactive scene is owing to coincidence. At the time of (rather than *by*) the "Look" and "Touch," the girls fell into fits. Brattle refutes the relationship between the two sensory phenomena that the trials take as their essential evidentiary premise. He further explains that the magistrates cultivated this theory of sensory interaction from a misuse of Cartesian philosophy.

58. Mather, *Wonders of the Invisible World,* 80; Boyer and Nissenbaum, eds., *Salem Witchcraft Papers,* I, 185.

59. Brattle, Letter, in Burr, ed., *Narratives of the Witchcraft Cases,* 170–171.

> The Salem Justices, at least some of them, do assert, that the cure of the afflicted persons is a natural effect of this touch; and they are so well instructed in the Cartesian philosophy, and in the doctrine of *effluvia*, that they undertake to give a demonstration how this touch does cure the afflicted persons; and the account they give of it is this; that by this touch, the venemous and malignant particles, that were ejected from the eye, do, by this means, return to the body whence they came, and so leave the afflicted persons pure and whole. I must confesse to you, that I am no small admirer of the Cartesian philosophy; but yet I have not so learned it. Certainly this is a strain that it will by no means allow of.[60]

Through his critique of method, Brattle exposes a discourse of philosophical expertise operative in courtroom scenes that traces the presence of an invisible witness as evidence of witchcraft. The "doctrine of *effluvia*" refers to a component of corpuscular philosophy in which the corruption of bodies is made by outward agents. Brattle admits to finding no philosophical basis for this use of corpuscular philosophy in Cartesian philosophy. This is not surprising, given that the magistrates rooted their method of tracing material motion in the natural philosophical tradition that developed in explicit response to Descartes as Boyle, Baxter, More, and Glanvill employed empirical data and sensory techniques in order to prove not only the existence but also the mappable, manifest presence of the demonic. Through the repetition of Carrier's "Look" and "Touch," which left the afflicted girls for dead, the specter of demonic truth emerged from the interstices of possessed performance. The trial becomes a quest for "Satan in signs," an attempt to assemble a science of demonic presence from the "book" and "black man" that appear as invisible forms in the trial through the girls' entranced state.[61]

The most notorious image of the specter within the trials appears "in the Shape of a black Man," its elusive status directly opposing the notion of a visible saint. Black signifies the darkness of the devil's work and the specter of hypocrisy condensed within the figure of the demonic. The irreducible condition of Reformation theology haunts the genre of testimony with its own epistemological limits. Carrier announces the black man's ontological status through her statement, "It is false the Devil is a liar." The devil displays the continuously transforming essence described in Boyle's cor-

60. Ibid.

61. Ibid.; Boyle, *Origine of Formes and Qualities*, 58.

puscular philosophy in supernatural form. Because the devil had no real, verifiable, or tangible essence, his appearance in 1692 called into question the very notion that evidence of grace could be deduced from human souls or that such evidence might contribute to an inductive understanding of divine secrets within the natural world. The pregnant absence of an invisible witness in the trials became a spectral correspondent to Spinoza's denial of the epistemological validity of the devil or to Locke's denial of "real" essence.[62]

Toward the end of the trial's opening scene, the afflicted girls conclude their sensory performance by proclaiming, "There is the black man wispering in [Martha Carrier's] ear." Like the Puritans, we desire to know what this specter symbolizes. What is the essence that the "black man" names? The Salem transcripts come closest to uncovering an essence beneath the name in the trial of George Burroughs, whose hearing and conviction occurred on the same day as Martha Carrier's. In Burroughs's trial, the testifiers describe a system of signs from the unseen region of the invisible world that begins to congeal into a coherent narrative form. Burroughs practiced witchcraft on the "two Successive Wives" that he kept "in a strange kind of Slavery," asking them to "Write, Sign, Seal, and Swear a Covenant, never to reveal any of his Secrets."[63]

From this history, we learn that the black man is the withholder of secrets who acquaints "dear neighbors" and wives with "hidden things." Enslaved under Burroughs's spell, his wives confide to neighbors their fear of "Apparitions of Evil Spirits with which their House was sometimes infested." The presence of these apparitions displays how demonic agency preys upon women within their domestic role, manipulating the *"Plastic Spirit* of the World" to infiltrate the core of pious and covenantal hope within a post–Halfway Covenant world. Burroughs, it turns out, is the visible manifestation of the black man present in the other trials who "put on his *Invisibility"* as a *"Fascinating Mist,"* which gives him the preternatural ability to "render many other things utterly *Invisible."* This is the most terrifying essence of the specter for Cotton Mather: the "Trick which the

62. Glanvill, *Sadducismus Triumphatus,* in Kors and Peters, eds., *Witchcraft in Europe,* 377. Elizabeth Reis discusses black as a marker of witchcraft as "the underside of covenant theology." Reis, "Witches, Sinners, and the Underside of Covenant Theology," Essex Institute, *Historical Collections,* CXXIX (1993), 103–118.

63. Boyer and Nissenbaum,, eds., *Salem Witchcraft Papers,* I, 169, 186; Mather, *Wonders of the Invisible World,* 64.

witches have to render . . . their Tools Invisible" and to "invisibliz[e] . . . the Grossest Bodies." It is frightening because it mimics precisely the failure rooted in the perpetually limiting condition of hypocrisy and immanent in the genre of testimony. Invisibility suggests that the specter has no real essence. If this was the case, grace could never become manifest, at least as a traceable and categorical divine phenomenon within the natural world. Partaking of the same semiotic configuration as spectral haunting, albeit manifest in a radically different way, grace created a world of divine shadows, the Lockean idea that conveys divine form but never its essence.[64]

In addition to signifying epistemological limits, the specter's blackness also seemed to carry with it a radical connotation (most persuasively interpreted by Mary Beth Norton as an inwardly directed communal index of the violent presence of Native Americans on the New England frontier). Indeed, a telling conflation of the black man and the Indian appears in Mather's retelling of George Burroughs's trial transcript. Thomas Greenslit, a neighbor of Burroughs from Casco Bay, recalls that during King Philip's War "he Saw Mr. George Burroughs lift and hold Out a gunn of Six foot barrel or thereabouts" and shoot "a full barrel of Malasses." Mather cites Burroughs's rebuttal that "an Indian was there," which Mather immediately interprets as a spectral presence. He explains that none of the witnesses "ever saw any such *Indian;* but they supposed the Black Man, (as the Witches call the Devil; and they generally say he resembles an *Indian*)."[65]

As we have seen, Martha's Vineyard had come to be positioned as an experimental site by the early eighteenth century. Just as Eliot found the Sandwich Islands of Plymouth Colony a safe enclave for witnessing more authentic signs of grace in a more remote missionary frontier, Daniel Gookin and Thomas Mayhew would also view the islands of Nantucket, Martha's Vineyard, and Cape Cod as reclusive sites on the eastern seaboard of New England. The display of grace among the island Indians functioned as proof within the missionary writings of these authors that not all Indians were the hopeless, fallen savages taking up arms against the English. Docile Christian Indians still existed, and their evidence of grace within these

64. Mather to John Richards, May 31, 1692, in Silverman, ed., *Select Letters of Mather,* 37; Mather, *Wonders of the Invisible World,* 44, 64, 65.

65. Mather, *Wonders of the Invisible World,* 63; Boyer and Nissenbaum, eds., *Salem Witchcraft Papers,* 160. For another interpretation of spectral blackness and Native Amerians, see Alfred A. Cave, "Indian Shamans and English Witches in Seventeenth-Century New England," Essex Institute, *Historical Collections,* CXXVII (1992), 239–254.

island repositories stood as evidence of their native sainthood and undying fidelity to the English. King Philip's War had given a political as well as epistemological meaning to the Native American hypocrite.[66]

The marked shift in colonial perceptions of the Native American in the wake of King Philip's War was largely what made the island home of docile Indians still displaying benevolent signs of grace so central to the few missionary writers still interested in justifying the Native American cause. This configuration of specific, contained, geographical sites, bordered on all sides by the Atlantic Ocean, removed the experimental practice of studying Indian souls from the turmoil of the war. However, the benevolent Praying Indians featured in tracts by Eliot, Gookin, and Mayhew could not erase the lingering and ominous presence of natives on the opposite side of the New England frontier, bordered by vast expanses of a howling wilderness rather than an ocean that symbolized baptismal rebirth and the journey home to England. The natives on this frontier embodied hypocritical politics as well as epistemology, both of which continually threatened the colony with violence. From the epistemological perspective of a viable soul science, these two disparate and opposing representations of the native frontier allegorized the kernel of Reformation optimism and its concomitant shadow of doubt. If the dying natives of Martha's Vineyard displayed their souls as sites of empirical plentitude, the natives who killed and maimed colonists on the rural Maine frontier must resemble their soulless counterpart, a people fallen beyond all hope of redemption whose presence would inevitably thwart the millennial goals of the missionary movement.

In the witchcraft trials, the black man looked like an Indian because the unseen regions of the demonic world resided "in places where the gospel is not preached," where "Daemons . . . infest the Gentiles and Heathens of old." The Maine frontier proved fertile ground for the residence of demons. The conflation of the black man and the Indian reflected the political as well as epistemological problem of a "heathen" presence in relation to the Puritan science of visible signs. If the Indian soul emerged as a space of indigenous, natural, and authentically American epistemological potential

66. According to Daniel Gookin, the evidence of grace displayed on Martha's Vineyard and Nantucket showed "their love to the English" and that "these Indians lived very soberly, and quietly, and industriously, and were all unarmed; neither could any of them be charged with any unfaithfulness to the English interest." "An Historical Account of the Doings and Sufferings of the Christian Indians in New England . . . ," in *Archaeologia Americana:* American Antiquarian Society, *Transactions and Collections,* II (1836), 486, 495.

by the late-seventeenth century, it carried with it the ever-present danger that such a science could fail. The Salem witchcraft crisis registered the shadow of doubt exposed by the specter of violence looming on the New England frontier, where, rather than repositories of special grace, Native Americans came to represent an irreversible descent to an even more debased heathen savagery.[67]

Mather concedes the problem that the heathenish and hypocritical specter posed for a science of divinity as that which "now Grievously Vexes us." He laments that there is nothing more "Unaccountabl[e]" than the witch's "Tools *Invisible*." This particular form of invisibility threatened to dissolve the natural philosophical system of visible signs. For Mather, witchcraft undermines the Neoplatonic approach to divine truth because witches have the "skill of Applying the *Plastic Spirit* of the World, unto some unlawful purposes, by means of a Confederacy with *Evil Spirits*." But, rather than achieve itself in epistemological crisis, the science of the soul emerges from the trials in a revised form. Mather uses the figure of the heathen Indian to contain the philosophical uncertainty that the specter encodes and makes manifest within the Puritan semiotic system. Ultimately, the conflation of the haunting black spectral presence and the Indian man receives the label of the "curiosity" in Mather's narrative of the trials.[68]

Mather compiles a list of curiosities in an attempt to catalog invisible malignity as he cataloged plants for the Royal Society. His first curiosity acknowledges the status of the devil as "an Impious and Impudent *imitation* of Divine Things." The heathen "Indian nation" represents the visible form of the devil's imitation of divinity. Mather situates his knowledge of this curious discovery alongside José de Acosta's *Natural and Moral History of the Indies,* which describes the overlapping presence of the devil in Mexico, where *"Six* Nations of Indians" have *"Discoursed* in Secret." Acosta's *History* reveals this secret discourse to Mather in the form of a desert *"Tabernacle* for their false god . . . placed [alongside] the *Ark* upon an *Alter*." In Mather's retelling, the "Devil in one Night, horribly kill'd" the Mexican Indians, a description that evokes the opening scene from Mary Rowlandson's narrative. Continuing to "imitate" the *"Church of the Old Testament,"* the devil moved to New England, persisting with this practice of "Bloody *Imitations*." Mather concludes this first curiosity with an expression of his insatiable desire to know how the devil works this imitative magic: "What is their

67. R. B., *Kingdom of Darkness,* preface.

68. Mather, *Wonders of the Invisible World,* 44.

Transportation thro' the *Air?*" "What is their Travelling in *Spirit,* while their Body is cast into a Trance?" "What is their Covering of themselves and their Instruments with *Invisibility?*" In his defense of the most infamous witch-hunt of his time, Mather reiterates his own empirical desire to see past the dark glass limiting human perception in order to transcend the double bind of human fallenness, described by Calvin as a longing to know the impossible. This series of questions suppresses the tragic irony that this very mode of inquiry produced the violent outcome that he is now called upon to defend.[69]

69. Ibid., 43, 44.

{6} Revivals

Evangelical Enlightenment

The dangers of misapplied science, fully realized in the Salem witchcraft trials and the events' larger context in Royal Society debates, put tremendous pressure on the science of the soul. By the late-seventeenth century, Harvard-trained ministers and Royal Society natural philosophers continued to advance their practices, though with renewed caution, aware both of mechanical philosophy's limitations and the parameters surrounding empirical inquiry—especially interdictions against studying preternatural phenomena. Yet neither New England ministers nor London natural philosophers abandoned efforts to apply empirical methods to advance knowledge of God. Rather, such theologians as Samuel Clarke recuperated the goal of knowing the essence of God in the Boyle lectures that he delivered in the early eighteenth century while also correcting the shortcomings of previous generations by making revelation subordinate to reason.

Clarke's contributions to both theology and philosophy reflect a larger pattern. As Protestant theologians continued their quest to move beyond the Calvinist crisis of limited knowledge of election, natural philosophers pursued the advancement of natural knowledge beyond the boundaries of mechanism. Clarke's *Discourse concerning the Being and Attributes of God* (1711) strongly refutes the materialism of Descartes and Spinoza by providing "certainty of the existence of God." Clarke identifies the problem of causality as his central evidence, explaining that it is not possible that "all Things have arisen out of Nothing, nor can they have depended one on another in an endless Succession." Indeed, according to Clarke, "there is Something in the Universe." The only way to advance knowledge of this something is by collecting more evidence of its "essence," a term that Clarke defines in the metaphysical sense as "that by which a thing is what it is." Interestingly, Clarke makes human testimony central to this quest for divine essence, placing the "testimony of [God's] followers at the center of the human rational capacity to judge whether revelation resembled certain evidence of God."[1]

What makes testimony—in particular the testimony of God's followers—capable of producing such certain evidence? Testifiers, according to Clarke, have the advantage of "Revelation," which, when paired with "the Judgment of right and sober Reason, appears even of it self highly credible and probable; and abundantly recommends it self in its native Simplicity." Clarke identifies the simplicity of human testimony as his most convincing evidence of how to identify God's essence within the world.[2]

What use did a natural philosopher have for the testimony of faith in the early decades of the eighteenth century? By this time, the testimony of faith had both persisted and transformed, caught in the currents of a transatlantic Enlightenment, the theological transformations that precipitated revivalism, and the geographic specificity of New England's entrenched Calvinist roots. I propose that the testimonies uttered during the revivals preceding Northampton's had absorbed the effects of Enlightenment thought. What has long been perceived as a split between reason and revelation, between Enlightenment elitism and an increasingly popular religious landscape of evangelicalism, obfuscates more deeply ingrained continuities. The science of the soul that bound empiricism to Calvinism and experimental religion to experimental science occurred within a finite historical period; the much-contested and elusive genre ebbed and flowed according to the tenability of the evidence of grace produced. The historical significance of the Great Awakening as a transitional event that marks a new phase of American religion has long been established. This chapter adds an account of how the particular New England history of soul science as well as its correspondent transatlantic currents of Enlightenment thought facilitated this transformation.[3]

The science of the soul answers an explicit philosophical concern for Clarke, which he revisited only four years later in his correspondence

1. Samuel Clarke, "The Evidences of Natural and Revealed Religion," in *A Discourse concerning the Being and Attributes of God* . . . (London, 1711), A2, 12–13, 21 (irregular pagination).

2. Ibid.

3. This literature is, of course, vast. Some of the better-known arguments include Sacvan Bercovitch, *The Rites of Assent: Transformations in the Symbolic Construction of America* (New York, 1993); Alan Heimert, *Religion and the American Mind: From the Great Awakening to the Revolution* (Cambridge, 1966); Susan Juster, *Doomsayers: Anglo-American Prophecy in the Age of Revolution* (Philadelphia, 2003); Ann Taves, *Fits, Trances, and Visions: Experiencing Religion and Explaining Experience from Wesley to James* (Princeton, N.J., 1999). Notable departures from this traditional narrative include Frank Lambert, *Inventing the "Great Awakening"* (Princeton, N.J., 1999).

with Leibniz over a similar question of how to accrue more knowledge of God's essence. In his fourth paper, Clarke references the experimental approach of German scientist Otto von Guericke's experiments with the vacuum pump and Italian physicist and mathematician Evangelista Torricelli's work with mercury barometers, both of which, Clarke feels, offer conclusive proof of the existence of vacuums. Leibniz refutes this point on the grounds that a universe willed by God is infinite and fully filled. Minister and philosopher debate theories of space as a "sensorium of God" with the goal of refuting the writings of Sir Isaac Newton, which both contend "hurt" the "foundations of natural religion." As the conversation evolves, Clarke defends Locke and Boyle's discovery of the vacuum, explaining that only a gross materialism threatens religion and that by contrast these materialist contributions may in fact be used in ascertaining new knowledge of God. Leibniz remains skeptical of this use of Lockean methods throughout his correspondence and doubtful that the invisible soul can accurately perceive or record what happens in the body. Clarke responds that the soul is a microcosm of the divine, capable of perceiving God as well as the "images to which it is present." The proper use of an experimental approach and a materialist method is that both may be directed toward the philosophical goal of discovering God's essence. Despite the pressures that the science of the soul underwent at the turn of the eighteenth century, Clarke's declaration positions the testimony of faith as a mainstream feature of emergent currents of natural philosophy and a repository of revealed truths that could be directed toward the production of new divine knowledge.[4]

On the ministerial front during these decades, a transatlantic wave of moderation gradually replaced previous forms of Calvinist piety. Clarke, Benjamin Colman, and John Tillotson were among the influential ministers that consciously sought to incorporate natural philosophy more fully than ever before. They espoused a theology that accepted the rational order of a Newtonian universe and exhibited greater confidence in the human capacity to detect that order. Colman gathered a group of Boston luminaries such as Increase Mather, Joseph Sewall, Thomas Prince, and John Webb to preach *A Course of Sermons on Early Piety* (1721). The ministers expressed the conventional lament that New England had turned from religious to worldly interests—even as their sermons use a Lockean frame to chart the impressions of God on the souls of young converts and the experimental philosophy of Robert Boyle to record the "evidences for heaven" conveyed

4. H. G. Alexander, ed., *The Leibniz-Clarke Correspondence* (Manchester, 1998), 22.

in an individual's "dying hour." Enlightenment rationalism blended more seamlessly with Calvinism than in years past, restructuring the anxiety produced by the unknowable soul into increased confidence in the rational order of the universe and the human place within that order. The status of the soul became increasingly intelligible through the display of human affections, as emergent currents of moral philosophy transformed understanding of the human faculties.[5]

In New England, congregational admission practices long based in a theology of epistemological uncertainty gradually gave way to the purportedly more tolerant era of church admission instigated by Solomon Stoddard. The testimony of faith began to disappear from religious communities in New England and the farther reaches of the Anglo world. This pattern suggests a departure from tradition and the gradual erosion of Puritan orthodoxy in favor of more lenient practices. Yet the disappearance of the test of faith in the late-seventeenth and early eighteenth centuries also reflected the status of divine evidence as it changed to accord with larger philosophical currents. Just a few years before Salem, Harvard-trained ministers had adjusted to a new curriculum that broadened the scope of required reading beyond Reformed theology to include Anglican and latitudinarian writers who promoted a program of free intellectual inquiry. In subsequent generations, this curricular shift gradually contributed to a transformation in the validity that ministers accorded testimonial evidence. In such sermons as *The Defects of Preachers Reproved* (1723), Solomon Stoddard argued for the resurgence of an inwardly directed piety, proposing that, because outward signs could not be fully known, the church needed to rely on the individual's inward spiritual quest.[6]

But, despite the general disappearance of the testimony of faith, the genre persisted in small communities, often scattered well beyond the Boston metropolis. In Westborough, Massachusetts, Ebenezer Parkman started recording testimonies of faith in 1727, just five years after he preached a sermon to Boston ministers on how Lockean psychology could provide a more stable index for advancing knowledge of the human soul. In the same year, Harvard University instituted the Hollis Professorship of Mathematics and Natural Philosophy along with a published "Experimental Course

5. John Corrigan, *The Prism of Piety: Catholic Congregational Clergy at the Beginning of the Enlightenment* (New York, 1991), 5–18; [Benjamin] Colman, "The Nature of Early Piety as It Respects Men," in Eight Ministers, *A Course of Sermons on Early Piety* (Boston, 1721), 32 (2d pag.).

6. Corrigan, *Prism of Piety,* 10.

in Mechanical Philosophy." Preserved in manuscript church records in the early decades of the eighteenth century, collections of testimonies resemble the traditional test of faith as it developed in the first generations of the Massachusetts Bay Colony, yet with much variation in the standards of evidence applied to the soul. Testimonies collected at Sturbridge Village, for example, or in the later decades of Parkman's Westborough church reflect a changed world in which converts were no longer consumed by the epistemological crisis of Calvinism and consequently were less caught up in the anxious search for intelligible signs of God on their souls. The evidence narrated in these accounts is a product of a more confident, self-assured world, where belief in God leads more directly to faith in the status of one's soul as worthy and eligible for church membership. The testimony of faith survived by adapting to changes in congregational structure as well as philosophical shifts in human faculty psychology and natural history.[7]

The collections of testimonies recorded in the Connecticut River valley in the 1710s and 1720s under the particular circumstances of a series of "harvests" or minirevivals formally resemble those recorded in the mid-seventeenth century Massachusetts Bay Colony. Yet closer inspection reveals a different pattern of evidence. These testifiers use nature in new ways to intuit evidence of God on their souls. They display anomalous forms of religious affect and evidence of an individual relationship to God that overturns tradition and narrowly escapes the dangers of heresy. Although these collections of oral testimonies reflect Calvinist tradition in form and content, the evidence of grace transcribed therein anticipates the theological transformations that would take place during the transatlantic revivals of the 1730s and 1740s. The revivals made new use of the practice of testifying, reinventing the genre to make it a permanent component of evangelical practice, whether in the manuscript testimonies transcribed by William McCulloch in Scotland's Cambuslang revival (1742) or in such widely circulated collections as Jonathan Edwards's *Faithful Narrative of the Surprizing Work of God in the Conversion of Many Hundred Souls* (1737).[8]

7. See Kenneth P. Minkema, "The East Windsor Conversion Relations, 1700–1725," *Connecticut Historical Society Bulletin,* LI (1986), 9–63; Minkema, ed., "The Lynn End 'Earthquake' Relations of 1727," *New England Quarterly,* LXIX (1996), 473–499; Douglas A. Winiarski, "A Question of Plain Dealing: Josiah Cotton, Native Christians, and the Quest for Security in Eighteenth-Century Plymouth County," *New England Quarterly,* LXXVII (2004), 368–413; Samuel Eliot Morison, *Three Centuries of Harvard, 1636–1936* (Cambridge, Mass., 1946), 79.

8. For an analysis of how the transatlantic revivals transformed the spiritual autobiography in the mid-eighteenth century, see D. Bruce Hindmarsh, *The Evangelical Conversion Narrative: Spiritual Autobiography in Early Modern England* (New York, 2005), 61–88, 193–226.

In September 1725, members of a small congregational community in East Windsor, Connecticut, assembled to hear the conversion testimony of twenty-four-year-old Abigail Strong. A small, intimate group of visible saints comprising her audience sat attentively while their minister, Timothy Edwards, sat with pen poised above his notebook, ready to record the evidence of God's grace upon her soul that Abigail was about to supply. Abigail stands and opens her mouth to speak, as instructed by Paul to "confess with thy mouth . . . believe in thine heart" and "be saved." Upon a first reading of this testimony, it seems that not much had changed in this small, provincial, Calvinist enclave hidden in the hills of the river valley. This was a community ostensibly protected from the not-too-distant Salem witchcraft trials that had exploded the very semiotic configuration that produced the testimony of faith, unaware that the testimony had been all but abandoned by such eminent ministers as Solomon Stoddard (who was Timothy Edwards's father-in-law nonetheless)—and forgotten by the philosophical circles of the Royal Society, including Isaac Newton, John Locke, John Webster, and others, whose perspective on the material world as a finite domain of mechanistic interactions was putting tremendous pressure on the Puritan belief that signs of grace could be applied toward the advancement of divine knowledge. For these reasons, the collection of conversion testimonies of which Abigail Strong's is a part seems to reflect an atavistic form of Calvinism. These were a people for whom the theological transformations taking place in Boston and the subsequent changes in ecclesiastical policy seemed not to matter.[9]

Yet woven into Abigail's testimony is an intricate juxtaposition of darkness and light, figured as day and night, spiritual sight and sinful blindness, the inner light of awakening and the recurrent shadows of doubt. This is a new imagistic development within the congregational testimony of faith, one that partakes of "an era concerned with enlightenment." From preaching to philosophy, light functioned as a portable image in the mid-eighteenth century, signaling the revelatory power of objects and methods from divinity to nature to human intellect. Light travels from the experiments published in Newton's *Opticks* (1704) to the small window at the

9. Quoted in Isaac Watts, *The Rational Foundation of a Christian Church, and the Terms of Christian Communion* (London, 1747), 113. The verse is from Romans 10:9–10: "That if thou shalt confess with thy mouth the Lord Jesus, and shalt believe in thine heart that God hath raised him from the dead, thou shalt be saved."

Conversion testimonies (including Abigail Strong's) in Minkema, "East Windsor Conversion Relations, 1700–1725," *Connecticut Historical Society Bulletin,* LI (1986), 9–63.

top of a Martha's Vineyard wigwam (see Chapter 4) that became a kind of camera obscura for taking a snapshot of a dying Indian's soul. The images of darkness and light that contour the vicissitudes of Abigail Strong's soul as the fleeting and elusive presence of moving grace partake in this fascination with light.[10]

Abigail's testimony captures what I call the Evangelical Enlightenment. Uttered at the brink of an evangelical revolution that would transform early American religious thought and practice through the concept of the indwelling light of Christ, the testimony also locates this indwelling light within the Lockean limits ascribed to human understanding and the conditions of uncertainty that the Enlightenment would inherit from the Reformation as a sustaining trait of experimental knowledge. The Evangelical Enlightenment appears in microcosmic form in Abigail's testimony, consisting of the theological ascendancy of the indwelling light, its proliferation in manifest forms upon the souls of lay converts, the attempt to develop inner light into certain knowledge of God, and the subsequent use of this knowledge to resolve the central philosophical dilemmas of the eighteenth century.

This Evangelical Enlightenment also builds upon modern work that has revised our traditional understanding of the Enlightenment as a temporal process, composed of universal and universally applicable laws that unfold over time. Such work on the Enlightenment recasts this philosophical frame as having spatial as well as temporal dimensions in order to investigate how knowledge circulates as ideas cross the Atlantic as well as how these ideas transform through their implementation in New World settings. We have learned how indigenous inhabitants and New World nature added to new knowledge, how this knowledge infiltrated its European purveyors individually as well as institutionally, how the hierarchies of colonial power operated within natural history, and how these hierarchies were both maintained and subverted across a disparate range of European national identities and imperial pursuits. Through its geographic specificity, I propose that the Northampton revival interacted with a broader transatlantic audience to partake in this new history of the Enlightenment. Jonathan Edwards implements a technique of observing souls in their native habitat and culling empirical data across a diverse human population

10. George M. Marsden, *Jonathan Edwards: A Life* (New Haven, Conn., 2003), 54–55. For a discussion of light, reverberating out from Newton's optical experiments to disparate cultural registers, see Mordechai Feingold, *The Newtonian Moment: Isaac Newton and the Making of Modern Culture* (New York, 2004).

parallel to that practiced by such natural philosophers as Mark Catesby or Hans Sloane. Just as these philosophers looked to indigenous or African sources to expand their own knowledge of the natural world (as in Susan Scott Parrish's revelatory example of the African Magi), Jonathan Edwards and his protégé David Brainerd discover true knowledge of divinity in the testimonies of Sarah Edwards, Phoebe Bartlett, and, eventually, the Mahican Indians on the New Jersey and Pennsylvania frontier.[11]

Listening to Abigail Strong

Many of these spiritual testifiers, including Abigail Strong in 1725, would not have been aware of the significance of her uttered words. She did not realize that, if she just managed to encase a discourse of divine light in a Lockean frame of experiential knowledge, she might advance the cause of Reformed theology within Europe's philosophical circles, which were becoming increasingly disillusioned with the prospect that the human senses might lead to knowledge of the invisible world or that this knowledge could be translated into a rational framework. Ten years after Abigail testified, Ebenezer Hunt was converted under Edwards's preaching. He replicates this imagery of divine light as a way of understanding his conversion experience. Hunt recalls a state of uncertainty that lasted about three weeks until he "began to indulge the hope that he knew something about spiritual light." A week later, he "records a prayer for spiritual light and knowledge," and, finally, two weeks after that, his soul was "filled with a sense of God's sovereign rich free grace and mercy." Hunt's library included "Locke on Education" and "Edwards Sermon on Divine Light."[12]

11. The classic example of the Enlightenment understood as a temporal process is Max Horkheimer and Theodor W. Adorno, *Dialectic of Enlightenment,* trans. John Cumming (New York, 1972). "Cultural geography" most succinctly summarizes current practices of reading the spatiality of Enlightenment culture. The term was coined by Ralph Bauer in *The Cultural Geography of Colonial American Literatures: Empire, Travel, Modernity* (Cambridge, 2003). See also Jorge Cañizares-Esguerra, *How to Write the History of the New World: Histories, Epistemologies, and Identities in the Eighteenth-Century Atlantic World* (Stanford, Calif., 2001); Susan Scott Parrish, *American Curiosity: Cultures of Natural History in the Colonial British Atlantic World* (Chapel Hill, N.C., 2006). For a similar critical framing of Enlightenment geographies as they pertain to Europe rather than the New World, see Jonathan I. Israel, *Radical Enlightenment and the Making of Modernity, 1650–1750* (New York, 2001).

12. Sylvester Judd, "Abstracts and Extracts from Deac. Ebenezer Hunts Journal," Sylvester Judd MSS (vol. I, Massachusetts Series), 27–28, Forbes Library, Northampton, Mass. This is Judd's synopsis of Hunt's description of his spiritual experience. I owe much thanks to Kenneth Minkema for this transcription and reference.

Abigail Strong's and Ebenezer Hunt's conversions reveal striking continuity over a ten-year period of lay converts who understood their spiritual experience through an image of divine light. This is the same light that features in Jonathan Edwards's famous sermon *Divine and Supernatural Light* (1734). This suggests that in 1725 at least one possible member of the audience listening to Strong's testimony would have been aware of the larger philosophical context surrounding her words. Jonathan Edwards was the son of Timothy. In 1725, Jonathan Edwards was a tutor at Yale and returning home in the summer to practice preaching. Although there is no way of knowing whether he was present at Strong's testimony per se, he was certainly aware of his father's commitment to the testimony of faith as a means of vetting congregants for their eligibility for church membership. What would the young Jonathan Edwards have heard in Strong's testimony?

Had Edwards been present, he would have heard evidence of an augmented sensory encounter upon hearing a "sermon" that "was very awakening." He would have heard instances in which the peaks and troughs of Strong's conversion experience exceeded, as if to spill over, the oscillating regularity of a traditional, conventional testimony of faith. Strong narrates an experience that is more individualistic than its seventeenth-century precedent. She remembers awakening one night in the dark and then seeing a vision of Christ. This vision recurs throughout, culminating with "a sight of the glory of God and the excellency of Christ." Strong sees her visions of Christ alone, at night, while engaged in "a solemn religious duty." These visions index reciprocity between her love for Christ and Christ's love for her, so that at times her love for God "acts" in her "in a more sensible feeling manner than what I commonly experience."[13]

Had he been in attendance in his father's church the day that Strong gave her relation, Edwards would have heard an account of a religious experience that was slowly, subtly, but unmistakably tearing at the seams of the testimony of faith, a genre whose measured and disciplined formality was designed to temper more emphatic, individualistic instances of religious experience. Edwards would have listened carefully to these moments of experiential rupture, because he, like Roger Williams, privileged the anomalous forms of gracious affections as the authentic utter formlessness of true divinity. He would have also heard his own spiritual experience—refracted

13. Quotes from Strong in Minkema, "East Windsor Conversion Relations, 1700–1725," *Connecticut Historical Society Bulletin,* LI (1986), 56–58.

through Strong's—of a personal piety that was superseding the forms available for its description.

The "sensible feeling" that Strong speaks of indexes an inwardly directed transformation that enables her to see the world in a new light. Her encounter with the natural world is transformed through this "sensible feeling" when she joins her experience of nature with her reading of scripture. While walking in the woods, she finds "a leaf of the Psalms . . . accidentally" upon the ground. Through this syntactical construction, "a leaf of the Psalms," she constructs a typological connection between reading scripture and experiencing nature. The material form of the leaf found in East Windsor in early autumn is a type of the leaf mentioned in Psalms; Strong's "accidental" discovery highlights the divine authority structuring this association. She removes her own agency from this discovery; the link between the leaf on the ground and its typological referent to Psalms is a literalization of a scriptural truth that is wholly authorized by God. Is it possible that Jonathan Edwards heard within Strong's discovery a solution to the Reformation and natural philosophical dilemma of unknowable knowledge?

Strong's testimony indeed foreshadows some of the primary theological transformations that took place under Edwards's ministry during the first Great Awakening. Psalms 90:5–6, the passage that Strong refers to through her leaf discovery, is the same passage that Edwards would preach on early in 1734 to the "many young people" in his congregation who were "much affected" by the "very sudden and awful death of a young man in the bloom of his youth." In addition to conveying the spiritual lesson to be gleaned from death, the sermon instructs parishioners of an intrinsic connection between the elements of nature and the elements of their own souls. This is the same connection that Strong observed in 1725 while walking through the woods of East Windsor after experiencing the new light of Christ. It is also the same connection that Edwards would eventually expand into an entire cosmological scheme that would culminate in the sermon series *History of the Work of Redemption,* written and preached in 1739 with the aim of revitalizing nature as a domain in which divine evidence could be gathered and where the human soul could be viewed as a microcosm of a larger cosmological structure.[14]

14. Quote is from *A Faithful Narrative of the Surprizing Work of God* (1737), quoted in Marsden, *Jonathan Edwards,* 153. *A History of the Work of Redemption* was not published during Edwards's lifetime; John Erskine published the sermon series in Edinburgh in 1774.

The "new light" that permitted Strong to see the leaf upon the ground as the scriptural leaf of Psalms is the same new light that Edwards would develop as the "indwelling principle" in *Divine and Supernatural Light* (1734) and then fully defend as the ontological state of the "spiritual man" in his *Treatise concerning Religious Affections* (1746). Strong's more ecstatic divine encounter not only carried resonance with Edwards's personal experience but also anticipated the epistemological potential that Edwards would recognize in certain case studies of female piety, most importantly of his wife, Sarah. Finally, Strong articulates a link between hearing a sermon and an "awakening" feeling that would have appealed greatly to Edwards's own sensibility as a minister in training.

Although sermons throughout the Puritan tradition often incited humiliation as the first phase of conversion, the connection expressed by Strong indexes a subtle shift toward an era of increased pastoral care, when sermons came to claim a greater capacity to elicit the appropriate feeling of grace from the convert. Edwards was pivotal to this transformation, one that would pick up from the very kind of sermonic awakening that Strong describes in 1725. It is through this combined nexus of influences from Enlightenment philosophy to New Light theology to the knowledge supplied by lay testimonies such as Strong's, that Edwards produced a New England version of the evangelical Enlightenment that would constitute the last phase of the science of the soul: an empirically discernible sign of regeneration, more certain than the evidentiary criteria to come before.[15]

There has been a persistent tension between viewing Edwards as an isolated genius versus a product of his time, and how American he was versus how embedded in transatlantic culture. Was he a universal thinker, or an integral though historically bound participant in mid-eighteenth-century evangelical and Enlightenment culture? Are we to understand Edwards within a transatlantic framework of Anglo revivals and Newtonian physics, or within the American exceptionalist paradigm of religious culture? Without privileging one perspective over the other, I position Edwards as each of these things. He certainly drew on the philosophical currents of his time but also transformed them in unprecedented ways, hence making substantial interventions into a long history of Calvinist and natural philo-

15. E. Brooks Holifield refers to this as the "Baconian style" of "evidential Christianity," which emerges through the nexus of a fragmented Calvinism and the rise of deism. *Theology in America: Christian Thought from the Age of the Puritans to the Civil War* (New Haven, Conn., 2003), 197–272.

sophical problems of knowledge. Revivalism was a transatlantic phenomenon, taking root throughout the mid-eighteenth-century Anglo world.[16]

Yet, to decouple Edwards too readily from a larger context of American exceptionalism would be a mistake, for he himself saw his theological interventions as geographically bound, first to New England and then to America as a particular spiritual space. We need not look further than Edwards's own work to see that he himself resisted these categorizations, to see that his revision of Locke bespeaks both his contribution to a transatlantic Enlightenment and his effectiveness as an American evangelical who used Lockean faculty psychology to inculcate a sensory experience among his listeners in order to produce a new performative religious affect that would be sustainable across disparate populations, denominations, and centuries. One need not look farther than *History of the Work of Redemption*, a grand narrative of evangelical thought, to see that Edwards himself both embraced an exceptionalist perspective on what made America spiritually unique and attempted to rewrite world history.[17]

World history unfolds within this series of thirty-nine sermons not only without precluding God's "prior actuality" but also by making "theological epistemology" dependent upon knowing the "temporal and spiritual world." This is a profoundly important philosophical claim and addresses the deepest schisms and crises of knowing that Edwards faced in the 1730s. But what is fascinating as well about this sermon series is that Edwards preached it in Northampton to his congregation. He did not publish it for Newton's descendants on the other side of the Atlantic, but, rather, left it to reverberate throughout the inwardly directed world of laypeople in the Connecticut River valley. Edwards understood how some of his best epistemological interventions could come only by studying the effects of grace upon his congregants, that the visible manifestations of divinity displayed upon them were in fact the key to the reconciliation of science and religion,

16. These perspectives have been summarized in Harry S. Stout, Kenneth P. Minkema, and Caleb J. D. Maskell, eds., *Jonathan Edwards at Three Hundred: Essays on the Tercentenary of His Birth* (Lanham, Md., 2005). Mark A. Noll, David W. Bebbington, and George A. Rawlyk, eds., *Evangelicalism: Comparative Studies of Popular Protestantism in North America, the British Isles, and Beyond, 1700–1990* (New York, 1994); Phyllis Mack, *Heart Religion in the British Enlightenment: Gender and Emotion in Early Methodism* (Cambridge, 2008).

17. This reference to Edwards as an American evangelical comes from Philip Gura's biography of Edwards: *Jonathan Edwards: America's Evangelical* (New York, 2005). And see Amy Plantinga Pauw, "Edwards as American Theologian," in Stout, Minkema, and Maskell, eds., *Jonathan Edwards at Three Hundred*, 14–24.

or of "cracking the code" to the "mystery" of how one might "reconcile freedom of the will with belief in an almighty God."[18]

Edwards preached *History of Redemption* to his congregants not merely to bring grandiose theological, historical, and biblical concepts to laypeople but also to gather information from them on how history might be understood and experienced through an "emanationist language" of divinity. Epistemology works in a reciprocal fashion in Edwards, mediating between his status as great thinker and evangelical preacher, between his status as an American evangelical and an Enlightenment philosopher, unraveling a spiral of intertwined thought just as *History of Redemption* entices its listeners to imagine infinite space despite the limits that John Locke placed on the human capacity to do so.[19]

By positioning Jonathan Edwards as the primary capstone to a history of the science of the soul, I recognize that I am largely underscoring what scholars have long recognized as his work's exceptional innovations and theological transformations. I do concur that Edwards was somewhat anomalous in imagining the most complete morphology of access to the soul's empirical potential. Although natural theologians such as John Ray, William Derham, and John Williams as well as Puritan ministers Samuel Clarke and Benjamin Colman had done much to advance the cause of soul science, their metaphysical inquiry was repeatedly curtailed by the caution that the human soul was the "least known" "part of Man" and nature within the divine cosmos. Edwards's philosophical and theological innovation culminated in his turn to the Puritan testimony of faith as the genre capable of eliciting these unknown realms of knowledge. He had learned from his grandfather, Stoddard, about the transforming genre of the sermon and its more productive use to induce a feeling of the effects of God's grace upon the souls of his converts as well as planned a study of this feeling to accord with what Derham called a "copy" of the divine. But, unlike his grandfather, Edwards could not abandon a long colonial history of attempts to request that lay converts perform this evidence of grace so that it might build upon a history of ministerial expertise generated through an empiricism of the human soul.[20]

18. Pauw, "Edwards as American Theologian," in Stout, Minkema, and Maskell, eds., *Jonathan Edwards at Three Hundred,* 11. Quotes from Gura, *Jonathan Edwards,* 188.

19. For an expanded version of this argument, see my review essay, "What We Can Learn from Jonathan Edwards," *Early American Literature,* XLIV (1999), 423–432.

20. W[illiam] Derham, *Physico-Theology; or, A Demonstration of the Being and Attributes of God, from His Works of Creation . . .* (London, 1714), 272–273.

Jonathan Edwards followed his own father, Timothy, in his attempt to revive the testimony of faith and invest the genre with renewed hope for the science of the soul. Disillusioned with the aftermath of the Great Awakening, which seemed to authorize a proliferation of rhetorical performances of conversion without the accompanying verification of gracious authenticity, Edwards decided to return to congregational tradition. He reintroduced the testimony of faith among lay converts—Anglo men as well as women, Native Americans as well as young girls—in an effort to claim the science of the soul as New England's own through the display of religious affections as occurring with pertinacity, regularity, and certainty on the Anglo-American frontier.

Extensive and verifiable evidence of Edwards's use of the testimony of faith does not exist. Two testimonies are quoted at the beginning of his "Farewell Sermon" (1750), and more were prepared as manuscript evidence for the committee examining the communion controversy at Northampton. Nonetheless, the prevalence and purpose with which Edwards used this genre are debatable. While recognizing the difficulty of identifying his use of these testimonies with historical precision, I suggest that Edwards integrated his theological concept of the indwelling light into the testimonial form as part of an experimental effort to salvage both the genre and its epistemic contribution to a worthwhile philosophical endeavor—one that could supply unparalleled and otherwise unavailable data on the workings of God's presence throughout the world.[21]

His failure to make the indwelling light epistemologically relevant as a certain sign of regeneration also marks the failure of the Puritan errand as a political and colonial enterprise. The "Farewell Sermon" concludes both an era of soul science as a philosophically sustainable endeavor and the status of New England as a unique, geographical setting in which such souls are produced. Once dismissed from Northampton, Edwards moves to Stockbridge, where he lives out his remaining years in a state of philosophical and ministerial exile. He still continues to write, preach, and convert, of course, but he does so without the ambitions of the historical and geographical specificities of the colonial endeavor or the conviction that philosophy might learn from the larger cosmic meaning behind the Puritan project.

21. For this collection of testimonies of faith, I am indebted to Kenneth Minkema, who not only told me about their existence but quite generously supplied me with his own transcriptions of the manuscripts. This transcription is from the "Drafts of Professions of Faith," Jonathan Edwards Collection, fol. 1245, Beinecke Rare Book Library, Yale University, New Haven, Conn.

Semiotics of Grace

By the time Abigail Strong stood before the East Windsor congregation in 1725, the very nature of the evidence that she hoped to supply was under extreme contestation, which had been building since the Halfway Covenant of 1662. The Puritan concept of the visible saint began as an attempt to render the numinous evidence of God's workings upon the human soul—the kernel of Reformation optimism—as empirical knowledge of the invisible world, even though doing so was also a violation of theological precedent. Consequently, the ecclesiastical history of colonial New England repeatedly warned against this effort, forcing the practice to expand into new sites of witnessing such as the deathbed, the missionary community, or the courtroom drama enacted under the investigation of specters at Salem.

Since at least the mid-seventeenth century, critics of the visible church model had been writing extensively against this practice. For example, Samuel Hudson argued that there was a danger of confusing the visible with the actual, that appearances of saintliness are all that religious communities really have on earth and appearances are not enough to constitute any kind of assurance. In *Certain Disputations of Right to Sacraments* (1657), Richard Baxter, who had been a longtime opponent of the practice but eventually embraced it, argued that the profession of faith was a useful measure of sincerity only if the audience recognized that there was no way to get beyond the verbal statement to some kind of secure knowledge of God. Some late-seventeenth-century ministers guarded the sanctity of the test as a true measure of eligibility for admittance to the Lord's Supper, but by the early eighteenth century Solomon Stoddard challenged the efficacy of the test even for this.[22]

Solomon Stoddard's 1723 sermon, *The Defects of Preachers Reproved,* calls for a resurgence of the brand of piety first espoused by William Perkins,

22. In his thorough introduction to the communion controversy, David D. Hall explains how a number of critics insisted that the practice of individual covenants threatened to abolish the distinction between visible and invisible churches. Samuel Hudson was a key critic, who wrote *A Vindication of the Essence and Unity of the Church Catholike Visible* (1650), which argues that "visible" should not be confused with "actual." To make this conflation is to violate a theologically grounded understanding of the semiotics of grace. Jonathan Edwards, *Ecclesiastical Writing,* ed. David D. Hall, The Works of Jonathan Edwards (New Haven, Conn., 1957–2008), XII, 22.

Hall reads Richard Baxter's *Certain Disputations* as favoring a middle way between the extremes of taking the sincerity of the profession seriously and acknowledging the risk of hypocrisy. Baxter resolved this problem by proposing that ministers and congregations accept the evidence of a profession at face value. Edwards, *Ecclesiastical Writing,* ed. Hall, Works, XII, 30.

though this time facilitated by discerning preachers. *The Defects of Preachers Reproved* functions as a ministerial manual like others in the period such as Cotton Mather's *Manuductio ad Ministerium,* published only two years later to aid in the training of the ministry at Harvard. Stoddard's sermon seeks to make ministers better preachers, but it also advocates a transformation in the minister's role in identifying evidence of grace. Integral to the early-eighteenth-century genre of the minister's advice manual was an effort to augment professional expertise. Religious communities relied less on the discerning authority of the congregants and more on the minister's capacity to correctly discern for them. Because lay audiences depended increasingly on the minister's ability to correctly identify signs of grace, Stoddard emphasizes that ministers must have a clear sense of the evidence of their own souls in order to intuit it upon the souls of others.

According to the model proposed in this sermon, there are two kinds of "Signs of Grace": the false and the probable. It is up to the minister to tell them apart. Stoddard describes the possibility of gradually increasing certainty as a compilation of the probable "may make the thing more Probable." However, he also warns that it will not make things "Certain." Stoddard emphasizes: "There is no infallible Sign of Grace, but Grace. Grace is known only by intuition: All the External Effects of Grace may flow from other Causes." Typical of the paradoxical nature of Puritan knowledge, a sermon designed to rejuvenate the ministry's awareness of its divine call does so by issuing a reminder of the epistemological limits surrounding any attempt to gather evidence of conversion as empirical data of God's place within the world.[23]

If it is the minister's job to identify "probable" signs of grace as presented through testimonies within religious communities, what is the role of the convert? Stoddard anticipates this question in *The Defects of Preachers Reproved:* God can be known only through "Self-evidencing Light in the Works of God" (based on 1 Cor. 2:9). Alongside the "probable" signs that the minister learns to intuit with increasing expertise, the convert undergoes an inward soul-searching, which, Stoddard affirms, is the only adequate proof of divine grace. Without this introspective quest, "the multiplying of ceremonies eats out the heart of Religion, and makes a People Degenerate."[24]

23. Solomon Stoddard, *The Defects of Preachers Reproved, in a Sermon Preached at Northampton, May 19th, 1723* (New London, Conn., 1724), 17–18.

24. Ibid., 19, 22–23.

The Nature of Saving Conversion, a complementary tract also written by Stoddard in 1719, advances this principle; it effectively remaps the morphology of conversion, promoted previously in such texts as Shepard's *Sincere Convert* (1641) and Baxter's *Call to the Unconverted* (1663), which ascribed a homologous condition of unknowing to the experience of conversion. The inward quest leads to signs that could not fully be trusted, and the outward representation of this quest replicated its equivocal pattern by producing a narrative sign of what grace felt like that carried with it an implicit understanding of the limits surrounding this representational form. This commensurability between inward feeling and outward sign defined the long colonial history of attempts to capture grace within its verbal, narrative form, from the 1630s up to when Stoddard wrote this tract. Stoddard transforms this tradition, not through an innovation of, but rather by returning to the notion of a discrepancy between inward feeling and outward sign that was promoted by William Perkins. The Perkinsian model taught that the only full knowledge of assurance came through an individualized awareness of the soul's evidence, an encounter with forms of divine evidence that could not be externalized. Post-Perkinsian theologians attempted to externalize the inner evidence as their collective effort toward resolving this epistemological problem.[25]

Stoddard both augments the level of ministerial expertise and encourages a return to a Perkinsian model of inward faith among potential converts. Reason cannot discern divine truth, he explains. No taxonomy or structure of externalized faith can supply an adequate logic for intuiting the workings of God throughout the world. Stoddard's theological contribution to the long-standing Reformation project of reading contestable signs as evidence of inaccessible knowledge of God was to decouple the phenomenal feeling of grace from its embeddedness within nominalistic structures. Returning to a Calvinist and Perkinsian notion of grace as the only form of true assurance, Stoddard gears his theological project toward reminding ministers and laity that inward revelation must be separated from reason and the narrative structures that seek to delineate how grace works in human terms. Stoddard's treatise quotes Hebrews 11:1, not trans-

25. In *The Covenant Sealed: The Development of Puritan Sacramental Theology in Old and New England, 1570–1720* (New Haven Conn., 1974), E. Brooks Holifield explains that about the turn of the century sacramental signs were becoming increasingly separate from the inward spiritual meaning to which they were supposed to refer. I agree with Holifield's reading. As the spiritual meaning of external signs diminished, the inward turn toward an individually realized faith became more important.

forming it as Baxter does in his *Call,* but rather as a conservative interpretation of the scriptural lesson: "Natural reason *doth not discover the certainty of divine things.* . . . Heb. 11.1. *Faith is the evidence of things not seen.*" Baxter quoted this passage in order to transform it into a new form of knowledge that could be supplied from the effects of saving grace. Stoddard uses the quote as a reminder of the epistemological limits surrounding grace and the inability of reason to act as a mediating force for discernment. Stoddard agrees with Perkins's notion of assurance as coming exclusively from an inwardly directed journey while also heightening his emphasis on the role of the preacher as a provider of pastoral care and on the sermon as a vehicle for eliciting conversion.[26]

Material Impressions of Spiritual Things

By the time Jonathan Edwards heard Abigail Strong and others testify to her sights of Christ, he would have been fully aware of his grandfather's warnings of the pitfalls of observing evidence of grace in this manner. By the early eighteenth century, the ontological elevation of the natural world as a site of collecting, taxonomizing, and collating data into new epistemic structures had converged irrevocably with a brand of Protestantism already transformed by these philosophical innovations. Repeatedly and collectively, philosophers and ministers grappled with the problem of uncertainty, a deep and ensconced awareness that their techniques of observation could extend to a world just beyond the natural but that the divine knowledge produced therein might also extend beyond revealed truths and scriptural law. This dangerous and uncharted territory meant that new knowledge carried with it the threat of repeating original sin.

While consumed with this dilemma (as were many of his contemporaries), Edwards was also, perhaps, unique in his attempt to definitely resolve this dilemma of unknowable knowledge by emphasizing rather than ignoring or carefully skirting around the consequences of the Fall. In his writings and in his ministerial practice, Edwards sought to propose a program with the specific purpose of achieving secure religious truth and philosophical accuracy without decoupling philosophical inquiry from its

26. Solomon Stoddard, *The Nature of Saving Conversion, and the Way Wherein It Is Wrought,* 2d ed. (Boston, [1770]), 66. Richard Baxter paraphrases this verse as "faith is a kind of sight; it is the eye of the soul: the Evidence of things not seen." Baxter, *Call to the Unconverted to Turn and Live, and Accept Mercy* . . . (London, 1663), 48.

entrenchment in the story of Genesis. New England was the laboratory in which he hoped to do this, even though his philosophical goals were transatlantic—even global. He remapped Northampton as the "city on a hill," both a theological beacon to the lapsed Christians of the world and a stage upon which natural philosophers could witness evidence of their highest ideal of scientific attainment: evidence of divine essence displayed clearly, empirically, and unequivocally upon the human soul. To accomplish this, Edwards looked to the testimony of faith, the very genre that had most haunted his New England Puritan predecessors with its slippery elusiveness, appearing sporadically and intermittently in the archive of election as the faintest visible evidence of divinity. This was the genre, as I have been arguing, that marked the most suspect yet tantalizing form of empirical evidence, evading the human capacity for visual sensory discernment but nonetheless supplying an array of divine evidence that was simply too valuable for natural philosophy to ignore.

Edwards devoted his life work to realizing a soul science that could revive the kernel of Reformation optimism and move past the resurgent shadows of doubt and uncertainty surrounding it. He fashioned a world in which each human soul was a microcosm of how divinity worked in nature, transforming religious experience as he simultaneously transformed the philosophy of Locke, Newton, and More. His record of the Northampton revival in *Some Thoughts concerning the Present Revival of Religion in New-England* (1742) and *A Faithful Narrative of the Surprizing Work of God* as well as his efforts to represent "true religion" in *An Account of the Life of the Late Reverend Mr David Brainerd* collectively demonstrate a design to render the signs of religious affection as a public and publicly discernible phenomenon. While these tracts make claims for the particularity of religious experience in New England, they are also directed toward and printed for a transatlantic audience of philosophers and ministers. Unlike Thomas Shepard, John Fiske, or Ebenezer Parkman, Edwards recorded the conversion narratives of his congregants with public consumption and circulation in mind. This goal aligned him with John Eliot and John Rogers, whose collected testimonies published in 1653 present evidence of the Algonquian and Irish religious experience to the larger Anglo World. Yet Edwards's tracts of the Northampton revival also speak more directly to the specific ways that Puritan conversion could respond to a philosophical question of how to find more evidence of God in nature.[27]

27. Lambert fully explores this idea of public circulation in *Inventing the Great Awakening*.

In his early philosophical manuscript "Of Being," Edwards demonstrates a proclivity toward abandoning this mechanical philosophical framework in order to ascribe a real ontological category to the notion of the spirit world. This notion of a divine spatiality counters the Newtonian framework of divinity as inaccessibly located outside a spatial and temporal reality. It is a "gross mistake," according to Edwards, to "think material beings the most substantial beings, and spirits more like a shadow; whereas spirits only are properly substance." The spirit's substance can be known and gained access to through this understanding of divine space as a real entity. Edwards denies the validity of proving divine existence through a dualistic Cartesian model with the maxim, "The mind can never . . . bring itself to conceive of a state of perfect nothing." If the reconfiguration of the human mind as an entity capable of understanding a real spiritual domain presented a workable resolution to the very inaccessibility of divine essence that natural philosophers had been struggling with, then Edwards had to come up with a scheme for perceiving this information. His essay "Of Being" does just this, positioning the mind as the essential human entity that could in fact supply entry to a spiritual realm. Giving form to the unseen, Edwards asks the reader to imagine "another universe only of bodies, created at a great distance from this, created in excellent order and harmonious motions, and a beautiful variety." Significantly, we are to imagine this other universe in material terms.[28]

"Of Being" marked the beginnings of a project to decouple the science of the soul from its embeddedness in nominalism. Ultimately, Edwards knew that the attempt to use evidence of the soul to trace divine secrets was futile. The soul could supply signs that were only faint shadows of the actual thing. Evidence of grace could never fully represent divinity itself, even though it not only inculcated but also increased empirical desire to further knowledge of God. Edwards resolved this semiotic problem by constructing a model of a human mind capable of penetrating an invisible world that he reconfigured as consisting of real substances. The effects of God's grace were no longer rendered by the convert as faint, representative shadows of the invisible. Rather, the mind conjoined worldly experience

28. Jonathan Edwards, "Of Being," in *Scientific and Philosophical Writings,* ed. Wallace E. Anderson, Works, VI, 202, 204, 206. Edwards's maxim comes from Locke's chapter "Of Infinity" in his *Essay concerning Human Understanding*. Edwards wants both to take Locke's claim seriously and to move beyond the limitations that it prescribes for divine comprehension. He does so here and elsewhere by refuting Descartes's resignation to the skeptics' stance of the unseen and invisible as inaccessible.

with the indwelling light of Christ. Edwards revived the practice of spiritual testimony, first as spiritual biography during his Northampton revival and then as a basis for church membership in Northampton, in order to cull the empirical data of indwelling Christly light from the souls of his visible saints.

The indwelling light of Christ constituted an ontological as well as theological shift in which divine essence opened the mind to "divine consciousness," which gathered knowledge of religious mystery from the innermost recesses of the self. In place of the universe of signs, the soul became the space where divinity resided. Edwards looked to the soul for a solution to the mechanical philosophical limits of nature. If decoupled from the nominalistic tendency of natural philosophy and observed as a spiritual entity unto itself, the "indwelling principle" might provide empirical access to the divine. Edwards proposes in his *Treatise concerning Religious Affections* that the transformative effects of this indwelling principle upon divine minds should be foregrounded as the object of empirical inquiry. This, rather than an external divine form, would expand spiritual understanding.

"Christ, the Light of the World," one of Edwards's earliest extant sermons, puts his philosophical schema into practice, proposing the concept of the "indwelling light" of Christ for the first time and constructing a homology between the human soul and the natural world as a model for establishing how this indwelling light might be rendered usable. The "indwelling light" permits the "enlightened person" to become sensible of divine truth in a way that makes him or her no longer reliant upon the theoretical truth of religious affairs. The "indwelling light" dissolves the problem of uncertainty surrounding assurance, because it is no longer incumbent upon the convert to differentiate the real from the unreal. The danger of hypocrisy is thus partially resolved in this early sermon through the notion of the revelatory knowledge that comes from Christ's descending and indwelling light. That Edwards was both obsessed with the problem of hypocrisy and determined to propose a resolution is integral to his theology.[29]

Edwards's concept of the indwelling light was not only a means of rectifying a long-standing theological problem within Calvinism but also to make a philosophical intervention. Edwards's essay "Of Being" trans-

29. Jonathan Edwards, *Sermons and Discourses, 1720–1723*, ed. Wilson H. Kinmach, Works, X, 533–550; Marsden, *Jonathan Edwards;* Edwards, *Ecclesiastical Writings*, ed. Hall, Works, XII; Perry Miller, "Edwards, Locke, and the Rhetoric of Sensation," in Harry Levin, ed., *Perspectives of Criticism* (Cambridge, Mass., 1950), 103–123; Holifield, *Theology in America*.

formed the mind into a vehicle capable of entering the real dimensions of the invisible world; "Christ, the Light of the World" simply applied this logic to the transformation that the soul undergoes during conversion. Edwards conceived of the presence of an indwelling light within the soul of the convert and within the natural world as a Christian solution to the Enlightenment's most pressing epistemological problems. The reduction and purported elimination of uncertainty would ameliorate the frustrations and limitations surrounding mechanical philosophy at the same time that it would revive and further the endeavor toward a taxonomy of grace through which the human soul could be used to accumulate divine knowledge. Edwards instructed his converts in the phenomenon of Christ's indwelling presence in their souls; he did so with the intention of calling upon his lay converts to supply the necessary data to further a frustrated empirical inquiry into the essence of divinity.

The doctrine of "Christ, the Light of the World" comes from John 8:12: "I am the light of the world." Most literally and directly, the "I" refers to Christ, but, detached from its embeddedness within the book of John, it is also a free-floating pronominal referent that stands in for speaker, listener, and world within the context of the sermon. The detached "I" begins the sermon by introducing the lay audience to the collapse of semiotic distinctions that Edwards wishes to cultivate, theologically and philosophically. Here the distinction between self, Christ, minister, and world becomes blurred. Through this doctrinal injunction, introducing the concept of the indwelling light, the world created by the sermon enfolds the speaker in the complementary images of darkness and light. The listener learns of a "universal night of . . . darkness" that covered "the world before Christ came." Christ's light evaporated the darkness of the world, generating a spatial and temporal dimension of gradually increasing divine legibility. Edwards establishes a correspondence between the darkness of the world and the darkness of the human soul. He explains, "Our souls are naturally like a dark, hideous dungeon where the sun, moon, nor stars never found an entrance for their beams." The dark state of the human soul left humans within a completely depleted sensory state: "Their eyes are closed that they cannot see, and their ears heavy like the deaf adder that cannot hear the voice of charmers charming never so wisely."[30]

The structuring of the sermon sets up a homologous relationship be-

30. Edwards, *Sermons and Discourses, 1720–1723*, ed. Kinmach, Works, X, 537, 538.

tween the human soul and the natural world that renders the former a microcosm of the latter. Both domains were enshrouded in darkness during an eschatological period that preceded Christ's descending light. The sermon invites the listener to understand Christ's descendant light within a presentist temporal frame. A simultaneous illumination takes place within each human soul and throughout the natural world. To place oneself within the "I" of the doctrinal framework is to imagine one's elect status as united to both Christ and the world. To be the "light of the world" is to enter into a state of universal connections where the discrepancy between the self, the world, and Christ dissolves.

Edwards explains that, "when a man is enlightened savingly by Christ,"

> he sees then that he was in darkness before; though he was not sensible of it till now, he is like one that was born and brought up in a cave, where is nothing but darkness, but now is brought out into the lightsome world, enlightened by the beams of the sun, and greatly admires and wonders at those things which he never saw before, looks and gazes with sweet astonishment on the pleasing variety of things that are discovered by the light unto him. He now sees things in their true shapes and colors that he never saw before: how he sees his own vileness and filthiness, which he had often heard of before but never believed.[31]

In a clear reference to the allegory of the cave from book 7 of *The Republic*, Edwards takes from Plato the emblematic, allegorical shift from "light to darkness or from darkness to light." He takes as well the framework of a story in which revelation comes through a gradual realization of the subtle discrepancy between "reality" and the "shadows" of "artificial objects," between "the scene-shifting periact in the theatre" and the ultimate "contemplation of essence and the brightest region of being." For Edwards as for Plato, the soul is the site of fruition in this allegorical description. Darkness and light oscillate between the cave and the sun to mirror the "intelligible region" of divine truth displayed upon the human soul. Plato describes this process as "a conversion and turning about of the soul from a day whose light is darkness to the veritable day—that ascension to reality of our parable which we will affirm to be true philosophy." Edwards incorporates this Platonic notion of conversion into the Puritan morphology as revised in "Christ, the Light" to consist of, following Plato, a telos of transformation.

31. Ibid., 539.

Prior to Edwards, Puritan conversion was dialectical with a promised but unrealized resolution. The movement between darkness and light, anxiety and assurance, meant that the convert would never arrive at that "veritable day" of "true philosophy"; to do so would be a mark of hypocrisy.[32]

In "Christ, the Light," Edwards revisits Christianity's Platonic tradition in order to borrow from the gradual Platonic ascendancy toward philosophical truth through the soul's conversion. Neoplatonism was important across a diverse range of Christian scholastic thought in the sixteenth and seventeenth centuries because Plato supplied a philosophical framework for contemplating knowledge of essence. Medieval Artistotelians, Ramist theorists, and aesthetic philosophers such as Sir Philip Sidney depended on Plato to parse the relationship between "intelligible and sensible forms" in order to transform sensory data into a "direct apprehension" of the divine. Philosophically, this was a difficult move to make while maintaining the integrity of the formal structures of reason designed by Aristotle and Ramus as the logical mechanisms for arriving at truth. Like Plato, the Puritans sought to reconcile the idea of reason as the source or container of truth with the clarity to be produced through revelation.[33]

Unlike the Christian Platonists who came before him, Edwards renders Plato's cave allegory immanently realizable within a Christian eschatological frame, geared toward philosophical truth as much as millennial fulfillment. The light of Christ is both transcendent and natural, imparting within the enlightened being a renewed sensibility to perceive the true form of colors and shapes that had previously been encased in the limits of one's own depleted sensory capacity. As the darkness of the world retreats and lightness unfolds, the veil produced by original sin, like Plato's cave, recedes. No longer an omnipresent haze clouding and obscuring the perception of true shapes within the world, sin congeals into an objectifiable, nameable site. The convert emerges from a world encased in darkness as the man in Plato's allegory emerges from the cave. The dissolution of darkness correspondingly evaporates the limits of perception. Within this new world of light, the convert enters into a new state of epistemological security whereby he or she no longer needs to guess at the accuracy of sensory discernment. As sin dissolves through the infusion of divine grace, the

32. Paul Shorey explains that a "periact" was the "triangular prisms on each side of the stage. They revolved on an axis and had different scenes painted on their three faces" (II, 134). Plato, *The Republic*, trans. Paul Shorey (Cambridge, Mass., 1935), II, 123, 131, 133, 135, 147.

33. Perry Miller, *The New England Mind: The Seventeenth Century* (New York, 1938).

"I" ceases to be attached to a specific embodied location. Instead, the "I" becomes increasingly transparent—setting the historical stage for the dissolution of the body and the ego in Ralph Waldo Emerson's famous "Preface" to "Nature" (1836). Just as Emerson's eyeball becomes a host for the Oversoul, the "I" becomes increasingly transparent and capable of acting like a cipher or lens through which humans can gain access to the invisible world. Humans no longer simply carry with them the faint evidences of God's grace. Instead, they act as a specific kind of spiritual host.

Quite far from the infamous image of Jonathan Edwards as the hellfire and brimstone preacher of *Sinners in the Hands of an Angry God,* "Christ, the Light of the World" could not present a more comforting description of the salvivfic effects of grace unfolding within the context of this new era. Christ's love awakens sinners from a deep sleep in which "all their affections are dead, dull and lifeless; their understandings are darkened with the dark shades of spiritual night." Awakening and movement come from the manifestations of Christ's love that Edwards describes through the warming effects of summer flowers dancing across the wintry darkness of the human soul. "As when the sun returns in the spring, the frozen earth is opened, mollified and softened, so by the beams of the Sun of Righteousness the stony, rocky, adamantine hearts of men are thawed, mellowed and softened, and made fit to receive the seeds of grace." The use of a seasonal metaphor describing the workings of grace both harkens back to Perkins's writings on conversion and signals Edwards's gradual incorporation and transformation of natural philosophy as an integral component of his theological innovations. Christ infuses the "seeds of grace" within the soul of the expectant convert through the sexual encounter described in Canticles 2:10–13. The "graces springing forth" are the offspring of this encounter, described as "sweet flowers that adorn the face of the earth" and "the sweet melody of singing birds."[34]

Edwards repeatedly uses whimsical—even sentimental—images from nature that produce a benevolent, comforting, and seductive description of the workings of grace. This highly descriptive language works to narrate for the converts what religious affections would feel like moving and dancing across their souls in an attempt to initiate the bridge between Edwards's theological and natural philosophical endeavors, between the human soul as a microcosm of the divine and the natural world as a macrocosmic map

34. Edwards,, *Sermons and Discourses, 1720–1723,* ed. Kinmach, Works, X, 540, 541.

of God's omnipotent presence. Edwards's oeuvre culminates in an explicit and direct link between the ministerial expertise gathering around human souls and the scientific expertise culled from the natural world.

By the early eighteenth century, however, this effort to generate epistemological continuity around the science of the soul faced an additional challenge through an emergent split between the methods of sensory learning and the structures of reason. On one side of this divide, Spinoza proposed to eliminate sensory data and rely completely on reason. Newton agreed, arguing for rational structures to map the pattern of the natural world and resigning himself to the limited scope that such knowledge could reveal—the world reflected the causal effects of divine existence but could not lead to knowledge beyond that. On the other side of the divide, Locke remained committed to the senses, proposing a precise and narrow window of information that could be acquired from sensory data. This approach produced the opposite dilemma as Locke grappled repeatedly with the question of what the senses could actually tell us—what larger rational structures might sensory data reveal? Edwards embraced the methods but eschewed the limitations imposed by each of these perspectives. He wanted to eliminate the mechanism generated by human reason and extract the full extent of information that could be derived from the human senses. He intended to accomplish this by elevating the ontological status of the senses, expanding the various forms of information that they could yield through a notion of an indwelling principle of Christ that transformed the limited scope of human capacity from a fallen to a redeemed condition of rational comprehension.[35]

Alongside his contemporaries, Edwards incorporated Locke into a Calvinist framework in an effort to reduce the condition of uncertainty inherent in the problem of grace. Locke both presented a usable method for and a direct challenge to the cultivation of a properly disciplined divine mind that could perceive this spatial entity. While he transformatively explained how experience worked, he also warned that his concept of experience could not be adapted to the divine. The *Essay concerning Human Understanding* (1689) quite carefully delineates a discrepancy between the physical event that sets various bodies in motion and the human perception about that event. Through this discrepancy, Locke's empirical model keys in on the problem of discerning between the real and the unreal. He wants to maintain that the senses receive impressions from "external Objects" and

35. Richard Mason, *The God of Spinoza: A Philosophical Study* (Cambridge, 2001).

that these "external Objects" then "convey into the mind what produces there those Perceptions."[36]

These external objects existed only and exclusively in a material realm that Locke defined quite narrowly, hence reinforcing rather than disrupting the binary between visible and invisible realms that presented such a challenge to mechanical philosophers interested in tracing divine evidence. The limitations that Locke establishes for his empirical program throughout the *Essay* present a tremendous obstacle for those ministers who wanted to capitalize on his insights in order to augment knowledge of religious experience. Lockean limitations on sensory data also challenged natural philosophers, who were either curtailed in their efforts to apply the mechanisms of sensory knowledge to a world of causal and mechanistic events or confined to the material domain as the ultimate source of knowledge without an adequate model of how to expand this knowledge by channeling human perceptive capacities.

Locke did proclaim that the testimony of God "cannot deceive, nor be deceived." An accurate, precise empiricism of the senses "absolutely determines [in] our Minds" the validity of "divine revelation." The correct application of these principles guards against the "Extravagancy of Enthusiasm." This premise had great appeal to Edwards and other ministers who were increasingly troubled by the detrimental enthusiasm unavoidably harnessed to the proliferation of revivalist practice. But Locke also thwarted the religious appeal of his sensory program, explaining that the essence of "immaterial Beings" could not be examined with this certainty because "humans Senses" would allow access only to individual experience. Divine revelation could be studied empirically only as a "simple idea" guided by subjective experience. Edwards revised Lockean sensationalism as he incorporated it into his study of religious affections. He eliminated the problem of hypocrisy in the sequence of perceptive faculties, and he located divine essence in human souls, where it could be rendered objectively and approached through "evidence and probability."[37]

36. Leon Chai, *Jonathan Edwards and the Limits of Enlightenment Philosophy* (New York, 1998), 13.

37. John Locke, *An Essay concerning Human Understanding*, ed. Peter H. Nidditch (New York, 1979), 663, 665, 667. Richard Ashcraft argues that Locke's critics were concerned that his denial of our knowledge of the "real essence" of sensual objects threatened the existence of God; thus, he was accused of atheism by his critics when this was in fact not the case. Richard Ashcraft, "Faith and Knowledge in Locke's Philosophy," in John W. Yolton, ed., *John Locke: Problems and Perspectives* (Cambridge, 1969), 194–223.

According to Perry Miller, Locke taught Edwards how to apply words to experience, rather than to the thing itself. In my reading, Edwards was more troubled by experience than Miller's argu-

Applying Locke toward a sequential understanding of human perceptions of the divine, Edwards reworked the mind as a site that could capture the reality of divine essence through the emergent concept of the "indwelling principle" as the place within the soul where Christ resides. These contributions toward a philosophical intervention in the philosophy of conversion can perhaps be most clearly traced in his essay "The Mind." An early section, the "Place of Minds," builds directly on Locke's *Essay*. Striving to correct Locke's notion of the shadowy immateriality of the spiritual, Edwards writes:

> Our common way of conceiving of what is spiritual is very gross and shadowy and corporeal, with dimensions and figure, etc.; though it be supposed to be very clear, so that we can see through it. If we would get a right notion of what is spiritual, we must think of thought or inclination or delight. How large is that thing in the mind which they call thought? Is love square or round? Is the surface of hatred rough or smooth? Is joy an inch, or a foot in diameter? These are spiritual things. And why should we then form such a ridiculous idea of spirits, as to think them so long, so thick, or so wide; or to think there is a necessity of their being square or round or some other certain figure?[38]

What is the relationship between "thought" and "spiritual things" in the above passage? The mind collects and records impressions of spiritual things as they enter into the soul of the convert. The epistemological challenge posed to Baconian empiricists and practitioners of Puritan piety alike is how to discern these impressions or evidences of grace accurately. The perception of sensory data depends upon human reason, and human reason, as most early modern philosophers recognized, was flawed owing to the Fall. Locke's *Essay* proposes a way out of this dilemma by charting an empirical model that privileges the data of the senses over reason. Specifically, Locke proposes that the inaccuracy of the rational mind could be reduced by increased philosophical focus on how sensible objects enter the mind and stake their impressionistic place. But Locke proposes this schema by limiting and circumscribing the types of data that can be interpreted and recorded by the mind. Only a very narrow domain of sensible objects

ment leads us to believe. Miller, "Edwards, Locke, and the Rhetoric of Sensation," in Levin, ed., *Perspectives of Criticism*, 103–123.

38. Edwards, "The Mind," in *Scientific and Philosophical Writings*, ed. Anderson, Works, VI, 338.

can be rendered accessible through this formulation. Edwards's work in the cited passage from "The Mind" and elsewhere in his writing on the phenomenal sensory experience of conversion would be to expand the range of sensory data, such that "spiritual things" might be subjected to the same empirical revision. If simple ideas enter the mind to create impressions of external things, the passage seems to ask, couldn't ideas of spiritual things also perform this kind of a transformation from the sensible to the visible, material impression left on the mind?

Thoughts, purified through the Lockean empirical model, transform the sensory impressions of spiritual things into real knowledge of the unseen. "The seat of the soul," writes Edwards, is "in the heart or the affections," the "immediate operations [of the soul] are there also." The "head" is where the ideas and thoughts about these operations "are constituted." Edwards explains that divine sense accumulates in the soul, becomes manifest in gracious affections, and then conveys to the mind an understanding of this process. This model of conversion almost directly applies Locke's perceptual model to the mind and the soul, the understanding and the heart. Yet Edwards also saw the problem that hypocrisy posed for the Lockean method. The heart collects "affections" that have "some sensible effect on the body," but these effects are not adequate proof of spiritual substance. If, as Edwards makes clear throughout his *Religious Affections,* the mind is the proper vehicle for distinguishing true religion from the pitfalls of enthusiastic hypocrisy, how does the mind determine truly gracious affections? Locke's sequential ordering of sensory reception and cognitive perception presented an empirical problem: the danger of hypocrisy made the differentiation between divine truth and a sense of divinity nearly impossible to achieve.[39]

Harvesting the New Science

Edwards not only drew upon his theological training but also looked to the phenomenon of conversion among his congregants in order to address this problem of hypocrisy head-on in the hopes of eventually overcoming the limitations of Lockean knowledge. In "The Heart of Man Is Exceedingly Deceitful," a sermon preached before 1733, Edwards explains to his congregants that the "heart is exceedingly prone to delusions and deceits about its

39. Ibid., 352; Jonathan Edwards, *Religious Affections,* ed. John E. Smith, Works, II, 23, 103.

own knowledge." Humans cannot be trusted to self-knowledge, because they inevitably mistake "common illuminations and affections for saving grace." Hypocrisy centers on this problem of false evidence, or the misinterpretation of sensory perceptions. The problem with Locke's perceptual sequence for the science of the soul was also interpretive. Descartes, Locke, Edwards, and others agreed that senses themselves do not deceive; rather, deception comes from assigning a rash or undisciplined interpretive framework to them. Hypocrisy thus proceeds from "judgments," not from the sensory "experience" itself. Erroneous judgments are so pervasive, according to Edwards, that they almost negate the validity of any profession of saving faith.[40]

In the years leading up to the Northampton revival, Edwards worked to link this theological problem of hypocrisy to the larger problems of piety plaguing the community locally as well as within the larger narrative of the New England mission. The problems of hypocrisy within each individual soul become the problem of Northampton, reconceptualized as the new "city on a hill." In the 1730s Edwards's sermons worked to regenerate the importance of this mission and to place piety firmly back within the public domain. Preaching to his Northampton congregation in 1736, Edwards explained that Satan's work had been "to hinder the credit" of divine revelations by "filling the country with innumerable false and groundless reports." In reference to the most recent wave of revivals, Edwards expressed his "astonishment" in "seeing how many strange and ridiculous stories have been carried abroad that have no foundation; and how those things that have been in part true, that have been a blemish to religion amongst us, have been magnified and added to." Establishing Northampton as a "city on a hill," Edwards explains that, alongside the revivals' display of divinity's "native beauty and excellency," comes its inevitably marring "defilement." Now that the "eyes" of the world are turned to Northampton, this

40. Jonathan Edwards, "The Heart of Man Is Exceedingly Deceitful," in William C. Nichol, ed., *Knowing the Heart: Jonathan Edwards on True and False Conversion* (Ames, Iowa, 2003), 21. We don't know the exact date on this sermon, but, according to the editor, undated sermons were preached before 1733.

Edwards expands upon this idea in his essay "The Mind": "Indeed, in some things our senses make no difference in the representation, where there is a difference in the things; but in those things our experience by our senses will lead us not to judge at all, and so they will not deceive. We are in danger of being deceived by our senses in judging of appearances by our experience in different things, or by judging where we have had no experience of the like." Edwards, "The Mind," in *Scientific and Philosophical Writings*, ed. Anderson, Works, VI, 369–370.

defilement has intensified and threatens to completely negate the "extraordinary light" of divine knowledge expressed through the dramatic increase in "professions of special experiences." Edwards effectively describes an epistemological crisis in this sermon. He had learned from these early "harvests" in Northampton the difficulty of implementing the philosophical endeavor outlined in his essays "Of Being" and "The Mind." Human sensory perceptions and interpretations of divine phenomena continued to thwart the effort to discover the soul as a spatial entity of divinity.[41]

In an effort to amend the philosophical problem posed through the unreliability of human sensory interpretation, Edwards draws the scrutinizing "eyes of all the land" inward to sharpen the community's own observational capacities. Like Natick, Cambridge, and other scenes that implemented natural philosophical methods to discover grace on human souls, Northampton became Edwards's laboratory for discerning Christianity in its "native beauty and excellency." This claim for a "native beauty" echoes claims made by Roger Williams as he advocated a detached, ethnographic stance as the best way to capture spiritual essence. Moving between the scriptural discussions of spiritual knowledge and the forms of spiritual biography recorded in such collections as *A Faithful Narrative of the Surprizing Work of God,* Edwards began to compile "Directions for Judging of Persons' Experiences." He urged "persons" to resist concluding that they are "converted" as soon as they "experience" something "remarkable," for the experience likely is an illusion or a figment of the imagination that will subside with time.[42]

Through this idea, Edwards built upon the writings of his grandfather who, in his *Nature of Saving Conversion,* encouraged converts to look for patterns of repetition. But, where Stoddard would argue that these repeated patterns could offer only inwardly directed assurance to the individual, Edwards wanted to propose a plan for extrapolating this inward knowledge, rendering it as legible evidence for others. Divine "discoveries and illuminations" came from experience, not as "superficial pangs, flashes, imagination, [and] freaks," but rather as "solid, substantial, deep, [and] inwrought" experiences that seep gradually "into the frame and temper of

41. Jonathan Edwards, "A City on a Hill," in *Sermons and Discourses, 1734–1738,* ed. M. X. Lesser, Works, XIX, 550.

42. Ibid., 549; Jonathan Edwards, "Directions for Judging of Persons' Experiences," in *Writings on the Trinity, Grace, and Faith,* ed. Sang Hyun Lee, Works, XXI, 523; Edwards, "Heart of Man," in Nichols, ed., *Knowing the Heart,* 39.

[the convert's] mind." He urged his congregants to resist interpretive form and to remain completely passive through the sensory encounter. This passivity would allow the sensory data to slowly infiltrate the mind of the convert—leaving impressions that, unmarred by human fallibility, congealed into an indwelling principle.[43]

Edwards arrived at this philosophical and theological concept of the indwelling principle through accumulating a particular kind of spiritual expertise, acquired by observing the various forms of spiritual expression generated in the Northampton revival and its aftermath. The observational techniques that Edwards brought to these scenes share epistemological continuity with the ethnographic techniques employed by Roger Williams nearly a century before. As discussed earlier (Chapter 4), Williams departed markedly from his more orthodox Massachusetts brethren by arguing for a specific method of discovering divinity's rarities within the natural world in their pure and unadulterated state. In *The Examiner Defended* (1652), Williams identifies New England as a unique geographic and spiritual space full of a certain kind of spiritual plentitude. "Extraordinary, supernatural, and miraculous" forms of God's presence reside in New England, the New World site of colonial settlement and religious fulfillment that Williams typologically links to the "New Canaan." The land and its inhabitants embodied tremendous spiritual potential. However, Williams recognized, along with his Puritan brethren and the natural and mechanical philosophers that would follow in subsequent decades, the inadequacies of the human capacity for discernment.[44]

To compensate for these inadequacies, Williams adopted a particular form of spiritual, ethnographic witnessing that Edwards then took up in his writings on the Northampton revival. Williams advocated the study of grace among "the meanest and lowest *Earthen Vessels*" who are more often chosen as repositories. The spiritual knowledge ascertained through "*Impressions* from *Heaven*" can be "distinguished," according to what Williams describes as "a more tender and observant Eye." The "Eye" must be trained with the knowledge that "*Impressions*" often come in the form of "*Incivilities*" within the "*Light* of Nature." Resisting the tendency of members of the Massachusetts orthodoxy such as Shepard, Hooker, and Fiske,

43. Jonathan Edwards, "Directions for Judging of Persons' Experiences," in *Writings on the Trinity, Grace, and Faith,* ed. Sang Hyun Lee, Works, XXI, 523.

44. Roger Williams, *The Examiner Defended, in a Fair and Sober Answer to the Two and Twenty Questions Which Largely Examined the Author of "Zeal Examined,"* in *The Complete Writings of Roger Williams,* VII, ed. Perry Miller (New York, 1963), 251.

to look for intelligible signs, Williams insisted on the pure formlessness of true divinity. His intervention into techniques of witnessing grace was twofold. He insisted that ministers pay special attention to the spirituality encoded in Anglo women and Narragansett Indian converts, and he urged ministers to hone and expand their observational capacities, namely by looking toward the anomalous forms of divinity displayed among these populations. Whether a result of ministerial solicitation or their own less orthodox relationship to pious practice, these populations tended to display grace in forms that were more surprising than conventional. These less conventional displays of religious affect are what drew ministers' attention as they recognized that their expertise could only be increased through an expanded notion of how evidence of grace appeared, filtered through the farthest reaches of the natural world.[45]

The observational techniques that Edwards brought to the Northampton revival (and its published narrative account) took part in this long colonial tradition that began with Roger Williams and extended through the anomalous forms of grace witnessed in deathbed scenes. Edwards made these observational practices more mainstream, bringing them to bear on the more public testimonial accounts that proliferated through his town in the 1730s. Like Williams, Edwards sought the essence of true divinity in a purer and less adulterated form, but, to address his own concern with hypocrisy, he also sought to link this formless manifestation of grace into a highly taxonomized and rigidly structured formula for religious affections. Edwards used this formula to implement a schema of religious practice that was designed to drastically reduce the problem of uncertainty surrounding election. His sermons sought to elicit the experience of the indwelling principle among his lay converts, who would then perform this indwelling principle as certain knowledge of God's presence within their souls.

Edwards's study of true divinity supplied the raw material for not only a new morphology of conversion but also a larger philosophical project in which he sought to map the truth of God upon the exemplary human soul as evidence for the way that God works throughout the world. Nowhere is this dual goal more evident than in his study of the conversion experience of his wife, Sarah. Sarah's conversion experience occurred in three installments in 1738, 1740, and 1742 (the year that it was first recorded, at Edwards's request). A subsequent edition appears under Edwards's authorship as "An Example of Evangelical Piety" in *Some Thoughts concerning the*

45. Ibid., 224, 241, 244.

Revival of Religion, where Sarah's conversion narrative is published with a deliberate dissociation of the experience from the testifier's identity and location in time and place. This deliberate dissociation functions to further abstract the evidentiary data recorded in the testimony from the subjective interpretation of the testifier.

The first recorded narrative of Sarah's conversion (the original transcription has been lost, but a version remains in Sereno Edward Dwight's *Life of President Edwards*) emphasizes the anomalous status of Sarah's experience: "Her piety appears to have been in no ordinary degree pure, intense, and elevated, and that her view of spiritual and heavenly things were uncommonly clear and joyful." From the outset, we are told that the experience is uncommon, setting up the expectation that the divine knowledge encoded therein will require the privileged discernment of ministerial expertise. Sarah Edwards's conversion narrative responds to this expectation for a unique and unusual spiritual expression through a language of formlessness that charts an experience of being "swallowed up," "overwhelmed," and "unusually submissive."[46]

Such language characterizes the spiritual ideal of the Northampton revival. *A Faithful Narrative of the Surprizing Work of God* contains several conversion experiences supplying similar data of God's real and historically unprecedented residence in New England. The religious experience of four-year-old Phoebe Bartlett, for example, includes exclamatory sightings of God in the privacy of Phoebe's prayer closet. Edwards's record of Phoebe's conversion replicates the narrative structure of the deathbed testimonies (examined in Chapter 4) while also repeating and putting to very different use some of the major tropes of the Salem witchcraft trials. Phoebe's age, her repeated retreats to her prayer closet, her mother's witnessing of the intensity of her spiritual devotion therein, and her preparation for Christian death render her religious experience quite resonant with the children in Janeway's *Token for Children.*

Additionally and even more explicitly, Phoebe Bartlett seems to undergo a religious experience that has much in common with that of New England's own Cataret Rede, whose mother, Sarah, repeatedly hears her five-year-old daughter's prayer closet exclamations and often finds her in tears. As with Cataret Rede, whose "Experience of the wonderful workings of the Spirit of God" was "faithfully taken from her own Mouth by her Mother," the authenticity of Phoebe's account depends upon parental

46. S[ereno] E[dwards] Dwight, *The Life of President Edwards* (New York, 1830), 171–186.

credibility. Repeating the trope of the mouth as the vehicle of authentic aural truth, Edwards tell us that he learns of Phoebe's experience directly "from the mouths of her parents, whose veracity, none that know them doubt." The account that we get is primarily related through the words of Phoebe's mother, who, like Sarah Rede, finds her four-year-old daughter in tears after her prayer closet devotions until Phoebe's exclamation: "Mother, the kingdom of heaven is come to me!" In each case, the witnessing authority of pious, upstanding mothers takes evidence of God directly from the mouths of their children and offers it up for the minister's and the public's scrutiny. Additionally, Phoebe's spiritual experience transforms some of the key dilemmas of the Salem witchcraft trials into evidence of God's residence in New England, which has become in Edwards's hands "A city that is set upon a hill [and] cannot be hid."[47]

Through his description of Phoebe's husking Indian corn with other children, Edwards constructs an allusion to the Native American presence on the frontier that recalls the tobacco that keeps reappearing in Mercy Short's demonic possession as a psychological link to the Indian wars. A neighbor's cow dies, an incident that would have been read immediately as maleficium in 1692 but that in 1735 merely testifies to Phoebe's pious virtue as she pleads with her father to supply the neighbor with a new cow. The incorporation of the prayer closet evidence of dying Christian children as a central trope of conversion and the revision of the signs of Salem from God's displeasure to divine favor make Phoebe Bartlett an exemplary case of revivalist piety in *A Faithful Narrative of the Surprizing Work of God.* From the original awakenings at Northampton in 1734 through Edwards's second redaction of it in *Some Thoughts concerning the Revival of Religion,* Edwards assigns privileged epistemic authority to children and young Anglo women that is much in keeping with the relegation of this population to a kind of tablua rasa purity in the deathbed tradition.[48]

Witnessing the dying testimonies of women, children, and Native Americans, ministers and community members would gather around the bed in order to observe evidence of the invisible world from the dying saints' final moments. In Northampton, the same scene occurred in the Edwardses' home during the three installments of Sarah's conversion ex-

47. Sarah Rede, *A Token for Youth . . .* (Boston, 1729); Jonathan Edwards, *The Great Awakening: A Faithful Narrative . . .* , ed. C. C. Goen, Works, IV, 199, 200.

48. Janice Knight develops this reading in "Telling It Slant: The Testimony of Mercy Short," *Early American Literature,* XXXVII (2002), 36–69.

perience. In her orally recited and ministerially recorded narrative of this event, Sarah recalls: "When I came home, I found Mr. Buell, Mr. Christophers, Mr. Hopkins, Mrs. Eleanor Dwight, the wife of Mr. Joseph Allen, and Mr. Job Strong, at the house." The gathered presence of community members, including the expertise of the local ministry, infuses the scene of spiritual testimony with theological legitimacy. Sarah's assurance increases through the approving observational stance of those present: "Seeing and conversing with them on the Divine goodness, renewed my former feelings, and filled me with an intense desire that we might all arise, and with an active, flowing and fervent heart give glory to God." Her own inwardly directed spiritual experience extends to a desire that the community collectively bear witness to God's presence. Once she has acquired the legitimacy derived from the approval of certain elect and expert witnesses within the community, Sarah then testifies to the evidence of God upon her soul through an interactive performance that exceeds and surpasses, rather than merely confirms, what her audience has been trained and accustomed to perceiving as adequate evidence of election.[49]

Sarah's conversion narrative invokes a new language in the history of the testimony of faith, because she supplies the "certeine signe" promised to dying saints in 1 John 3:14 but then lives to carry out the saintly life accorded a lay recipient of such a high degree of saving faith. Sarah's retrospective narrative of her conversion registers the experience as historically unprecedented through a language of inexpressibility: "I cannot find language to express, how *certain* this appeared—the everlasting mountains and hills were but shadows to it." The inability to "find language to express" marks a gap between the event of Sarah's conversion and her narrative about the event that exceeds Locke's notion of the event as exclusively knowable through the nominal impressions that it leaves on the mind. Her experience surpasses adequate representation through the conventions of a spiritual relation, Sarah tells us.[50]

Throughout her narrative, Sarah compensates for linguistic inadequacy by performing the truth of God's presence upon her soul through an enhanced sensory encounter. She hears Mr. Buell read the "melting hymn of Dr. Watts' concerning the loveliness of Christ, the enjoyments and employments of heaven, and the christian's earnest desire of heavenly things." The hymn elicits within Sarah a spiritual transformation that is both ren-

49. Dwight, *The Life of President Edwards,* 176.
50. Ibid., 173.

dered in a Lockean epistemic frame and amplified to a spiritual potential that exceeds the limitations of Locke through the simultaneous presence of the indwelling presence of Christ. Sarah explains that the hymn "made so strong an impression on my mind" that her "soul was drawn so powerfully towards Christ and heaven, that [she] leaped unconsciously from her chair." The impression on her mind is Lockean, but its effect can be the consequence only of the immanent, indwelling light of Christ. The experiential feeling of the body drawing upward and a compulsory ascension toward heaven establishes Sarah as near death within this moment. As such, her conversion narrative refers retrospectively to the deathbed testimonies to come before, condensed with the most vividly immanent true evidence of grace just in the moments before death.[51]

Jonathan Edwards encapsulates these anomalous elements in his redaction of Sarah's conversion narrative in *Some Thoughts concerning the Present Revival of Religion:* its formal proximity to the deathbed scene, the language of inexpressibility paired with increased certainty, and the uniqueness of a divine sensory encounter that includes heavenly visions and bodily movements. Capitalizing on its epistemological potential of Sarah's conversion, her testimony becomes the evidentiary fulcrum of a tract designed to legitimate the revival through an unequivocal claim for the presence of God in New England. Sarah stands as a test case both for an exemplary model of piety and for establishing a macrocosmic connection to the particular forms of evidence displayed on individual souls in Edwards's revival as well as the larger unfolding of the redemption story that Edwards foretold in his *History of the Work of Redemption,* preached as a sermon series in 1739, exactly one year after the first manifestations of God's presence upon the soul of his wife.[52]

Sarah's conversion testimony as it was originally performed and then redacted by Edwards as an exemplary model of piety marks the culmination of Edwards's philosophical and ministerial goal of using the "indwelling

51. Ibid., 177.

52. Edwards works to establish this connection between the individual soul, the revival, and the larger, cosmic structure of nature and the universe through scripture from John 3:8 that begins *Some Thoughts concerning the Revival: "The Wind bloweth where it listeth, and thou hearest the Sound thereof; but canst not tell whence it cometh, and whither it goeth.* We hear the Sound, we perceive the Effect, and from thence we judge that the Wind does indeed blow; without waiting, before we pass this Judgment, first to be satisfied what should be the Cause of the Wind's blowing from such a Part of the Heavens, and how it should come to pass that it should blow in such a Manner, at sch a Time." Jonathan Edwards, *Some Thoughts concerning the Present Revival of Religion in New-England . . .* (Boston, 1743), 2.

principle" to counter Locke's epistemological limits while still incorporating Locke's method for the empirical use of sensory data. Rather than constitute a threat through a radical claim to a kind of spiritual authority, Sarah's conversion narrative represented a moment of tremendous opportunity. Edwards's absence from the occasions of Sarah's progressive conversion generates an aura of authenticity around the event, reflected in Edwards's choice to displace Sarah chronologically as well as spatially, away from Northampton and away from the sway of his ministerial power. His revision of Locke can be sustained only by the autonomy of his lay converts who speak for themselves, through the indwelling presence of the light of Christ. Sarah's conversion was not influenced by ministers of the likes of "Whitefield" and "Tennet." Rather, it marks an exclusive communion between Christ and her soul. Sarah herself is not even present in the moment of her conversion. Picking up on the language of ascension in his wife's account, Edwards explains that Sarah's soul seems to leave her body in the most intense moments of her conversion. This produces an experience of being

> perfectly overwhelmed, and swallowed up with Light and Love, and a sweet Solace, Rest and Joy of Soul, that was altogether unspeakable; and more than once continuing for five or six hours together, without any Interruption, in that clear and lively View or Sense of the infinite Beauty and Amiableness of Christ's Person, and the heavenly Sweetness of his excellent and transcendent Love.[53]

Edwards buttresses the authenticity of this experiential description by explaining that these are the "Person's own expressions." Through an inwardly directed and phenomenal experience that can be known only by the recipient of Christ's indwelling light, Sarah's soul became immersed in a total sensory encounter with the divine. This passage, directly paraphrased from Sarah's testimony, produces what Henry More describes as the journey of the soul after death in *Immortality of the Soul.* More describes the soul's passing out of the body and merging with the "vast Ocean" and "sensible Spirit" of the infinite world. But where More could identify this only as the postmortem journey of the soul, so as not to further offend his contemporaries, Sarah Edwards successfully renders this theory of the soul as an integral and authentic part of one's spiritual experience on earth. Edwards utilizes the separation of Sarah's soul from her body to generate his

53. Ibid., 36.

model in *Some Thoughts concerning the Revival of Religion* of the ideal of an "indwelling principle" without a subjective self, even though her subjective self persists long after this "indwelling principle" gets extracted and recorded.[54]

This transformation of Sarah's conversion experience from a marginal and anomalous form of spiritual experience to the evidentiary crux of Edwards's philosophical and theological program recuperates the failures of witnessing with which the genre began: the failure, that is, of the testimonial form to bear witness to the evidence of grace recorded on female souls, to produce the patterns of silence, reservedness, and reluctance as in Thomas Shepard's "Confessions" (traced in Chapter 1). In contrast to the male witnessing subject entrusted by the audience to guide it through an inward spiritual journey, Shepard's "Confessions" displaced the female testifier from her position as reliable witness to the evidence of her soul. Consequently, women relied heavily upon the discerning authority of the minister who teased out the signs of grace that had gradually accumulated within the soul yet still remained unformed. Mary Angier, for example, testified to her own malleability after learning from a neighbor that "she was God's clay," a metaphor from Isaiah 64:8; Isabell Jackson finds comfort in "a godly minister's instruction" that teaches her that she is the "bruised reed" from Isaiah 42:3. This record of female souls containing the sense of divinity but lacking perceptual form and interpretive structure was what Edwards would ultimately perceive as a solution to the problem of hypocrisy. The patterns of reservedness and reluctance in Shepard's records of female testifiers might have modeled for Edwards the suitability of women and younger girls for his empirical study of the soul. Whereas Thomas Shepard built his empirical project around a masculine norm of the ideal witnessing subject, Edwards did the reverse, uncoupling the subjective experience of the witnessing subject from the soul's evidence of grace.[55]

A Divine Typology of the Natural World

The initial phase of Sarah's conversion experience, in 1758, came on the heels of Edwards's publication of *A Faithful Narrative of the Surprizing Work of God,* which established him as a successful revivalist preacher with an in-

54. Ibid.; Henry More, *The Immortality of the Soul . . .* (London, 1659), 276.

55. Michael McGiffert, ed., *God's Plot: Puritan Spirituality in Thomas Shepard's Cambridge* (Amherst, Mass., 1994), 170, 213.

ternational reputation. The final phase occurred just before *Some Thoughts concerning the Revival of Religion,* which Edwards penned to garner support for the Northampton revival despite mounting criticism and charges of enthusiasm. In the time between the initial awakening of Christ within his wife's soul and the culmination of her saintly experience in 1742, Edwards also preached his most ambitious sermon series, *A History of the Work of Redemption.* In thirty sermons, preached in a newly built meetinghouse in Northampton from March to August 1739, Edwards performed a sustained exegetical reading of Isaiah 51:8: "For the moth shall eat them up like a garment, and the worm shall eat them like wool: but my righteousness shall be for ever, and my salvation from generation to generation." His goal was nothing short of proving that the work of redemption, providentially ordained to occur between the fall of man and the end of the world, was taking place in New England with the Northampton awakenings as its most compelling evidence.[56]

The series works typologically by particularizing New England as representative of the Old Testament transition from Moses to David, beginning with the redemption out of Egypt. Starting with the renewal of the covenant between God and the elect in Deuteronomy 29, the Bible foretells how God preserves his church and "wonderfully possessed his people of this land." Through a classic typological hermeneutic of rendering present historical events as literal fulfillments of the narrative types recounted in the Old Testament, Edwards assigns New England the privileged geographical and spiritual position within this historical overview. New England is where the third phase of millennial fulfillment—from the Fall to the Resurrection to the end of the world—is providentially scripted to take place.[57]

But, unlike earlier examples of this mode of typological reading, such as Winthrop's "Modell of Christian Charity" or John Cotton's *Gods Promise to His Plantation,* Edwards supplements his typology with empirical claims of a microcosmic rendering of this history within the soul of each individual convert. After establishing the series's central doctrine, "The Work of Redemption is a work that God carries on from the fall of man to the end of the world," the first sermon in Edwards's *History* explains this connection between the typological and the empirical. The work of redemption carries on in two locations, according to Edwards: "with respect to the effect

56. Lambert, *Inventing the "Great Awakening,"* 143–150.

57. Jonathan Edwards, *A History of the Work of Redemption,* ed. John F. Wilson, Works, IX, 193.

wrought on the souls of the redeemed" and "with respect to the grand design." This proclaimed correspondence sets out to establish a connection between the empirical project of observing the effects of God's grace on human souls and the exegetical practice of reading worldly events within an eschatological frame. For Edwards, the evidence of grace displayed on human souls constitutes a microcosmic representation of the grand story of divinity's manifest presence throughout the natural world.[58]

Edwards adapts traditional Puritan typology to incorporate a broader canopy of natural forms. This epistemic shift in Puritan typological practice was integral to Edwards's attempt to incorporate the observational methods coming out of the natural philosophy of Newton and others into a more comprehensive hermeneutic practice. The link expressed in *A History of the Work of Redemption,* between the empirical and the typological, the individual soul and the grand design of nature, reflects an expansion of this epistemic model in an effort to call upon the full epistemological potential encoded within the science of the soul. If, as Edwards tells us, the work of redemption is most clearly displayed in the thousands of souls manifesting "conversion, justification, sanctification, and glorification," then these souls can be mobilized as the most effective evidence toward the discovery of divine essence within the natural world.[59]

Written in the accessible style of a sermon series that is designed to popularize theology among lay preachers and lay converts, *A History of the Work of Redemption* imagines the resolution of the Calvinist problem of knowledge through a full reversal of the effects of the Fall realized on earth. Redemption, he tell us, moves the Christian community collectively toward the end goal of Christ's Resurrection as well as toward the removal of the veil covering the eyes of the elect and eluding human optic capacities. Redemption comes in the third phase of an eschatological frame that is unfolding in the present. The reversal of the limiting conditions that ensued from the Fall comes in the form of an "indwelling principle" that infuses the minds and souls of redeemed saints with the effects of Christ's Second Coming. This theological transformation prompted the shift from the doctrine of election to a more expansive, democratized Christianity.

58. Ibid., 116, 120–121.

59. For an account of how Edwards expands typology to include manifestations of God within the natural world, see Janice Knight, "Learning the Language of God: Jonathan Edwards and the Typology of Nature," *William and Mary Quarterly,* 3d Ser., XLVIII (1991), 531–551; Joan Richardson, *A Natural History of Pragmatism: The Fact of Feeling from Jonathan Edwards to Gertrude Stein* (Cambridge, 2007).

It marked a movement away from the problems and shortcomings of recognizing assurance toward autonomous spiritual selves with less equivocal and then unequivocal knowledge of the status of their souls.

A History of the Work of Redemption carries forth the kernel of Calvinist Reformation optimism through the new concept of the indwelling light. Edwards's goal for the indwelling light was not only to foreground the inner light's intrinsic relationship to the story of millennial fulfillment but also to entice increasingly secular natural philosophical perspectives that faith had empirical potential. Edwards was the progenitor of modern American Christianity, but he was also deeply embedded in the context of his time. The relationship between his brand of evangelicalism and Enlightenment philosophy is what starkly distinguishes his theology from more modern forms. As it is most ambitiously and optimistically presented in *A History of the Work of Redemption,* Edwards's theological project was not simply to salvage the errand for New England but in fact to recuperate Puritan New England's relevancy for an increasingly enlightened world.

This idea had historical precedent. Ministers such as Samuel Clarke and Benjamin Colman as well as natural theologians such as John Ray, William Derham, and John Williams had each tried to develop new techniques for gaining access to divine knowledge. Ray, Derham, and Williams preached and published tracts on the origins of the world that drew from scientific discoveries such as the telescope, from biblical eschatology, and from the "testimonies" of "heathen philosophers" such as Cicero, Seneca, and Plato. Yet, their study was careful not to disrupt the perspective of "mechanism," as each natural theologian sought to extend natural philosophical insights to a philosophy of the supernatural.[60]

Williams explains that "certainty" of supernatural knowledge comes by way of "declaration" and that revelation coupled with "Human Testimonies" gives "unquestionable Evidence" of God. He repeats this language of certain evidence throughout the sermon, attempting to produce a rhetorical terrain to match a natural one in which God's existence is not only unquestionable but the information conveyed via divine manifestations is accessible and accurate. This rhetorical construction happens formally as well as through the repeated trope of certain evidence. At the beginning, Derham references Robert Boyle's recent lectures and explains that his treatise, like Boyle's lectures, is "sermons," designed to be preached from a pulpit. Natural theologians draw upon the authority of the pulpit to attribute di-

60. Derham, *Physico-Theology,* introduction (unpag.).

vine authority to their oratorical performance and to encode within the natural philosophical landscape the expansion and refraction that the sermonic form allows. Delivered by Royal Society members at the turn of the eighteenth century, lectures on natural theology were tailored to the formal characteristics of the sermon series, the essential ministerial genre of revivalism. Rearticulated through the sermonic form, natural philosophy took on new resonance, claiming to wield essential knowledge of God.[61]

The sermons on natural theology preached around the turn of the eighteenth century present the human soul as a site of new epistemological possibility. Derham explains, for example, that the soul contains a "copy" of divinity and that, when we "contemplate" this copy, we arrive at an understanding of "Divine Nature and Operations" that exceeds the "reach and compass" of human understanding. However, he quickly retreats from the implications of this statement with the dismissive proclamation that he "shall not dwell on this" because, even though the soul is the "superior part of Man," it is "the least known." Williams compares the human soul to a "flower," the "choicest part of the visible creation." This effort to encase the soul in an empirical potentiality rooted in its very status as a terra incognita of the various forms of divinity on earth accords with late-seventeenth-century tracts such as Henry More's *Immortality of the Soul* (1659). But natural theology also yokes this conception of the soul to an entire cosmological scheme, designed to squelch perceived atheistic tendencies more thoroughly and aggressively than ever before. As such, the soul becomes the fulcrum of a renewed investigation by natural theologians and revivalists over the first half of the eighteenth century.[62]

A History of the Work of Redemption picks up where natural theology left off, performing a single biblical verse over the time span of an hour or more and then further explicating the meaning of this verse over several months as each sermon links to the previous. Each sermon unfolds a spatial dimension of the Old Testament journey from the Fall in the Garden of Eden to the dispersal of nations to redemption out of Egypt to the parting of the Red Sea to the coming of Christ. The sermons render these events in an ever-expanding geographical scale that makes New England an explicit type to this ancient story. *A History of the Work of Redemption* thus generates meaning from a single scriptural verse that proliferates, seemingly

61. John Williams, *The Possibility, Expediency, and Necessity of Divine Revelation: A Sermon Preached . . .* (London, 1694), 3, 39.

62. Derham, *Physico-Theology*, 272–273; Williams, *Possibility, Expediency, and Necessity*, 2.

infinitely, over time and across space. The promise of Isaiah 51:8 expands through the course of the sermon so that the audience of listeners as well as readers comes to experience and bear witness to the infinite durability encoded in this phrase. Edwards conveys the meaning of this passage through the form as well as the content of his sermon series, moving from the contained singularity of the doctrine to the eternal dimensions of the message encoded therein. The witnessing of this work carries forth from one sermon to the next over the time of several weeks and months; the breadth of the work spreads like the dispersal of nations following the collapse of Babel from the voice of God in the garden to "remote parts of the world."[63]

Through this accordionlike frame of tersely encapsulated scriptural meaning and rhetorical expansion, Edwards uses the sermon series to address the central Lockean limit on the human capacity to comprehend the divine. In his chapter "The Infinite," Locke explains that, even though the mind is well equipped to receive ideas of infinity from sensation and reflection, there is always a contradiction between the mind's idea of the infinite and the infinite itself. As soon as the received sensations and reflections on the infinite solidify into an idea, the idea takes on a finite quality, thus running counter to the eternal expansion of the infinite. The infinite thus transforms into the finite at the moment of human comprehension. Locke identifies expansion and duration, time and space, as the two components of perception that both allow the mind to receive sensations that gesture toward the infinite and transform these sensations into a finite idea. Our idea of infinite space, for example, comes from our own sense that space is actually boundless, "to which," Locke explains, "Imagination, the *Idea* of Space or Expansion of it self naturally leads us."[64]

But, as soon as the imagination leads us to an idea of what infinite space might be, the idea solidifies according to the finite terms of human comprehension. For Locke, space and time, expansion and duration, link the infinite to the finite, permitting some understanding of the former while also issuing a reminder that we are inevitably bound to the latter. Given the structure of this epistemological limit, expressed at precisely the point in his *Essay concerning Human Understanding* that Lockean empiricism might be employed to intuit divine forms, it is not surprising that Edwards makes duration and expansion key narrative strategies for elaborating meaning in

63. Edwards, *A History of the Work of Redemption,* ed. Wilson, Works, IX, 120, 155.

64. Locke, *Essay concerning Human Understanding,* ed. Nidditch, 211.

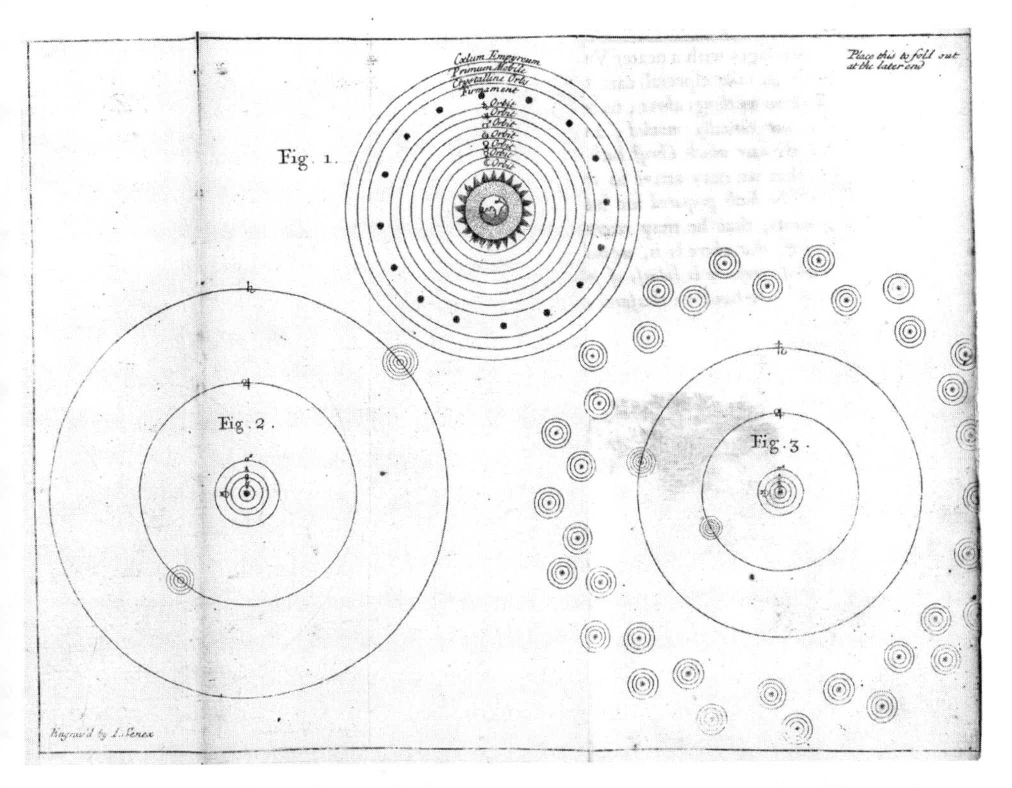

FIGURE 18 • *Foldout, William Derham,* Astro-Theology; or, A Demonstration of the Being and Attibutes of God, from a Survey of the Heavens *(London, 1715). Huntington Library, San Marino, Calif.* Derham uses this image to show how the telescope has amplified the works of God. The image is supposed to demonstrate a rational organization of the heavens and the universe and to expand knowledge far beyond Ptolemy and Copernicus

A History of the Work of Redemption. The genre of the sermon series itself formally resolves the problem set forth by Locke. The idea of the infinite—dually expressed in scriptural verse and doctrine of the sermon—does not have the chance to congeal into a finite idea, for Edwards continues to perform its meaning. Each of the thirty sermons becomes a concentric circle, overlaid upon the one coming before and always returning to the same circumferential point: "My righteousness shall be for ever"; the work of redemption carries on to the end of the world. As such, the sermon series performs the infinite expandability of the cosmic circles featured in the copperplate illustration from William Derham's *Astro-Theology* (1715), transcending the limits of mechanical philosophy through the sermonic form. If the mind of the audience rests or closes in upon this meaning, tending toward the very fixity that Locke argues eludes infinite comprehension, the next sermon opens it again, expanding verse and doctrine across space as well as time, from Eden to Egypt to America.

Of course, the history of redemption does end with a final sermon on the end of the world, preached in Northampton, August 1739. But, by this time, the mental conception of the infinite has transformed into the indwelling feeling of Christ within the souls of Edwards's audience. The indwelling light of Christ produced a sensation of the infinite within the soul that could not be reduced to a finite, mental idea of the infinite because, as Edwards observed and rendered exemplary in his wife's conversion narrative, this sensation eludes expression. It engenders an experience that cannot be reduced to the finite dimensions of Lockean terms. The indwelling light allows the convert to live within the end times forecast by the end of *A History of the Work of Redemption* as well as within the historical present. It allows the immanent realization of the scriptural within the natural world. This was Edwards's way of bridging the ever-widening schism between the material and the spiritual, the visible and the invisible, the finite and the infinite, that was rapidly creating an irreparable divide between early modern natural philosophy and theology.

Through his renowned effectiveness as a revivalist preacher and through the relative spiritual autonomy and assurance of salvation provided through this theological concept, Edwards secured the place of the indwelling principle across religious circles, from popular preachers, to lay converts, to ministerial elites. However, he also sought to integrate the empirical potential of the indwelling principle among philosophical circles, such that followers of Locke and Newton as well as the growing circle of Continental philosophers committed to revitalizing nature as a site of divine discovery might incorporate New England's science of the soul as a concomitant and integral site for pursuing this form of inquiry.[65]

True Religion

Edwards's *Treatise concerning Religious Affections* attempts to develop a technique of philosophical and ministerial discernment for recognizing signs of grace with increasing and unprecedented certainty. Initially, Edwards

65. According to Peter Hanns Reill's account, the emergence of a new natural vitalism was inaugurated by Buffon and followed by such philosophers as Alexander von Humboldt. Reill, *Vitalizing Nature in the Enlightenment* (Berkeley, Calif., 2005).

preached *Religious Affections* as a sermon series. This was not simply a gesture of translating philosophy into a popular form. Rather, Edwards both believed and depended upon the testimonies of his lay converts for answers to the most pressing contemporary philosophical problems.

Lay testimonies, such as performed by Abigail Strong in 1725 or reported by his wife, Sarah, in 1738 and 1742, supplied the data for inductively resolving the post-Reformation crisis of knowledge, catalyzed by a late-seventeenth-century mechanical philosophical frame. Edwards presented his philosophical and theological scheme for arriving at a certain sign of regeneration directly to his congregants, for this was the population that he hoped would supply him with the necessary data for advancing philosophical inquiry. The published treatise *Religious Affections* revises and expands the ideas developed in this initial setting such that they might reach a broader audience of philosophers, skeptical ministers, and theologians. The treatise encodes the minister's oral performance of the expansion and duration of the Lockean infinite within its formal structure. A taxonomy of grace unfolds that explains the impossible: how to extract a certain sign of God from the human soul without neglecting a rigorous adherence to the consequences of the Fall, namely the dual problem of hypocrisy and the condition of unknowable knowledge that, since Augustine's revival in Calvin, had rendered such knowledge desirable but unattainable.

Edwards's commitment to these dual consequences of the Fall explains why Thomas Shepard is the most-often-quoted theologian or minister in *Religious Affections.* A passage from Shepard's *Sound Beleever; or, A Treatise of Evangelicall Conversion* (1645) emphasizes Edwards's and Shepard's shared sense that individuals "cannot so clearly discern and feel" evidence of what is "wrought in the soul." *Religious Affections* references Shepard most extensively on hypocrisy while expanding upon Shepard's caution that the interpretive framework surrounding sensory experience cannot be relied upon. Shepard despises "the Formalist who contents himself with some holinesse, as much as will credit him," and warns against a single individual's ability to corrupt the entire church. While Shepard addressed this danger by instituting a pedagogy of transparency and warning his congregants to "feare to sinne . . . in secret, unlesse thou canst find out some darke hole where the eye of God cannot discerne thee," Edwards supplements these warnings with a plan to methodologically eliminate subjective feeling and false discernment from the public performance of grace. He builds into these public performances a long history of warnings issued to potential converts, such as we see in Timothy Edwards's sermon on the practice, where he instructs his congregants in the proper frame that they should bring to the scene of testifying.[66]

The objective of *Religious Affections* is to arrive at a form of religious truth that could still be mapped and taxonomized according to the cautious discernment that Edwards learned from Shepard. Edwards uses Shepard's critique of formalism as an appropriate corrective to the depleted and formulaic status of religious testimony among many of his contemporaries. But he also augmented the technique of empiricism inaugurated by Shepard by doing more to preserve the authentic experience and secure more certain evidence of God. Edwards celebrated the anomalous moments in lay conversion while Shepard privileged more conventional forms of intelligibility. This division explains why Edwards's women occupy quite different epistemological positions in *A Faithful Narrative of the Surprizing Work of God* and *Some Thoughts concerning the Revival of Religion* than they do in Shepard's "Confessions." Edwards sought to capture rather than silence the anomalous forms of true grace.

Edwards writes that the production of grace in the human soul is a "new inward perception or sensation of their minds." As manifestations of true religion, gracious affections are "entirely different in nature and kind from anything that ever [human minds] were the subjects of before they were sanctified." The goal of *Religious Affections* is to separate these manifestations from the persons in whom they are embodied, to read them as unadorned marks of divinity. *Religious Affections* completes Edwards's grandfather's project of developing a theory of the nature of saving conversion; but, where Stoddard maintained that full assurance was possible only as an individualized phenomenon, Edwards writes *Religious Affections* as the opus of true divinity, positing that an indwelling and native light, the essence of true religion, can supply epistemic certainty despite the risk of hypocrisy. Edwards takes the opposite approach from his grandfather by isolating the objective quality of this light rather than the subjective authority of each testifier to intuit it accurately. This has the consequence of reclaiming pious practice from an inwardly directed and secretive domain in order to relocate the display of gracious affections within the public arena.[67]

According to Edwards, the testifier receives the generative principle of

66. Edwards, *Religious Affections,* ed. Smith, Works, II, 163; Thomas Shepard, *The Sincere Convert: Discovering the Paucity of True Beleevers and the Great Difficulty of Saving Conversion* (London, 1641), 20–21, 34.

67. Edwards, *Religious Affections,* ed. Smith, Works, II, 205.

grace in the soul as Christ was conceived in the womb. This analogy highlights the notion of passivity as the crucial ingredient to the fruition of an indwelling principle through which divinity becomes actualized in the soul. The soul becomes the "seat" of spiritual properties, or "signs" of grace that have "carnal" and "fleshy" manifestations. Once this indwelling principle actualizes divinity, the convert undergoes an ontological transformation. He or she develops a "new nature" that is discernible and verifiable to external witnesses. The fusion of the "indwelling principle" within the soul itself marks the closest form of epistemic access to divine essence.[68]

By reintroducing a version of the practice of public profession in his Northampton congregation, Edwards attempted to cultivate this technique of recording indwelling formlessness among his congregants in order to observe the soul as a space of divinity. In doing so, Edwards separated the testifier's subjectivity from the objective records of gracious manifestations as a general technique for recording evidences of his congregants' souls. He eliminates the discerning perceptive and witnessing capacity that casts the subjective experience of grace as the root of the risk of hypocrisy. This increases degrees of certainty surrounding evidence of grace while also increasing the need for the privileged observational discretion of a minister expert. Edwards's inducement of the indwelling principle among his converts greatly reduces their agency as well as their authority to know, decide, and display the contents of the soul. Subtly working beyond the testifier's interpretation of his or her own sensory perceptions, Edwards demanded of the genre a more certain epistemic access to divinity at the expense of the individual's autonomous narrational capacity to describe this form of evidence.

Emphasizing the necessity of audience to verify the legitimacy of the experience, Edwards cautions his congregation against testifiers who have become too adept at performing what the audience wants to hear. One should look for individuals who "tell of their experiences . . . not with such an air that you as it were feel that they expect to be admired and applauded, and won't be disappointed if they fail of discerning in you something of that nature; and shocked and displeased if they discover the contrary." Experiences should be unique so that the response of the audience cannot be anticipated. This antiformalist focus extends to converts as well as audience. The convert should resist placing any interpretive construct on the sensory encounter, while the witnessing audience should operate inductively, wait-

68. Ibid., 129.

ing for a singular expression of the indwelling principle. Edwards writes that the witnessing audience should recognize the "indwelling principle" through the testifier's expression of a "new inward perception or sensation" that is "entirely different in its nature and kind, from anything that ever their minds were the subjects of before they were sanctified."[69]

Ontologically different from anything that the testifier might have experienced or even conceived of before the conversion experience, the indwelling principle is Edwards's most complete and final attempt to develop a science for discerning true, "native" religion. Native religion was connected to the kernel of Reformation optimism, or the idea that the revelation of grace came exclusively from God, untouched by the human hand. Humans were powerless to effect or induce their own conversion; when authentic grace became manifest within the convert's soul, it reflected this untouched, unadorned, and pure quality.[70]

Edwards's notion of spiritual singularity draws upon the Baconian sui juris, or the idea that the mind must seek the anomalous in nature in order to gain access to truth. For Bacon, as for Edwards, the "singular" guards against false appearances. According to Bacon, false appearances can be "recalled to examination" or corrected through the continued contemplation of singular, "true" natural phenomena. Edwards applies the same logic to conversion by urging his congregants to continually reexamine what their sensory encounter appears to be until their soul becomes the seat of the "indwelling principle," the singular spiritual phenomenon that offers access to divine truth.[71]

Through his disciplined rules for "judging of persons' experiences," Edwards rejects the mechanical philosophical and Lockean denial of human access to divine essence by rendering the soul the spatial entity of divin-

69. Jonathan Edwards, "Directions for Judging of Persons' Experiences," in *Writings on the Trinity, Grace, and Faith*, ed. Lee, Works, XXI, 524. "Indwelling principle" quoted in Chai, *Jonathan Edwards*, 24. This is the idea behind the notion of the "surprising" conversion and, to a large extent, what Edwards finds so compelling about the conversion narrative of his wife, Sarah. Edwards remarked that her "constant Assurance" was different and more intense than "any he ever saw appearance of, in any person." Dwight, *The Life of President Edwards*, 186, 188.

70. Both Edwards and Brainerd privilege a concept of "native" religion. The term appears first in Edwards's "City on a Hill" sermon; it is also used by Brainerd in his preface to Thomas Shepard, *Meditations and Spiritual Experiences of Mr. Thomas Shepard* (Boston, 1747). Brainerd refers to Shepard's "native Excellency" (i).

71. Mary Poovey, *A History of the Modern Fact: Problems of Knowledge in the Sciences of Wealth and Society* (Chicago, 1998).

Bacon discusses this concept of sui juris in *The Advancement of Learning* (1607); see Markku Peltonen, ed., *The Cambridge Companion to Bacon* (Cambridge, 1996), 251.

ity (described in his essay "Of Being"). This rendering of the soul as the visible ontological immanence of divine essence required such scrupulous attention to the evidence of divine grace that the fallacy as well as the autonomy of the testifier was completely eliminated from the witnessing practice. Sermons were about how to isolate this spiritual soul and make it useful. Working toward this project of isolation, Edwards transformed the genre of the conversion narrative into the spiritual biographies found in such tracts as *A Faithful Narrative of the Surprizing Work of God* and *Some Thoughts concerning the Revival.* Edwards then attempted to transfer the "indwelling principle" performed by the converted subjects of these tracts back into the testimony of faith. Working within the formal conventions while also transforming the genre that he heard in his father's church, Edwards attempted to isolate the evidence conveyed through the indwelling principle and preserve its epistemic certainty through the testimonial form.

Up to the time that Edwards revived it, the testimony of faith had modeled evidence of grace that was necessarily uncertain owing to its fleeting, numinous, and barely traceable quality. Edwards proposed an alternate route of access to this "hidden" and "subtle spirit." His notion of an "indwelling principle" transformed the soul into a space where divine matter became actualized. The indwelling principle illuminated within human souls a science of signs that would lead to the discovery of "true Religion" through the phenomenal transformation of the converted "divine mind." Through the indwelling principle, the divine mind could conceive of an essence existing beyond epistemological limits. In his philosophical writing and practices of observing conversion, Edwards attempted to bring testimony to a state of epistemic certainty through the notion of an indwelling principle as an entity that could be abstracted from the testifier's subjective experience. Following the goal of arriving at a true religion that could respond to the quandaries of modern science, the testimony of faith developed at Northampton dissolved the problem of hypocrisy by replacing subjective epistemological limits with objectified scientific accuracy. The human soul, like the natural world, gets infused with Christ's light, establishing a micro-macrocosmic connection coded as divine truth.[72]

Ultimately Edwards's attempt to attain religious certainty through testimony proved to be impossible. The scientific methods that Edwards in-

72. Edwards discusses the indwelling principle at length in *Religious Affections.* It might be summarized as the implanting of a "spiritual sense" that leads to new sight, a "spiritual opening of the eyes in conversion." Edwards, *Religious Affections,* ed. Smith, Works, II, 219.

stituted exceeded the limits of the genre: the integral status of the speaker's subjectivity as well as the inevitable dilemma of human fallacy could not be abstracted from the testimonial practice. The testimony of faith could not, in other words, wield the objective evidence of the soul that Edwards sought, even though the genre of the conversion narrative would transform through the course of the first Great Awakening, eventuating in new forms that permitted the formal representation of religious certainty at the dawn of a new era of American Christianity. Like Shepard in Cambridge, Eliot in Natick, Williams in Rhode Island, and Cotton Mather in Salem, Edwards fails in his empirical quest to transform Northampton into a laboratory of grace. His dismissal from the pulpit marks the end of the scientific uses of the genre. Given the prevalence with which the testimony of faith was practiced elsewhere, in Boxford (1738–1752), Westborough (1727–1771), and Sturbridge Village (1757–1762), for example, it was not the conservatism of the practice itself that bothered Edwards's critics, but rather his attempt to render the practice certain.[73]

In the preface to the "Farewell Sermon," preached in Northampton on June 22, 1750, Edwards suggests that his demand for testimonial certainty was one of the primary reasons for his dismissal from the pulpit. Defending himself against the "slanderous representations" of his ministry, he begins the preface by cataloging the accusations of misconduct that his congregation has brought against him. This controversy over church communion is commonly read as a reaction to a conservative theology that Edwards assumes toward the end of his career, often interpreted as a frustrated attempt to counteract the proliferation of enthusiasm surrounding him. Congregational critique came from a religious community whose experience of grace had moved past more antiquated forms of New England orthodoxy. Clerical resistance to Edwards's theological transformations came in part from a growing sense among the religious that, in a world transforming through rapid economic growth and the ascendancy of more secular concerns, ministers had to practice more inclusive methods of admitting church members.[74]

As part of his own defense, Edwards denies that he has "undertaken to

73. The Westborough Relations are at the American Antiquarian Society, and the other two collections are housed in each town's respective historical society. Thanks to Catherine Brekus for sharing her copies of the Sturbridge Village "Experimental Relations" with me.

74. Jonathan Edwards, *The Works of President Edwards* (Worcester, Mass., 1808–1809), I, [103]–141. On controversy over communion, see Marsden, *Jonathan Edwards;* Edwards, *Ecclesiastical Writing,* ed. Hall, Works, XII.

set up a pure church, and to make an *exact and certain* distinction between saints and hypocrites, by a pretended infallible discerning of the state of men's souls" (my emphasis). But this is, in fact, precisely what he does. He tries to deflect attention from the accusation that his test of faith demands epistemic certainty by quoting directly from professions in his "Farewell Sermon" that were deemed passable even though clearly lacking exact and certain evidence of faith. One simply and succinctly states, "I hope I truly find in my Heart a willingness to comply with God." Although he found this testimony lacking in the "fuller and more particular" evidence that he sought, Edwards explains that he accepted it rather than "break" with his people. The "Farewell Sermon" clearly and repeatedly refutes the accusation that Edwards demanded certain evidence of grace from his testifiers.[75]

As in his rendition of Sarah Edwards's testimony in *Some Thoughts concerning the Revival,* the short testimonies of faith included in the preface to the "Farewell Sermon" are anonymous. While we expect from the context of his sermon that the testimonies would be from two Anglo congregants in Northampton, they in fact appear to be from two Mahicans in his Stockbridge mission, Cornelius and Mary Munnewaumummuh. Edwards suggests their authorial identity in the "drafts of the professions of faith" that he prepared for the Northampton church committee during the controversy. He inscribes their names with the longest edition of profession drafts prepared for this manuscript collection. The testimonies summarized and published in the "Farewell Sermon" are brief and conventional enough to make their authorial identity debatable, but there is much overlap in the language between the published passages and the manuscript drafts ascribed to Cornelius and Mary Munnewaumummuh.

Reading the Mahican testimonies alongside the other anonymous (presumably Anglo) testimonies shows uneven displays of spiritual evidence. The Anglo testimony is an exemplar of the genre. In a short, succinct, and poignant form, it encapsulates the story of Adam's fall "by eating the forbidden Fruit" from the "Tree of Knowledge," the redemptive outcome despite this unforgivable transgression through the covenant of grace, the "free grace" dispensed upon the soul of the testifying saint, and the testimonial rendering of this kernel as a social offering as the speaker "desires publicly to Join my self to the People of Christ." This is the ideal and long-

75. Jonathan Edwards, "Two Fragments of Preface to Farewell Sermon, 1750," Jonathan Edwards Collection, fol. 1244, Beinecke Rare Book Library, Yale University; Edwards, *Works of President Edwards,* I, [103].

idealized testimony of faith. It supplies a microcosmic version of *A History of the Work of Redemption* and the certain evidence of "free grace" that can be extracted and extrapolated to fulfill empirical desire and the covenantal promise of the "city on a hill" simultaneously.[76]

The anonymous testifier supplies the evidence that Edwards sought as an experiential resolution to the growing division between theology and natural philosophy. Given the implications of this evidentiary rendering, it is no wonder that Edwards decided to publish the much more subdued testimonies of two Native Americans in place of this one. In a publication decision that recalls Thomas Shepard's careful guarding of the testimonies of faith in a small church notebook, Edwards offers the Anglo testimony anonymously and only in manuscript form, intent upon protecting this form of evidence from more skeptical attack or from inappropriate use and display. He supplies this testimony to the Northampton church committee in a final attempt to demonstrate the effectiveness of this testimonial technique in culling more-certain-than-ever evidence of God from human souls.

For his record of this event, published for posterity as well as for his congregation and the members of the church committee, Edwards includes the testimonies of two Native American converts, whose words had been heavily edited by Edwards. The fragmented record of either Cornelius or Mary Munnewaumummuh (authorial identity is blurred) contains the following revisions: Edwards reveals his editorial hand in tailoring the expressions to more accurately suit the "sentiments and experience" of the professor's heart according to his own discerning expertise. The crossed-out lines and overwrites—"It is in my Heart I have been enabled to yield my self to God"—intensify at the moments in which Cornelius supplies extra evidence of her inward searching and self-examination. The testimony becomes a record of the evidence itself, selected and refined at the transcriber's prompting while detached from Cornelius's subjective identification with inward signs.

> I hope I do truly find a Heart to give up my self wholly to God according to the Tenor of that Covenant of Grace which was Sealed in my Baptism and to walk in a way of that Obedience to all the Commandments of God which the Covenant of Grace requires as long as I live.[77]

76. "Drafts of Professions of Faith," Edwards Collection, fol. 1245.

77. "Two Fragments of Preface to Farewell Sermon, 1750," Edwards Collection, fol. 1244; and see also Edwards, *Works of President Edwards,* I, 106. According to Kenneth Minkema, who has edited and described the "Drafts of Professions of Faith" as well as generously shared his transcription with me, Edwards composed these sample professions for the benefit of the Northampton church

The published preface to the "Farewell Sermon" mutes and neutralizes these signs of grace. Edwards supplies a rationale for taking the "Liberty" to interpret testimonial utterance: the compelling need to distinguish true religion from hypocrisy. To prevent the "rashness" of "profession making" that leads to "hypocritical and deceived" accounts of faith, Edwards requested that each testifier more "seriously examine and search his own Heart." This request marked an attempt to institute the conversion process described in "Directions for Judging of Persons' Experiences." By resisting a quick interpretation of the inward evidence of the soul, the testifier dwells in the sensory encounter, passively absorbing sensory data. The "indwelling principle" has time to congeal and form within the soul. The empirical discovery of this true, "native" religion became Edwards's primary reason for reintroducing the practice within his congregation. While on the one hand explaining to the council that he has "ever declared against insisting on a Relation of Experiences" for "Admission into the Church," Edwards also insists that the "Relation of Experiences" offers an occasion to witness "the great Things wrought wherein true Grace and the essential Acts and Habits of Holiness consist."[78]

True religion consists in the sensory encounter itself, as grace becomes actualized as the immanent "indwelling principle" of the human soul. The "Farewell Sermon" concludes by returning to an idea that he most likely heard in the moving sermon that his father gave in East Windsor in 1720. Conveying his theory of public profession, Edwards explains that he did not mean to demand certainty of grace as a condition of church membership. Rather, what he discovered through the original genre of church membership was a mechanism for discerning true grace.[79]

---

committee. "During this same time or shortly after—perhaps at the beginning of the Stockbridge period—he drew up lengthier drafts, one in its entirety, followed by four revised or added portions. . . . After the first and longest addition Edwards inscribed the names of two Indians, apparently to indicate that this was part of the profession they submitted or agreed to. Whether they had any part in composing the text is a matter of speculation, though the changes in this version are minimal."

78. Edwards, *Works of President Edwards,* I, [107]; Edwards, "Two Fragments of Preface to Farewell Sermon, 1750," Edwards Collection, f. 1244.

79. In a sermon that explains his rationale for and pedagogy of the testimony of faith, Timothy Edwards writes: "When This is done upon a weighty and Solemn Occasion, as to give God the Glory of his Grace, and Gain the Charity of his People it is that which is very pleasing and acceptable to the Godly, and them that are Serious and well Minded Among us; and which the Church hath Many a Time given a public Testimony of their Liking and approbation of" (21). My thanks again to Kenneth Minkema for his transcription and for sending me this manuscript. "Timothy Edwards Manuscript, c. 1720–25," Edwards Collection.

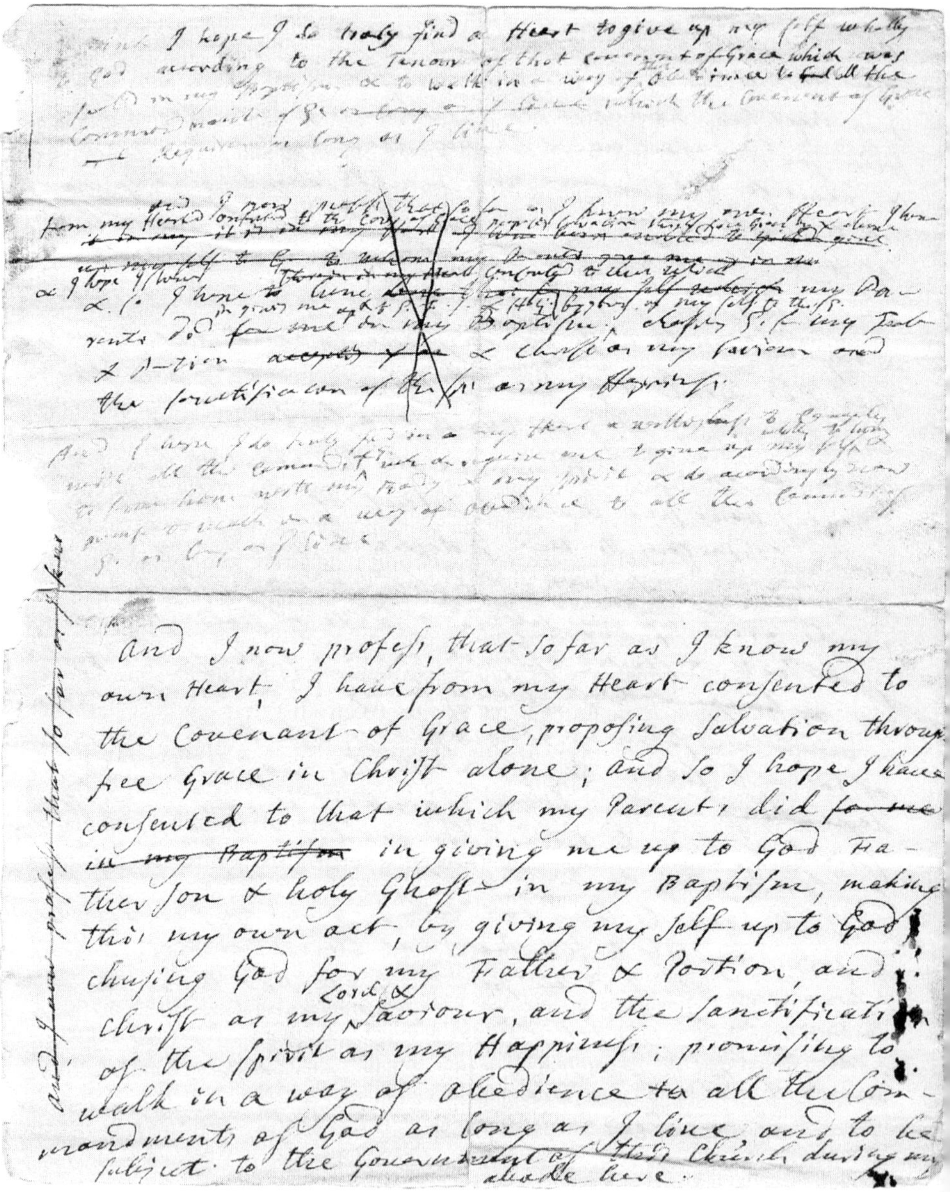

And I now profess, that so far as I know my own Heart, I have from my Heart consented to the Covenant of Grace, proposing Salvation throu[gh] free Grace in Christ alone; and so I hope I have consented to that which my Parents did ~~for me in my Baptism~~ in giving me up to God Father Son & holy Ghost in my Baptism, making this my own act, by giving my self up to God, chusing God for my Father & Portion, and Christ as my Lord & Saviour, and the sanctification of the Spirit as my Happiness; promising to walk in a way of obedience to all the Commandments of God as long as I live, and to be subject to the Government of this Church during my abode here.

FIGURE 19 • *Testimony of Cornelius or Mary Munnewaumummuh. 1740s. Jonathan Edwards Collection, General Collection, Beinecke Rare Book and Manuscript Library, Yale University, New Haven, Conn.*

and I do now appear before God & his People solemnly to give up my self to God to whom my Parents gave me upon my Baptism having so far as I know my own Heart chosen Him for my Portion & set my Heart on Him as my greatest & sweetest Good. and now would solemnly give up my self to Christ ~~whom~~ having as I hope seen my need of Him ~~to~~ being sensible of my sin & misery as I am in my self the insufficiency of my own Righteousness & my unworthiness of any mercy and my deserving that God should cast me off forever. and also seen the sufficiency of Jesus Christ as a saviour. I now also appear ~~to~~ openly to Renounce all the ways of sin which I hope I have seen the Hatefulness of sin ~~having~~ being made burdensome to me I desire to spend my Life in watching striving & fighting against it. and would give up my self to a Life of holiness earnestly to seek & strive after it as what I chuse & delight in ~~& hunger thirst~~ having hungered & thirsted after it & would now solemnly give up my Life to the service of God & to follow Christ in all his ways & ordinances ~~humbly depending on his H~~ & to give up my self to the Holy Spirit of God to follow his Leading & Guidance humbly depending on his Influences to ~~enable~~ sanctify me & enable me to live an holy Life. And ~~as~~ I now desire to join my self to the People of Christ as those whom I hope my Heart is ~~ess~~ peculiarly united to as ~~the~~ my Brethren in Christ. I profess universal forgiveness & good will to mankind & promise to be subject to the Government of this Church during my abode here.

Cornelius / Mary. [illegible]

The sermon that Edwards preached in 1750 issues a farewell to more than his Northampton congregation. Recognizing that his long-standing attempt to achieve epistemic certainty in the testimony of faith was also his own undoing, the two professions formally announce that Edwards recognizes the futility of his own attempt to offer evidence of grace to natural philosophy as unambiguous evidence of God. The attempt to render the human soul a "certain and exact" space of scientific inquiry failed because it appeared at a time when natural philosophers were becoming comfortable with the condition of uncertainty's surrounding modern inquiry. This was not simply because science was becoming more modern and secular, but rather because the condition of uncertainty surrounding knowledge was becoming a condition of a rapidly modernizing and professionalizing scientific world, not an obstacle to overcome through cosmic reconciliation.

Protestantism, conversely, shed the humility of its post-Reformation epistemological frame, gradually moving away from Calvinism's equivocal, hesitant, and uncertain forms of knowledge toward a religious landscape of increasing religious truth. This historical transformation has been understood by scholars of Christian history in America as an incremental transition from a doctrine of election to one of universal salvation as the New Lights replaced the Old Lights with the concept of Christ's descending spirit, eschatologically capable of bringing more believing Christians into the fold. This transformation depended on a correspondent rise in degrees of religious conviction; believers had to be certain that their faith was true. Sincerity and authenticity—rather than humility and doubt—became the hallmarks of rhetorical persuasion.

I have been proposing a direct link between Edwards's capacity as a minister and as a philosopher. His break from the Northampton congregation ensued from his attempt to import a framework of certainty in order to salvage the validity of the testimony of faith for a Puritan and natural philosophical inquiry into divine origins. This attempt failed precisely because of what ultimately became too much of a forced link between Lockean empiricism and evidence of grace. Edwards's exile to Stockbridge signaled a deep epistemological split that marked the end of the science of the soul, recounted in this study as the neat interdependency between a particular intellectual history and a history of religious practice. It is not mere coincidence that prompted Edwards to continue his *History of the Work of Redemption* in an unfinished manuscript, written across the pages to a published copy of the "Farewell Sermon." *A History of the Work of Redemption* was not published in Edwards's lifetime. Though he would not abandon the project of marrying certain signs of grace to a natural philosophical frame, recast within a biblical topos, he knew that the enlightened world was no longer receptive to it. He saw his dismissal as the end of a philosophical as well as an ecclesiastical era.

The record of the manuscript written across the printed page of the "Farewell Sermon" is a fitting palimpsestic representation of the conclusion to an era of soul science, for *A History of the Work of Redemption* might be thought of as a culminating and grandiose moment in the history of Reformation optimism. It is a history in which the virtues gained by Milton's Eve as she tastes the forbidden fruit promise to become sustainable traits of the postmillennial convert rather than small and temporary openings into the "things visible in heaven" but fleeting and punishable for the transgressive act of inquiry that preceded them. One of the central philosophical questions of Milton's epic—Do the rewards gained by transgression outweigh the consequences of the act?—is, as discussed in the Introduction, one of the main philosophical questions of the seventeenth century. In his most ambitious sermon series, Edwards seeks to address this question, not as one of balanced proportionality, but rather as a dilemma in need of millennial resolution. *A History of the Work of Redemption* transcends the limits that Locke applied to the infinite in order to capture God's promise for a certain kind of revealed truth. The unfinished manuscript copy, not to be published until 1774, sixteen years after Edwards's death and even then in Edinburgh, is faintly legible scribbled over the title page to Edwards's "Farewell Sermon."

But, even as the epistemological relevance of Edwards's project ceased to matter in a world of knowledge that focused increasingly and exclusively on the materiality of nature, the science of the soul did not end as a practice. Rather, it migrated to an increasingly marginalized domain of knowledge acquisition, represented literally through Edwards's exile to Stockbridge. It is not that God has stopped working, Edwards tells his correspondents in Scotland, but rather that God has stopped working through the nationalist and imperialist agenda of the "city on a hill" that could not only bear witness to God's residence among his elect but also supply evidence of God's essence for the Royal Society. "God is still carrying on his work," Edwards tells William McCulloch, "if not in one place, yet in another." Though his work seems to have ceased in the west of Scotland as it had in New England, it has "broken out" in "some parts of the United Netherlands." On the eastern seaboard of America, God's work is increasingly prevalent in "New Jersey, Pennsylvania, Maryland, and Virginia." The franchise of New Light

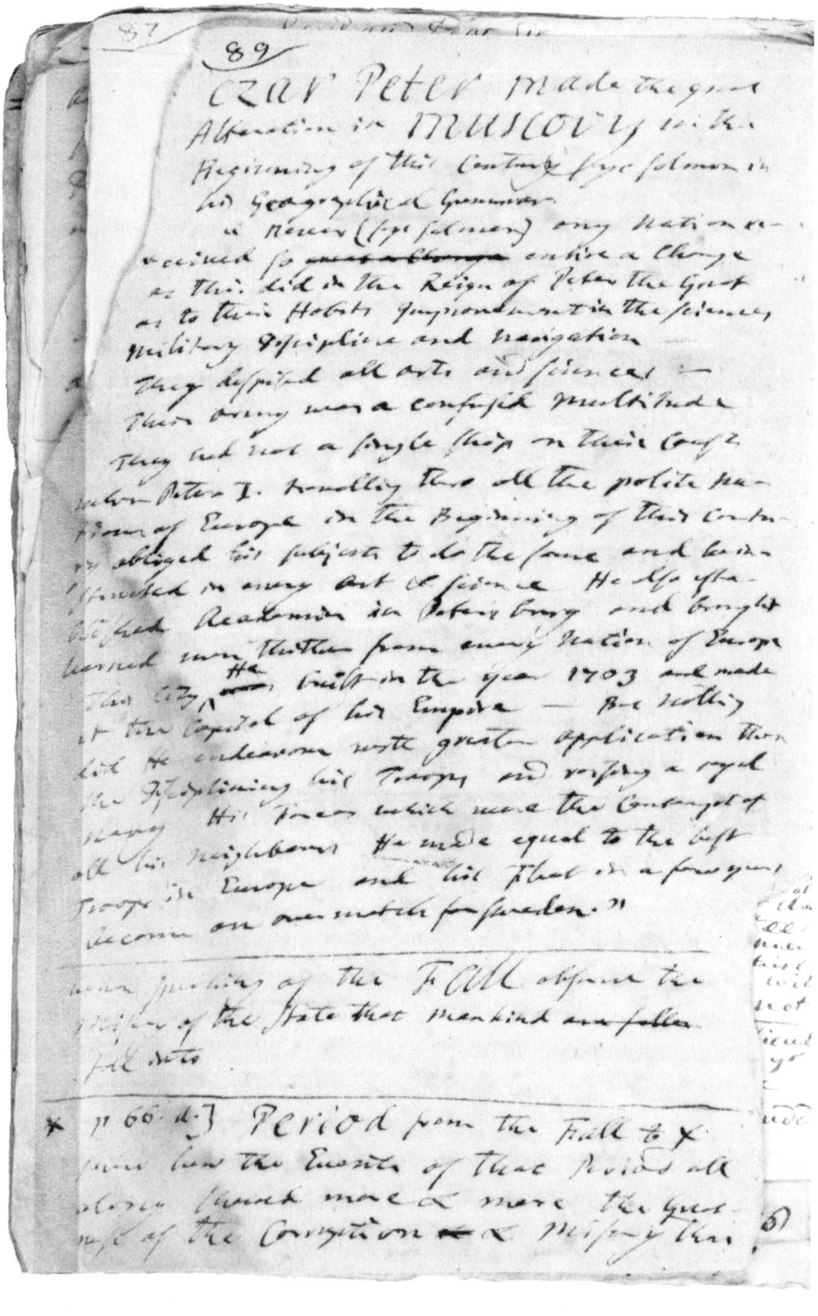

89

Czar Peter made the great Alteration in Muscovy in the Beginning of this Century says Salmon in his Geographical Grammar.

"Never (says Salmon) any Nation received so ~~great a change~~ entire a Change as this did in the Reign of Peter the Great as to their Habits Improvement in the Sciences, Military Discipline and Navigation — They despised all arts and Sciences — Their Army was a confused multitude. They had not a single Ship on their Coasts when Peter I. travelling thro all the polite Nations of Europe in the Beginning of this Century obliged his subjects to do the same and became finished in every Art & Science. He also established Academies in Petersburg and brought learned men thither from every Nation of Europe. This City He built in the year 1703 and made it the Capital of his Empire — But nothing did He endeavour with greater application than the Disciplining his Troops and raising a royal Navy. His Forces which were the Contempt of all his Neighbours He made equal to the best Troops in Europe and his Fleet in a few years become an over match for Sweden."

When speaking of the Fall observe the Misery of the State that mankind ~~are fallen~~ fell into

* p. 66. d.] Period from the Fall to X. [illegible] how the Events of that Period all along shewed more & more the Greatness of the Corruption & Misery they [illegible]

FIGURE 20 • *Notebook, Jonthan Edwards,* A History of the Work of Redemption. *Drafted on top of the* Farewel-Sermon. *Jonathan Edwards Collection, General Collection, Beinecke Rare Book and Manuscript Library, Yale University, New Haven, Conn.*

mankind had fallen into more & more manifested the depth of the abyss how total & desperate the Ruin how utterly helpless

The Period from [illegible] to the glorious times still further showed these Things

The Great Apostacy at the end of

Mr. *Edwards's*

of the C[illegible] preceding the most complete Redemption of all further manifested it preparing for complete Redemption of [illegible]

Farewel-SERMON.

and each of these preparations still the greater [illegible]

To his People at Northampton,

such a tendency or still under a tendency to prepare the [illegible] for that particular work of

June 22. 1750.

Redemption then to be accomplished. The state of the [illegible] before [illegible] came showed more especially the need of such things as were accomplished in the days of [illegible] and the Apostles and in the primitive ages of the Christian church. The Apostasy in the days of Antichrist &c. showed the need of such things as shall be in the glorious times & so on.

theology has simply expanded as compensation for the disintegration of New England as a laboratory of grace, no longer uniquely positioned to augment an imperial and empirical pursuit of divine existence.[80]

Native Religion Revealed

Beginning with Roger Williams and his linguistic inauguration of the Puritan missionary movement, the attempt to discover pure and authentic divinity upon human souls returned consistently and faithfully, though with varying degrees of success, to the Indian soul. As imagined within colonial discourse, the Native American represented a privileged status as a population discovered upon New World arrival as well as one believed to embody primitive sacred essence for those willing and attentive enough to recover it. Edwards did not write extensively about the potential that he discerned among his Stockbridge converts, yet he wrote three of his most famous philosophical treatises while living among them, including *Original Sin* (1758), *Freedom of the Will* (1765), and *The Nature of True Virtue* (1765). *Original Sin* contains a telling description of the natives, revealing the inspiration that Edwards gathered from them:

> I have sufficient reason, from what I know and have heard of the American Indians, to judge, that there are not many good philosophers among them; though the thoughts of their hearts, and the ideas and knowledge they have in their minds, are things invisible; and though I have never seen so much as the thousandth part of the Indians; and with respect to most of them, should not be able to pronounce peremptorily, concerning any one, that he was not very knowing in the nature of things, if all should singly pass before me.[81]

Edwards dismisses the American Indian's metaphysical capacity in this passage while highlighting the native heart and mind as a repository for promising forms of knowledge, of "things invisible" and of a "very knowing" intuition about the "nature of things." Alongside the presence of a true philosopher, such as himself, living among them, Edwards suggests that this native knowledge may come to fruition as an integral component of the advancement of religious and philosophical knowledge. If Edwards saw only this potential among one "thousandth" of the Indians, his dis-

80. Jonathan Edwards, *Letters and Personal Writings*, ed. George S. Claghorn, Works, XVI, 183–184.
81. Jonathan Edwards, *Original Sin*, ed. Clyde A. Holbrook, Works, III, 160.

ciple and colleague David Brainerd saw it among several hundreds more. Through their collective writing about the American Indian, a third definition for the term *native* arises. Native connotes a desire for access to divine essence "untouched" by the potentially flawed constructs of human sensory impression. Native religion can be discerned from the impressions of divinity that seep passively, almost unwittingly, into the convert's mind, where sensory data condenses into an inner light that evades the hypocritical and externally imposed constructs of human understanding. It stands for an ontological category of truth, mapped across a human population whose purported innocence and indigenousness most effectively facilitates the display of true religion.

Brainerd's *Mirabilia* (1748) exemplifies the continuing double play of the term *native* as both true and indigenous, as divinity becomes observable in Brainerd's ethnographic subjects. In his travels along the Pennsylvania and Delaware frontier, Brainerd observes, "There was not only a solemn Attention among them, but some considerable Impressions ('twas apparent) were made upon their Minds by divine Truths." Divine truth is always at the center of Brainerd's quest; it becomes something passively implanted yet discernible on "savage" minds and hearts where "some serious Impressions of a religious Nature" are recorded.[82]

Hypocrisy ceases to be an epistemological obstacle, as the method of recording no longer depends upon the testifier's perceptual capacity; instead, embodied manifestations of grace testify to the immanence of the indwelling principle in native subjects. Brainerd tells of an "Indian Woman" who experienced "her Soul's Salvation . . . like one pierced through with a Dart." Throughout the church meeting, "she could neither go nor stand, nor sit on her Seat without being held up." After the service concluded, "she lay flat on the Ground praying earnestly . . . *Gürtummáukálümméh wéchäuméh kméléh Ndah,*" which Brainerd interprets as, "Have Mercy on me, and help me to give you my Heart." This prayer remarkably constructs an image of the heart for which Edwards was searching, abstracted and available for ethnographic scrutiny. As in Experience Mayhew's *Indian Converts,* the Algonquian rendering of the prayer offers added rhetorical emphasis, creating a record of the anomalous ways in which native souls become vessels for

82. David Brainerd, *Mirabilia Dei inter Indicos; or, The Rise and Progress of a Remarkable Work of Grace amongst a Number of the Indians in the Provinces of New-Jersey and Pennsylvania, Justly Represented in a Journal Kept by Order of the Honourable Society (in Scotland) for Propagating Christian Knowledge* (Philadelphia, 1748), 2, 15.

the direct channeling of divine truth. Through the Sisyphean sequence of the native's "heathenish dance" and Christian "melting," Brainerd's ethnographic gaze becomes a lens for observing true divinity. He writes of one convert, "There was something in his Temper and Disposition that look'd more like true Religion than any Thing I ever observed amongst other *Heathens*."[83]

Given Brainerd's success in witnessing and recording "true religion" among the American Indians, it is unsurprising that Edwards finds his own most convincing and comforting evidence of God's continued work in the life and missionary efforts of David Brainerd. Writing to his colleague in Scotland, Edwards describes Brainerd as "a missionary employed by the Society in Scotland for Propagating Christian Knowledge to preach to the Indians, [who] has lately had more success than ever." Brainerd's success stands as testimony to Edwards's and McCulloch's theological achievement of formulating a soul science, because Brainerd is at work on the frontier putting this soul science into practice. Brainerd's untimely death in 1747, only a few years before the communion controversy that cost Edwards his pulpit at Northampton, was undoubtedly another factor that led Edwards into exile and a retreat to the Stockbridge mission in 1751.[84]

The fragmentation of Calvinism in the first half of the eighteenth century reflected not simply the collapse of a providentially ordained and outdated world but also the demise of a program designed to claim a place for the divine truths revealed to lay converts in a world increasingly driven by a rationalist framework and a Baconian engagement with the world of matter. As New England ceased to be the only place with the epistemic authority to produce evidentiary forms of divine grace, American Protestantism adapted its previous focus on geographic specificity into a new, nativist frame. The typological and ontological claim to the land as a New Canaan shifted from the model that we see in Williams—the New Canaan as the site for observing new and anomalous forms of grace—to the transmogrification of the "city on a hill" into a political trope built upon the vestiges of messianic promise.

83. Ibid., 24–25, 56.

84. Edwards, *Letters and Personal Writings*, ed. Claghorn, Works, XVI, 184–185.

During the Cambuslang revival in Scotland, William McCulloch transcribed several testimonies of faith. These testimonies reveal a parallel concern with charting evidence of divine grace upon the souls of Cambuslang's converts. The testimonies are carefully preserved in manuscript at the New School archives at the University of Edinburgh: "Examination of persons under Spiritual Concern at Cambuslang, during the Revival in 1741–42 By the Revered William Macculloch, Minister of Cambuslang with Marginal Notes by Dr. Webster and other Ministers."

As originally construed by Winthrop, the "Citty upon a Hill" promoted an integral relationship between the land and the evidence supplied by the souls of redeemed saints. Central to the Puritan errand was an attempt to capture this evidence, to display it as a beacon to the world, not only as a means of redeeming God for England or demonstrating the entitlement of a godly sanctioned community but also as a means of carving out a place that would recast and propagate the need for divine presence in a rapidly modernizing world. The "Citty upon a Hill" engendered a series of experiments—or laboratories of grace—delineated through the various communities and scenes of witnessing described earlier. Edwards sought to revive, in order to recuperate, a dying program that had accorded a historically unprecedented and subsequently unreplicable place for evidence of grace. His Northampton experiment was, not atavistic, but, rather, as deeply tied to paradoxically overlapping structures of primitive Christianity and scientific modernism as the communal experimental models that came before. Along with Edwards's retreat to Stockbridge came the end of a Puritan era that made such a compelling case for New England as a site of spiritual plentitude perpetuating forms of sensory interaction between its inhabitants and God. The formlessness of true divinity took root in New England, offering itself to the world as integral data in a deep quest for divine knowledge that was tragically and inevitably fraught with the simultaneous awareness of its utter impossibility. Edwards's exile marked the end of a quest for divine forms of knowledge that required a particular kind of continuity across the divergent and often competing domains of empiricism, Reformed theology, and natural philosophy, but it also suggested the beginning of a modern American era of new Protestant genres and forms of inquiry that have absorbed this history of the science of the soul.

Conversion in America

On Sunday, October 4, 1747, David Brainerd lay dying in Jonathan Edwards's home in Northampton. He had been ill for several years and had endured hardships as he traveled on horseback along the Delaware and Susquehanna Rivers in search of Indians to convert. His missionary journey, as Edwards would relate in his *Account of the Life of Brainerd,* based on Brainerd's journal, became a spiritual journey into the depths of his own heart. Manifestations of grace among Brainerd's native converts were a measure of his own. His illness and impending death indexed the depletion of religion as also expressed among his flock: "A view of their want of this was more afflictive to me, than all my bodily illness." Despite all that he endured, Brainerd pressed on well beyond what seems physically possible. His hope for the advancement of Christ's kingdom on earth, realized through each successful Indian conversion, sustained him. When bodily illness eventually overcame even Brainerd's capacity to promote the gospel, he retreated to Sarah and Jonathan Edwards's Northampton home. He bade goodbye to Jerusha Edwards, the Edwards's seventeen-year-old daughter who had also been appointed Brainerd's nurse, and called for his younger brother, John, who came to attend Brainerd's deathbed. Two days later, Brainerd died quietly and peacefully, attended by those who loved him and with the certainty that he and those closest to him were in the hands of God.[1]

Much affected by the death of his beloved disciple with a grief that was compounded a year later when Jerusha died "in the bloom of her youth," Edwards suspended all of his writing projects in order to focus entirely on his hagiographical *Account of the Life of Brainerd.* Edwards saw in Brainerd renewed hope for an international awakening, generated by the land and people of North America. Additionally, Edwards saw within Brainerd's

1. George Marsden, *Jonathan Edwards: A Life* (New Haven, Conn., 2003), 320–327; Jonathan Edwards, *An Account of the Life of the Late Reverend Mr David Brainerd, Minister of the Gospel* . . . (Edinburgh, 1765), 218, 230.

journal a means of discerning true religion. In the weeks leading up to his death, Brainerd proclaimed:

> I enjoyed as much serenity of mind, and clearness of thought, as perhaps I ever did in my life; and I think, my mind never penetrated with so much ease and freedom into divine things, as at this time; and I never felt so capable of demonstrating the truth of many important doctrines of the gospel as now. And as I saw clearly the truth of those great doctrines, which are justly stiled the *doctrines of grace;* so I saw with no less clearness, that the *essence of religion* consisted in the soul's *conformity to God,* and acting above all selfish views, for *his glory,* longing to be *for him,* to live *to him,* and please and honour *him* in all things: and this from a clear view of his infinite excellency and worthiness *in himself,* to be loved, adored, worshipped, and served by all intelligent creatures.[2]

Encoded within Brainerd's *Life* is a mutually dependent relationship between the evidence of the minister's soul and the evidence of God's presence among his flock. Brainerd looks to indigenous populations for a pure and authentic expression of grace. He does this, not with the detached form of ethnographic witnessing that we saw in John Eliot's Natick, but rather through the search for authentic religion within himself. Displayed within this passage is the collapse of an Evangelical Enlightenment into a new form of piety designed to withstand the pressures of a rapidly changing world. Empirical and natural philosophical frameworks of observing objects within the world have been incorporated into a unified theological frame such that Brainerd comes to know the "essence of religion" through "the soul's conformity to God." While reflecting the culmination of a long colonial history that uses the methods from the natural sciences in an effort to chart divine evidence upon human souls, Brainerd's near-death expression simultaneously reflects the transformation of religion into a more independent and self-sufficient phase. Religion claims an autonomous status to know through revealed truths, having absorbed the methods of the natural sciences to do so. By grasping this autonomy, pious certainty, and confidence in an immanent kingdom of God, Edwards scripted the modern American conversion narrative in *The Life of Brainerd.*[3]

2. Edwards, *Life of Brainerd,* 218, 230.

3. Mark Noll describes this as a phase of "epistemological self-sufficiency." *America's God: From Jonathan Edwards to Abraham Lincoln* (New York, 2002), 12.

The Life of Brainerd builds on precedent, recalling first-generation Puritan conversion testimonies and spiritual autobiographies that mapped the New World journey as an allegory for the soul's spiritual quest. Like the authors of these narratives, Brainerd's journey is not an easy one. He lives in a "most lonesome wilderness" that is even more lonely, since he knows only one other person who can speak English and "no fellow-Christian" to hear of his many troubles. His missionary work is discouraging. Brainerd enters into periods of utter despair when his only solace is to "spen[d] most of the day in study," such that he might begin to feel the "intense and passionate breathings of [his] soul after holiness" and regain some "bodily strength" to go back out into the cold, desolate, and lonely wilderness. Tracing this movement between the inner self and the outer world, the comforts of an enclosed but temporary spiritual community and the howling wilderness beyond, *The Life of Brainerd* replicates the archetypal oscillation of Thomas Shepard's "Autobiography," John Winthrop's *Christian Experience,* John Bunyan's *Pilgrim's Progress,* and Mary Rowlandson's captivity narrative.[4]

The Life of Brainerd also departs from precedent, for the narrative does not rest in this oscillation, but, rather, arrives at the certainty of true religion. Brainerd's frontier evolves from the site of merciful affliction that it was for Rowlandson into a place of renewal and expansion that permits a freer-flowing movement between the civilized and the savage, Christian and heathen communities, order and chaos, the primitive and the modern, the sick and the healthy, life and death, this world and the next. Where the Puritans understood conversion as an uncertain process whose truth could be understood only allegorically as a movement between the desolation of the world and the inner strength of the soul, between the outward and inward self, Brainerd's *Life* arrives at a distinctly different level of conviction. Brainerd's conversion comes through observation and dissolution. He sees the savage wilderness transforming into a spiritual landscape through the touch of divine grace on the souls of his native proselytes. The native religion displayed on the souls of Brainerd's converts dissolves into his own spiritual quest such that Brainerd discovers true divinity within himself. Through its celebration of the frontier as a site for the embrace of a primitive spirituality, a space on the margins where religion can be certain and where Christian values happen naturally, the *Life of Brainerd* exemplifies an emerging canon of modern American evangelical conversion narratives.

Brainerd enjoyed considerable success. Perhaps the scope of his evange-

4. Edwards, *Life of Brainerd,* 71, 77, 218.

lizing work would have rivaled that of his contemporary George Whitefield, had he lived longer. In his journal, Brainerd rejoices at reports of Whitefield's revivals, revealing an enthusiasm that Edwards chooses to leave out of *The Life of Brainerd*. Edwards's reasons for this omission might have been that he did not want to further inflame the already entrenched battles between Old Lights and New Lights. However, in other instances Edwards tamps down the more enthusiastic strains of Brainerd's piety, such as his reports of visions in the style of John Bunyan. Edwards clearly wanted to present Brainerd as a true saint, impervious to the theological disputes that had marred Brainerd's academic career at Yale and that were sweeping the colonies. However, these omissions also suggest that Edwards was not fully prepared to embrace the direction that American piety was headed in, even though he both recognized and celebrated Brainerd's pious *Life* as the exemplary model leading the way.[5]

Indeed, Brainerd can be aligned more clearly with the New Lights than the Old. He has much in common with Whitefield, whose *Journal of a Voyage from London to Savannah in Georgia* (1738) recounts commensurate degrees of physical illness, counteracted by the spiritual renewal that he derives from preaching. Like Brainerd, Whitefield is an observer of the spread of the gospel whose own soul maintains no distance from the converts that he observes. His *Journal* describes a new landscape of American spirituality, where thousands of people line Market Street in Philadelphia in order to hear him preach from the steps of the congregations that have banned him from their pulpits. The Whitefieldian revival bespeaks a fervor that is insatiable, that crowds streets and fields with an unchecked emotionalism that supplies catharsis and salvation to the poor, the orphaned, and the enslaved, who, forgotten by society, are beyond the pale of social redemption. This was a religious landscape that gave Jonathan Edwards great pause. He retreated, following the communion controversy that cost him his Northampton pulpit in 1751, to Stockbridge, where he discovered the conjoined projects of local missionary work and philosophical writing as the only salve for a religious and secular world that would no longer be the same.[6]

5. Marsden, *Jonathan Edwards,* 331. See also Jonathan Edwards, *The Life of David Brainerd,* ed. Norman Pettit, Works of Jonathan Edwards, VII (New Haven, Conn., 1985), 5–24.

6. George Whitefield, *A Journal of a Voyage from London to Savannah in Georgia* (London, 1738), and *A Continuation of the Reverend Mr. Whitefield's Journal, from His Arrival at Savannah to His Return to London* (London, 1739).

The 1750s bore witness to the rise of unchecked mercantile capitalism, the emergence of a powerful print public, the radical realignment of governmental authority, rapid urbanization on the eastern seaboard of North America, and the powerful, palpable force of an evangelical God. If these changes promoted Edwards to retreat, few individuals embraced them more fully than George Whitefield and Benjamin Franklin. While espousing a deist perspective, Franklin's *Autobiography* is highly conscious of the way that it builds upon as well as departs from the tradition of the Puritan spiritual autobiography. He recounts his European roots in a Nonconformist tradition, describing his ancestors' practice of hiding their Bible under a stool. He names John Bunyan's *Pilgrim's Progress* as his favorite book, invoking an implicit link between his own life journey and the trials and tribulations of Christian on his way to the Celestial City. Franklin is, as many have argued, Puritan in his demeanor and character even as the *Autobiography* replaces the struggle of the soul with a highly disciplined form of discursive rationalism that insists on its separation from revealed truths. Franklin reads the Boyle lectures and decides that he is a deist. He mentions having the near-opportunity to meet Isaac Newton but learns nothing from these great thinkers of natural theology and mechanical philosophy of their quest for discovering the essence of God in nature. This is an irrelevant question for Franklin, who systematically separates reason from revelation.[7]

Whitefield embraces the opposite perspective, upholding revelation as the most important category of knowledge. In his *Journals*, God's revealed truth comes entirely unmoored from the rational Christianity of Samuel Clarke, the natural theology of Boyle, or Edwards's careful condensation of Lockean faculty psychology into the indwelling principle. The intellectual link between evangelicalism and the Enlightenment underwent great challenge by the 1750s; as each side became aware of this challenge, evangelicals and deists became more self-consciously rooted in their own perspectives and less likely to try to imagine the continuities across an ever-widening divide between American revivalism and an American Enlightenment. Yet, while seemingly on opposed sides of the eighteenth-century Enlightenment, Franklin was captivated by Whitefield and in fact underwent his own form of a conversion experience upon hearing him preach.

7. Michael Warner develops this idea of discursive rationalism in his *Letters of the Republic: Publication and the Public Sphere in Eighteenth-Century America* (Cambridge, Mass., 1990), 79–80.

> It was wonderful to see the change soon made in the manners of our inhabitants; from being thoughtless or indifferent about religion, it seemed as if all the world were growing religious, so that one could not walk thro' the town in an evening without hearing psalms sung in different families of every street. . . .
>
> I had in my pocket a handful of copper money. . . . As he proceeded, I began to soften and concluded to give the coppers. Another stroke of his oratory made me ashamed of that and determined me to give the silver; and he finished so admirably that I emptied my pocket wholly into the collector's dish, gold and all.[8]

Formally, this passage follows a typical Protestant conversion narrative. Franklin begins with a closed pocket, his version of a hardened heart. The sound of Whitefield's voice, which Whitefield claimed to come from the power of God, softens the heart. Franklin relinquishes a little and then, upon feeling his shame, more and more. His shame in this scene is equivalent to a Protestant convert's realization of his or her own implication in original sin and the accompanying acceptance of responsibility for a debt that can never be repaid. At this point, as in a typical Protestant conversion narrative, Franklin submits completely. His heart-pocket is emptied, patiently awaiting the redemptive infusion of Christ's indwelling light. Except, of course, Franklin does not believe in revelation.

Franklin recognized an enterprising capitalist in George Whitefield. His donation to this field itinerant's cause demonstrates the shared system of economic exchange between a deist and an evangelical in the rapidly developing economy of the mid-eighteenth-century Philadelphia world. Indeed, there are many points of comparison between these two luminous mid-eighteenth-century contemporaries. In Whitefield's *Journal*s and Franklin's *Autobiography,* both men model a form of self-fashioning that invites emulation. Opening his *Journal of a Voyage* with a description of his own birth in an inn, Whitefield self-typologizes as Christ and then catalogs his sins as Franklin does his errata. Both men understand innovation as the key to

8. Benjamin Franklin, *The Autobiography and Other Writings,* ed. L. Jesse Lemisch (New York, 2001), 116–117. In *The Enlightenment in America* (New York, 1976), Henry F. May locates a potential split between "the Age of Reason and the Age of Enthusiasm" in the 1740s and 1750s. He proposes that movements become more self-conscious when they are challenged, so that the "partisans of order, balance, reason, and moderation were placed on the defensive by a revival of religious enthusiasm" (42).

mastering a successful public persona. Their humility thinly veils arrogant grandiosity. Each sacrificed the private self in the service of his respective public capacities. Each engaged in private war with the body in order to fashion a self that was larger than life. Franklin catalogs the food and drink that he refuses to put into his body, and Whitefield repeatedly tells us what he violently expels from his body by vomiting before delivering a sermon.

Forms of affect, personality, self-promotion, conversion, and replication represent striking continuity across the divide of reason and revelation, suturing the worlds of Franklin and Whitefield. In the *Autobiography*'s movement from sixteenth-century religious dissent to Whitefieldian revivalism, Franklin observes and records Protestantism's transformation from a religion rooted in profound Calvinist ambivalence and uncertainty to one that unabashedly proclaims privileged access to certain and revealed truths. The *Autobiography* indexes the mid-eighteenth-century American religion as discontinuous with the past and therefore representative of a specific form of American modernity.

Religion became more exceptionally American, as ministers from Edwards to Whitefield to Timothy Dwight to Charles Grandison Finney develop theology in ways that increasingly depart from European precedent. Religious practice becomes tied to the land, from the field gatherings of the Whitefieldian model to the camp revivals that became firmly ensconced in an American religious tradition by the Second Great Awakening. Although religious certainty may look like the vestige of an age in which divine authority reigned, or the very thing that marks the religious as pre-Enlightenment, religious certainty is in fact more a product of the Enlightenment epistemology that we are often too quick to label as its opposite. The certainty of religious conviction that the nonreligious find so puzzling and at times troubling does not reflect religion's rootedness in a historical past that believes wholly and unquestioningly in divine authority. Rather, religious certainty both derives from and marks a break with the long history of Protestantism's encounter with the structures of empirical practice and desire recounted in this study. The methods of empiricism and the techniques of observation, experiment, and verification that developed through the scientific revolution contributed to the rise of modern religious certainty.

This version of certainty is different from what we see in a book such as John Foxe's *Book of Martyrs* or in less mainstream prophetic traditions such as the Mormons and the Millerites in the nineteenth century. The certainty that we see in the mainstream Protestant evangelicalism that descended from the Whitefieldian revivals reflects a form of religious affect that had absorbed and transformed the long colonial history of the science of the soul. By the mid-eighteenth century, natural philosophers began to relinquish the idea that personal revelation could contribute to a rational plan for knowing God. Sensing the profound unmooring of their relevance to the world, theologians such as Edwards and his New Light contemporaries changed the terms of their contribution to human knowledge through the concept of an indwelling light that could, in fact, supply an unprecedented degree of certain knowledge to human souls. According to the renewed empirical potential supplied by the indwelling light, the cosmos need not be the "inexplicable mystery" that David Hume described in a mode of skeptical resignation to the limits of human understanding.

In explicit contrast to the certainty concurrently invented by contemporary theologians, Hume's *Dialogues concerning Natural Religion* (written in the 1750s though not printed until 1779) presents uncertainty as unavoidable and necessary to philosophical advancement. Like Locke, Hume denies sensory access to divine truth, yet for Hume this is a point of empirical liberation rather than limitation, for it frees the senses to function as vehicles for the pursuit of other truths. The existence of God remains an undeniable fact for Hume as he explains that humans can rest confident only in the "Being" of God; divine "nature" can never be known. The pursuit of divine nature, Hume explains, is futile, owing to the "infirmities of human understanding." Divine essence will forever remain "covered in a deep cloud from human curiosity." Hume writes: "It is profaneness to attempt penetrating through these sacred obscurities. And next to the impiety of denying his existence, is the temerity of prying into his nature and essence, decrees and attributes." In the decade surrounding Hume's *Dialogues,* natural philosophy accepted divine essence as mysterious rather than tried to solve the riddle or lament its inability to do so, as Isaac Newton had done at the end of the *Principia* (1687).[9]

Edwards's indwelling light and Hume's radical skepticism represent two distinct and opposed responses to the epistemological possibilities and limitations introduced by the watershed impact of Lockean sensational psychology on philosophy and religion. Edwards reinvests with the capacity for certain knowledge of God. Hume delineates a method of sensory discernment that must distinguish itself from the infinite mystery of God's existence in order to remain empirically sound. Each philosopher believed

9. David Hume, *Dialogues concerning Natural Religion,* ed. Martin Bell (New York, 1990), 51.

in the absolute certainty of God's existence, but Edwards wanted to salvage some notion of sensory access to divine essence via the manifestations of free grace upon the soul. The difference between their respective advancements of Lockean knowledge hinged on an essential disagreement over the use to which sensory data could be put. This disagreement features prominently in Edwards's *Freedom of the Will* (1754), which seeks to supplement Scottish Common Sense philosophy with a theory of the privileged moral capacity available through the sensory data that were uniquely available to spiritual men.

Despite these remarkably different ways of resolving the narrow window of Lockean psychology, Hume has his own version of the indwelling principle, represented through the position of Demea within the dialogues, who explains that "each man feels, in a manner, the truth of religion within his own breast." Where Edwards wanted to map and taxonomize this truth such that it remained integral to Enlightenment philosophy, Demea explains that humans know this truth through "a consciousness of his imbecility and misery, rather than from any reasoning." But, for Hume as well as Edwards, true religion within the heart constituted certain knowledge, and it is this very form of certain knowledge that becomes central to modern religion as it appears across the proliferation of Protestant conversion narratives in the late-eighteenth century. From John Marrant to Phillis Wheatley to Olaudah Equiano to John Woolman and Elizabeth Ashbridge, late-eighteenth-century conversion narratives abounded, as if to announce a kind of generic liberation from the carefully regulated and disciplined attention to the uncertain evidence of the soul imposed in Puritan testimonies prior to Edwards.[10]

The indwelling principle marked a point of philosophical departure, a distinct and ultimately irreconcilable difference between how Edwards and Hume responded to Locke as well as in the pursuit of philosophical truth represented by each thinker. But the indwelling principle also supplied the religious with a tool for surviving in the Age of Enlightenment, for moving adeptly between sacred and profane registers as Brainerd moves across the Pennsylvania and New Jersey frontier, and for making a truth claim based upon divine certainty in an era increasingly skeptical of such claims. These various uses of the indwelling principle come across clearly and auspiciously in Phillis Wheatley's letter to Samson Occom, written on February 11, 1774. The letter articulates a position on her own certain truth

10. Ibid., 103.

of religion that both builds from the Edwardsian and Humean notion of inner truth and transforms this certainty into a capacity to speak directly and authoritatively to new political purposes.

> In every human Breast, God has implanted a Principle, which we call Love of Freedom; it is impatient of Oppression, and pants for Deliverance; and by the Leave of our Modern Egyptians I will assert, that the same Principle lives in us.[11]

This notion of an indwelling principle references the influence of New Light theology on Wheatley. In her exchange with Occom, the "implanted principle" functions as certain proof of a millennial vision of a "divine Light . . . chasing away the thick Darkness which broods over the Land of Africa." While this letter, like Wheatley's better-known poem "On Being Brought from Africa to America," offers a highly controversial evangelist image of Christianizing Africa as part of God's master plan, Wheatley mobilizes this image to make a strategic claim for natural rights. She not only agrees with Occom that the "natural Rights" of Africans have been violated but also signs on to his plan of "Vindication." Wheatley expresses feeling deep justification in this plan for vindication precisely because the image of a rapidly evangelizing Africa is also one of a gradual conversion from chaos to order and from slavery to freedom. Wheatley's own individualized indwelling principle places her squarely within this millennial vision, giving her the privileged spiritual perspective of one who can discern this pattern of global conversion.[12]

From her position of spiritual authority, Wheatley equates "civil and religious Liberty," which she sees as "inseparably United." The effects of God's manifest grace within her own soul allow her to move freely between the religious and political, sacred and secular domains of a highly charged and permeable late-eighteenth-century discourse of liberty, freedom, and natural rights. As if responding directly to the exclusion of African peoples from the science of the soul discussed in the Introduction, Wheatley and other black Atlantic writers, such as John Marrant and Oladuah Equiano, used the indwelling principle to make explicit and certain claims for their status as Christians. Throughout their published works, the indwelling light reshapes the conversion narrative as a powerful genre within an Age

11. Julian D. Mason, Jr., ed., *The Poems of Phillis Wheatley,* rev. ed. (Chapel Hill, N.C., 1989), 203–204.

12. Ibid.

of Enlightenment, for they each claim religious conviction and certainty from positions of extreme social marginalization. Spiritual authority permits each of these authors to place a claim for natural rights alongside a broader secular discourse that did not need to depend, as Wheatley and Equiano depended, upon a topos of Christian virtue. Wheatley mobilizes religious certainty in order to present a powerful position of authority based on the undisputable presence of a conviction that comes from the transcendent and illuminating power of the indwelling light.[13]

American religious genres have adapted and thrived upon their colonial precedent. Their social and political function varies widely within any given historical moment as well as across centuries. The jeremiad has proved capable of addressing a political crisis of any scale and ideological position. The captivity narrative consistently offers a convention for transforming the unpredictable, chaotic, and unknown dimension of the American frontier—both westward in the nineteenth century and globally in the twentieth and twenty-first centuries—into a site of regenerative possibility. Religion seeks to recuperate an original primitivist dimension as the city upon the hill resurfaces through political sermons such as Samuel Sherwood's *Church's Flight into the Wilderness* (1776) and tracts such as Thomas Paine's *Common Sense* (1776). Such genres proliferate and adapt formally and thematically throughout late-eighteenth-century America, not only because religion maintained its stronghold despite the rise of secular values but also because religion evolves historically with the capacity to negotiate these values. Through the theological transformation of the doubt, hesitancy, and uncertainty that we have seen so intensely expressed in the Puritan testimonies into the certain claim to religious truth, the conversion narrative emerged as a distinct American genre. Certainty, as it has evolved from a long history of soul science, is the genre's hallmark, allowing the religious to speak powerfully, persuasively, and with much conviction within our shared secular age.

13. Ibid. Vincent Carretta has conveniently compiled these authors into a single anthology: *Unchained Voices: An Anthology of Black Authors in the English-Speaking World of the Eighteenth Century* (Lexington, Ky., 1996). On Equiano, see 185–318; on Marrant, 110–133.

Index